Length

$\frac{0.3048 \text{ m}}{1 \text{ ft}} = 1 \qquad \frac{2.54 \text{ cm}}{1 \text{ in.}} = 1$

Area

$\frac{9.29 \cdot 10^{-2} \text{ m}^2}{1 \text{ ft}^2} = 1 \qquad \frac{6.452 \text{ cm}}{1 \text{ in.}^2} = 1 \qquad \frac{}{1 \text{ acre}} = 1$

$\frac{640 \text{ acre}}{1 \text{ mile}^2} = 1 \qquad \frac{1 \cdot 10^6 \text{ m}^2}{1 \text{ km}^2} = 1 \qquad \frac{1 \cdot 10^4 \text{ m}^2}{1 \text{ m}^2} = 1 \qquad \frac{2.59 \text{ km}^2}{1 \text{ mile}^2} = 1$

Volume

$\frac{2.832 \cdot 10^{-2} \text{ m}^3}{1 \text{ ft}^3} = 1 \qquad \frac{1 \cdot 10^6 \text{ cm}^3}{1 \text{ m}^3} = 1 \qquad \frac{1 \cdot 10^3 \ \ell}{1 \text{ m}^3} = 1$

$\frac{7.4805 \text{ gal}}{1 \text{ ft}^3} = 1 \qquad \frac{3.7854 \ \ell}{1 \text{ gal}} = 1 \qquad \frac{43560 \text{ ft}^3}{1 \text{ acre-ft}} = 1$

$\frac{3.2585 \cdot 10^5 \text{ gal}}{1 \text{ acre-ft}} = 1 \qquad \frac{1233.48 \text{ m}^3}{1 \text{ acre-ft}} = 1$

Velocity

$\frac{0.3048 \text{ m/sec}}{1 \text{ ft/sec}} = 1 \qquad \frac{30.48 \text{ cm/sec}}{1 \text{ ft/sec}} = 1$

Discharge

$\frac{694.44 \text{ GPM}}{1 \text{ MGD}} = 1 \qquad \frac{448.83 \text{ GPM}}{1 \text{ ft}^3\text{/sec}} = 1 \qquad \frac{226.29 \text{ GPM}}{1 \text{ acre-ft/day}} = 1$

$\frac{15848.5 \text{ GPM}}{1 \text{ m}^3\text{/sec}} = 1 \qquad \frac{86400 \text{ m}^3\text{/day}}{1 \text{ m}^3\text{/sec}} = 1 \qquad \frac{3153.6 \text{ cm} \cdot \text{km}^2\text{/year}}{1 \text{ m}^3\text{/sec}} = 1$

$\frac{479.3 \text{ in./year/mile}^2}{1 \text{ m}^3\text{/sec}} = 1$

Mass

$\frac{14.594 \text{ kg}}{1 \text{ slug}} = 1 \qquad \frac{14594 \text{ g}}{1 \text{ slug}} = 1$

Density

$\frac{515.4 \text{ kg/m}^3}{1 \text{ slug/ft}^3} = 1 \qquad \frac{0.5154 \text{ g/cm}^3}{1 \text{ slug/ft}^3} = 1$

Force

$$\frac{4.4482\ \text{N}}{1\ \text{lb}} = 1 \qquad \frac{1 \cdot 10^5\ \text{dyn}}{1\ \text{N}} = 1$$

Pressure

$$\frac{6894.7\ \text{N/m}^2}{1\ \text{psi}} = 1 \qquad \frac{68947\ \text{dyn/cm}^2}{1\ \text{psi}} = 1 \qquad \frac{6.8011 \cdot 10^{-2}\ \text{atm}}{1\ \text{psi}} = 1$$

$$\frac{1\ \text{bar}}{1 \cdot 10^6\ \text{dyn/cm}^2} = 1$$

Temperature

$$(°\text{F}) = 1.8(°\text{C}) + 32, \text{ i.e., } 20°\text{C} \equiv 68°\text{F}$$

Energy

$$\frac{1\ \text{Joule}}{1\ \text{N}\cdot\text{m}} = 1 \qquad \frac{1 \cdot 10^7\ \text{erg}}{1\ \text{Joule}} = 1 \qquad \frac{1\ \text{erg}}{1\ \text{dyn}\cdot\text{cm}} = 1 \qquad \frac{0.23901\ \text{cal}}{1\ \text{Joule}} = 1$$

Viscosity

$$\frac{10\ \text{Poise}}{1\ \text{N}\cdot\text{s/m}^2} = 1 \qquad \frac{42.88\ \text{N}\cdot\text{s/m}^2}{1\ \text{lb}\cdot\text{s/ft}^2} = 1 \qquad \frac{100\ \text{cP}}{1\ \text{Poise}} = 1$$

Kinematic Viscosity

$$\frac{1 \cdot 10^4\ \text{Stoke}}{1\ \text{m}^2\text{/s}} = 1 \qquad \frac{9.294 \cdot 10^{-2}\ \text{m}^2\text{/s}}{1\ \text{ft}^2\text{/s}} = 1 \qquad \frac{1\ \text{cm}^2\text{/s}}{1\ \text{Stoke}} = 1$$

$$\frac{100\ \text{cS}}{1\ \text{Stoke}} = 1$$

Power

$$\frac{550\ \text{lb}\cdot\text{ft/s}}{1\ \text{hp}} = 1 \qquad \frac{745.7\ \text{N}\cdot\text{m/s}}{1\ \text{hp}} = 1 \qquad \frac{1.356\ \text{N}\cdot\text{m/s}}{1\ \text{lb}\cdot\text{ft/s}} = 1$$

$$\frac{1\ \text{W}}{1\ \text{Joule/s}} = 1 \qquad \frac{1\ \text{W}}{1\ \text{N}\cdot\text{m/s}} = 1$$

Second edition

FUNDAMENTALS OF HYDRAULIC ENGINEERING SYSTEMS

Ned H. C. Hwang
Carlos E. Hita

University of Houston
Houston, Texas

Prentice-Hall, Inc., Englewood Cliffs, New Jersey 07632

Library of Congress Cataloging-in-Publication Data

Hwang, Ned H. C. (date)
Fundamentals of hydraulic engineering systems.

Includes index.
1. Hydraulics. 2. Hydraulic engineering.
I. Hita, Carlos E., (date). II. Title.
TC160.H86 1986 627 86-4965
ISBN 0-13-340027-1

Editorial/production supervision and
interior design: *Gretchen K. Chenenko and Gloria Jordan*
Manufacturing buyer: *Rhett Conklin*

Printed in the United States of America

10 9 8 7 6 5

ISBN 0-13-340027-1 025

Prentice-Hall International (UK) Limited, *London*
Prentice-Hall of Australia Pty. Limited, *Sydney*
Prentice-Hall Canada Inc., *Toronto*
Prentice-Hall Hispanoamericana, S.A., *Mexico*
Prentice-Hall of India Private Limited, *New Delhi*
Prentice-Hall of Japan, Inc., *Tokyo*
Prentice-Hall of Southeast Asia Pte. Ltd., *Singapore*
Editora Prentice-Hall do Brasil, Ltda., *Rio de Janeiro*

DEDICATION

To our beloved families

Maria, Leon, and Leroy Hwang
and
Fatima, Danny, and Juliana Hita

CONTENTS

PREFACE

This book is a fundamental treatment of engineering hydraulics in S.I. units (Systéme International d'Unités). It is primarily intended to serve engineering students and their teachers as a textbook. It is also intended to serve practicing engineers who need a convenient reference to basic principles and their applications in hydraulic engineering problems.

Teaching hydraulics in engineering schools has changed tremendously during the last two decades. Such change was prompted by two concurrent developments. First, many problems in hydraulics that seemed to be computationally insurmountable previously are now being solved by high-speed computers. Second, reduced credit hours in the basic engineering curriculum have resulted in a more comprehensive but condensed program at the undergraduate level. Both of these developments emphasize the teaching of fundamentals.

Hydraulics is an area of fluid mechanics in which many empirical relationships are applied to achieve practical engineering solutions. Experience has shown that many engineering students, possibly well versed in basic fluid mechanics, were having trouble dealing with practical problems in hydraulics. For this reason, a good hydraulic textbook, presenting a systematic and comprehensive approach to practical problems, was needed.

This book contains ten chapters. They provide a broad coverage of most of the common topics in hydraulics systems. The first chapter discusses the fundamental properties of water as a fluid. The second chapter presents the concepts of water pressure and pressure force on surfaces in contact with water. In this chapter the basic differences between the S.I. system and the British units are discussed. Chapter 3 introduces the basic elements of water flow in pipes. These elements are applied to practical problems of pipelines and pipe networks in Chapter 4 with emphasis on system approach. Chapter 5 discusses the fundamentals of water pumps. The system concept is again emphasized with examples provided for pump selections as an integrated part of the pipeline system.

Water flow in open channels is presented in Chapter 6. Detailed discussions of wave phenomena in supercritical flow channels and design of supercritical transitions are included. The hydraulics of wells and seepage problems are treated in Chapter 7. This chapter also contains a section on sea water intrusion due to excessive pumping in coastal regions.

While detailed design procedures of hydraulic structures are this text's scope, Chapter 8 presents several types of common hydraulic structures to demonstrate the fundamental considerations in hydraulic structures. Chapter 9 discusses measurement of water pressure, velocity, and discharge in pipes and in open channels. The proper use of scaled models is an essential part of hydraulic engineering. Chapter 10 discusses the use of hydraulic models and the laws of engineering similitude.

This book is designed for a one-semester, three-hour-a-week course (16 lecture weeks) in the undergraduate curriculum. No prerequisite in basic fluid mechanics is needed. The book can also be used for the quarter-system academic program (12 weeks) by omitting a few special topics such as water pumps (Chapter 5), hydraulics of wells and seepage (Chapter 7), and hydraulic structures (Chapter 8). Examples and homework problems are provided for the students after every major section.

Computer programs are presented only where they are deemed necessary to emphasize a concept, such as the Hardy-Cross method for water distribution in pipe networks. In other areas in which computers are commonly used for heavy computations, such as open channel surface profiles, water hammer analysis, and flood hydrograph computations, digital programs are not included because we believe that simplified versions of these programs have little real significance and a complete treatment of these programs is beyond the scope of this book. Many of these programs (e.g., HEC-1, HEC-2*) are readily available to the public, and workshops are offered periodically by several universities and by the Hydrologic Center, U.S. Army Corps of Engineers.

Ned H. C. Hwang
Carlos E. Hita
Houston, Texas

*HEC-1 Flood Hydrograph Package, The Hydrologic Center, Corps of Engineers, U.S. Army.
HEC-2 Water Surface Profiles Package, The Hydrologic Center, Corps of Engineers, U.S. Army.

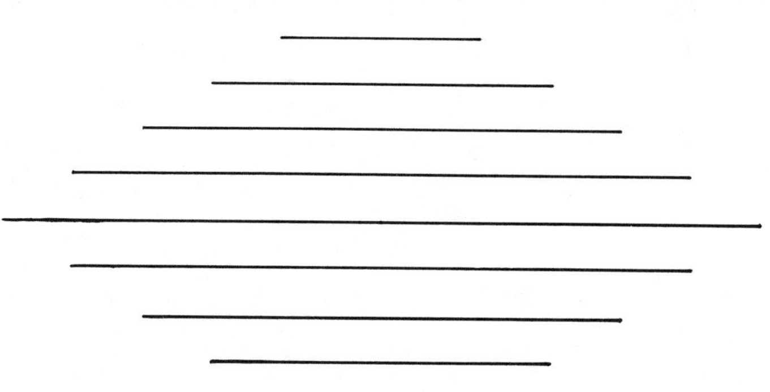

ACKNOWLEDGEMENTS

The first attempt to arrange my lecture notes into a textbook was made in 1974. My colleagues at the University of Houston, particularly Professors Fred W. Rankin, Jr. and Jerry Rogers, had provided valuable suggestions and assistance, which formed the foundation of this book. Dr. Rogers also carefully reviewed the first edition text after its completion.

I am also deeply indebted to my students, Drs. Travis T. Stripling, John T. Cox, James C. Chang, and Po-Ching Lu. All provided assistance during the various stages of preparation. Dr. Ahmed M. Sallam, who used a first draft of this book to lecture a course in hydraulics, provided many suggestions. Dr. Carlos E. Hita, who is coauthor of this current edition, has offered valuable suggestions, and also provided, with comprehension and competence, most of the sample problems used in the text.

My dear friend, Dr. David R. Gross, an ever-inquisitive physiologist (as was Jean Louise Poiseuille, 1799–1869) with a great interest in hydraulic matters, reviewed the first edition text and offered many irrefutable criticisms.

During the preparation of the first edition, I was ill for some time. The continuous encouragement, loyalty, and love of Maria, Leon, and Leroy kept me on track during the dark hours that have since passed. To them also, I dedicate this book.

Ned H. C. Hwang
Houston, Texas

INTRODUCTION

Hydraulic systems are designed to control, conserve, or transport water. The planning, construction, and operation of hydraulic systems involve the application of fundamental engineering principles in fluid mechanics, soil mechanics, structural analysis, engineering economics, and many other related fields.

Unlike most other branches of engineering, each hydraulic project encounters a unique set of physical conditions to which it must conform. There is no standard solution for which simple handbook answers can be assessed. Hydraulic engineering itself relies on fundamental knowledge that must be applied to meet the special conditions of each project.

The shape and dimensions of hydraulic systems may vary from a small flow meter a few centimeters in size to a levee several hundred kilometers long. Generally, however, hydraulic structures are relatively massive as compared with the productions of other engineering branches. For this reason, the design of large hydraulic systems is usually confined to the topography of the land. It is not always possible to select the most desirable location or material for a specific system. Commonly, a hydraulic system is designed to suit the local conditions including topographic, geological, ecological, and social considerations, and the availability of native materials.

The vital importance of water to human life is justified by the fact that hydraulic engineering is as old as civilization itself. There is much evidence that hydraulic systems of considerable magnitude existed several thousand years ago. For example, a large-scale drainage and irrigation system built in Egypt can be dated back to 3200 BC. Rather complex water supply systems, including several hundred kilometers of aqueducts, were constructed to bring water to ancient Rome. Dujonyen, a massive irrigation system in Siechuan, China, build nearly 2500 years ago, is still in effective use today. The abundant knowledge developed from these experiences has been found to be indispensable.

In addition to the analytical approach, modern hydraulic system design and operation depend heavily on many empirical formulas that produce excellent results in water works. No better replacement for these formulas can be expected in the near future. Unfortunately, most of these empirical formulas cannot be analyzed by modern methods in fluid mechanics. In general, they are dimensionally not homogenous. For this reason, the conversion of units from the British system to S.I. units and vice versa is more than just a matter of convenience. Sometimes for practical purposes the rigorous form (for example, the Parshall formulas for water flow measurements) must be maintained with its original units. In these cases, all quantities should be converted to the original units specified by the formula for computations.

For the most part, this book uses S.I. units. However, examples are provided when it is necessary to demonstrate the use of an empirical formula that is derived for the British units and the corresponding S.I. version is not commonly available. A detailed table of conversions is also provided on the inside front cover.

FUNDAMENTAL PROPERTIES OF WATER

The word hydraulic comes from the Greek word *hydraulikos,* meaning water. Hydraulic systems are designed to deal with water at rest and in motion. The *fundamentals in hydraulic engineering systems,* therefore, involve the application of engineering principles and methods to the planning, control, transportation, conservation, and utilization of water.

It is important that the reader understand the physical properties of water in order to properly solve the various problems in hydraulic engineering systems. For example, the density, the surface tension, and the viscosity of water all vary, in one way or another, with water temperature. Density is a fundamental property that directly relates to the operation of a large reservoir. Change of density with temperature causes lake water to stratify in summer with warmer water on the surface. During fall, the surface water temperature drops rapidly and sinks toward the lake bottom. The warmer water near the bottom rises to the surface resulting in fall overturn of the lake. In winter, when water temperature falls below 4°C (highest water density), lake surface freezes while warm water remains at the bottom. The winter stratification is followed by spring overturn of the lake. Similarly, variation of surface tension directly affects the evaporation loss from a large water body in storage; the variation of water viscosity with temperature is important to all problems involving water in motion.

This chapter discusses the fundamental physical properties of water that are important to problems in hydraulic engineering systems.

1.1 THE EARTH'S ATMOSPHERE AND ATMOSPHERIC PRESSURE

The earth's atmosphere is a thick layer (approximately 1500 km) of mixed gases. Nitrogen makes up approximately 78% of the atmosphere, oxygen makes up approximately 21%, and the remaining 1% consists mainly of water vapor, argon, and trace amounts of other gases. Each of the gases possesses a certain mass and consequently has a weight. The total weight of the atmospheric column exerts a pressure on every surface that it comes in contact with. At sea level, under normal conditions, the *atmospheric pressure* is approximately equal to $1.014 \cdot 10^5$ N/m^2, or 1 bar.* The pressure unit 1 N/m^2 is also known as 1 Pascal, named after the French mathematician Blaise Pascal (1623–1662).

Water surfaces that come in contact with the atmosphere are subjected to the atmospheric pressure. In the atmosphere, each gas exerts a partial pressure independent of the other gases. The partial pressure exerted by the water vapor in the atmosphere is called the *vapor pressure*.

1.2 THE THREE PHASES OF WATER

The water molecule is a stable chemical bond of oxygen and hydrogen atoms. The amount of energy holding the molecules together depends on the temperature and pressure present. Depending on its energy content, water may appear in either solid, liquid, or gaseous forms. Snow and ice are the solid forms of water; liquid is its most commonly recognized form; and moisture, water vapor in air, is water in its gaseous form. The three different forms of water are called its three *phases*.

To change water from one phase to another phase, energy must either be added or taken away from the water. The amount of energy required to change water from one phase to another is known as *latent energy*. This amount of energy may be in the form of heat or pressure. Heat energy is usually expressed in the unit of calories (cal). One cal is the energy required to increase the temperature of 1 gram of water, in liquid phase, 1°C. The amount of energy required to raise the temperature of a substance by 1°C is known as the *specific*

$$
\begin{aligned}
\text{* 1 atmospheric pressure} &= 1.014 \cdot 10^5 \text{ N/m}^2 = 1.014 \cdot 10^5 \text{ Pascals} \\
&= 1.014 \text{ bars} = 14.7 \text{ lbs/in.}^2 \\
&= 760 \text{ mmHg} = 10.33 \text{ m } H_2O
\end{aligned}
$$

heat of that substance. The latent heat and specific heat of all three phases of water are discussed next.

Under standard atmospheric pressure, the specific heat of water and ice are, respectively, 1.0 and 0.465 cal/g/°C. For water vapor, the specific heat under constant pressure is 0.432 cal/g/°C, and at constant volume it is 0.322 cal/gr/°C. These values may vary slightly depending on the purity of the water. To melt 1 gram of ice, changing water from its solid to liquid phase, requires the latent heat of 79.71 cal. To freeze water, the same amount of heat energy must be taken out of each gram of water, thus, the process is reversed. *Evaporation,* the changing of liquid-phase water into its gaseous phase, requires a latent heat of 597 cal/g.

Evaporation is a rather complex process. Under standard atmospheric pressure water boils at 100°C. At higher elevations, where the atmospheric pressure is less, water boils at temperatures lower than 100°C. This phenomenon may be explained best from a molecular-exchange viewpoint.

At the gas-liquid interface there is a continual interchange of molecules leaving the liquid to the gas and molecules entering the liquid from the gas atmosphere. Net evaporation occurs when more molecules are leaving than are entering the liquid; net *condensation* occurs when more molecules are entering than are leaving the liquid.

The continuous impingement of vapor molecules on the liquid surface creates pressure on the liquid surface known as the *vapor pressure*. This partial pressure combined with the partial pressures created by other gases in the atmosphere makes up the total atmospheric pressure.

As the temperature of the liquid increases the molecular energy is raised, which causes a large number of molecules to leave the liquid, which, in turn, increases the vapor pressure. Finally, the temperature reaches a point where the vapor pressure is equal to the ambient atmospheric pressure, evaporation increases drastically, boiling in liquid takes place. The temperature at which a liquid boils is commonly known as the *boiling point* of the liquid. For water at sea level, the boiling point is 100°C. The vapor pressure of water is shown in Table 1.1 on page 6.

In a closed system, for example, in pipelines or pumps, water vaporizes rapidly in regions where the pressure drops below the vapor pressure. This phenomenon is known as *cavitation*. The vapor bubbles formed in cavitation usually collapse in a violent manner, which may cause considerable damage to a system. Cavitation in a closed hydraulic system can be avoided by maintaining the pressure above the vapor pressure everywhere in the system.

PROBLEMS

1.2.1 How much heat energy must be removed from 500 ℓ of water at 80°C in order to freeze it?

1.2.2 A shallow pan contains 1200 g of water at 45°C. Compute the heat energy (in calories) required to evaporate this body of water under ambient pressure of 0.9 bar.

1.2.3 In a thermal container, 5 kg of ice at −4°C are mixed with 10 kg of water at 45°C. Determine whether all the ice will be melted or not and what the resultant temperature is.

1.2.4 Heat is added to a pan containing 3 kg of water at a rate of 500 cal/sec. If the initial temperature of the water is 22°C, determine the time it will take for one half of its mass to evaporate at standard atmospheric pressure.

1.2.5 At 0°C and at an absolute pressure of 911 N/m^2, 100 g of water, 100 g of vapor, and 100 g of ice are in equilibrium in a sealed container. Determine how much energy should be removed to freeze all water and vapor.

TABLE 1.1 Vapor Pressure of Water

Temperature	*Vapor Pressure*		*Temperature*	*Vapor Pressure*	
°C	*atm*	*N/m²*	*°C*	*atm*	*N/m²*
−5	0.004162	421	55	0.15531	15745
0	0.006027	611	60	0.19656	19924
5	0.008600	873	65	0.24679	25015
10	0.012102	1266	70	0.30752	31166
15	0.016804	1707	75	0.38043	38563
20	0.023042	2335	80	0.46740	47372
25	0.031222	3169	85	0.57047	57820
30	0.041831	4238	90	0.69192	70132
35	0.055446	5621	95	0.83421	84552
40	0.072747	7377	100	1.00000	101357
45	0.094526	9584	105	1.19220	120839
50	0.12170	12331	110	1.41390	143314

1.3 MASS (DENSITY) AND WEIGHT (SPECIFIC WEIGHT)

In the S.I. system, the mass of a substance is measured in the unit of gram (g) or kilogram (kg). The *density* of a substance is defined as the mass per unit volume. It is a property inherent to the molecular structure of the substance. This means that density not only depends on the size and weight of the molecules but also on the mechanisms by which the molecules are bonded together. The latter usually varies as a function of temperature and pressure. Because of its peculiar molecular structure, water is one of the few substances that expands when it freezes. The expansion of freezing water when contained in a closed container causes stresses on the container walls. These stresses are responsible for the bursting of frozen water pipes, chuck holes in pavement, and for the weathering of rocks in nature.

Water reaches a maximum density at 4°C. It becomes less dense when further chilled or heated. The density of water is shown as a function of temperature in Table 1.2. Note that the density of ice is different from that of liquid

TABLE 1.2 Density and Specific Weight of Water

Temperature (°C)	*Density* (ρ, kg/m^3)	*Specific Weight* (γ, N/m^3)
0° (ice)	917	8996
0° (water)	999	9800
4°	1000	9810
10°	999	9800
20°	998	9790
30°	996	9771
40°	992	9732
50°	988	9692
60°	983	9643
70°	978	9594
80°	972	9535
90°	965	9467
100°	958	9398

water at the same temperature. We observe this phenomenon when we see ice float on water.

Sea water contains salt, which increases the density of the sea water to about 4% more than that of fresh water.

In the S.I. system the weight of an object is defined by the product of its mass (m, in grams, kilograms, etc.), and the gravitational acceleration (g = 9.81 m/sec^2 on earth). The relation may be written as

$$W = m \cdot g \tag{1.1}^*$$

Weight in the S.I. system is usually expressed in the force units of *newton* (N). One newton is defined as the force required to accelerate 1 kg of mass at a rate of 1 m/sec^2. The specific weight (weight per unit volume) of water, γ, is thus the product of the density, ρ, and the gravitational acceleration, g. The specific weight of water is also shown as a function of temperature in Table 1.2.

The ratio of the specific weight of any liquid to that of water at 4°C is called the *specific gravity* of that liquid.

Example 1.1

A cylindrical water tank (Figure 1.1) is suspended vertically on the sides. The tank has a 2-m diameter and is filled with 40°C water to 1 m in height. Determine the force exerted on the tank bottom.

* In the British (Imperial) system the mass of an object is defined by its weight (ounce, pound, etc.) and the gravitational acceleration (g = 32.2 ft/sec^2 on earth). The relation may be written as

$$m = W/g \tag{1.1a}$$

Mass in the British system is expressed in the units of *slug*. One slug is defined as the mass of an object which requires 1 lb of force to reach an acceleration of 1 ft/sec^2.

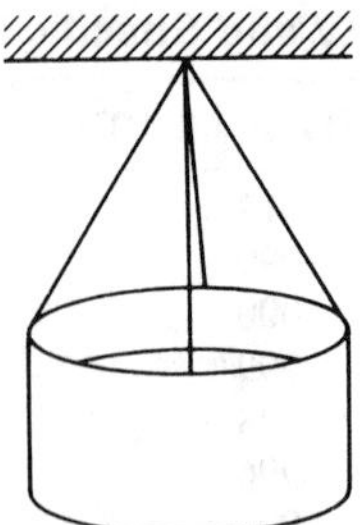

Figure 1.1

Solution

The force exerted on the tank bottom is equal to the weight of the water body.

$$F = W = m \cdot g = [\rho \cdot (\text{Volume})] \cdot (g)$$
$$= \left[992 \text{ kg/m}^3 \cdot \left(\frac{\pi \cdot 2^2}{4} \times 1 \text{ m}^3\right)\right](9.81 \text{ m/sec}^2)$$
$$= 30{,}572 \text{ kg-m/s}^2 = 30{,}572 \text{ N}$$

PROBLEMS

1.3.1 A container has a 5-m^3 volume capacity and weighs 1500 N when empty and 47,000 N when filled with a liquid. What is the mass density of the liquid?

1.3.2 A liquid has specific weight of 8000 N/m^3. What is the density of the liquid?

1.3.3 A steel tank 120 m^3 in volume is filled with water at atmospheric pressure. Find the weight of the water and its specific gravity if the temperature is 60°C.

1.3.4 A rocket carrying a tank of water weighing 9000 N lands on the moon where the gravitational acceleration is one sixth of that of the earth. Find the mass and the moon-weight of the water.

1.3.5 Determine the weight of a gallon of liquid if it has a mass of 0.258 slugs.

1.3.6 Determine the specific weight, the density, and the specific gravity of a liquid that occupies a volume of 200 ft^3 and weighing 11,023 lbs. Provide results in both the British system and S.I. units.

1.3.7 The unit of energy in the British system is foot-pound. Convert one unit of energy in this system to the S.I. system.

1.3.8 The unit of mass in the S.I. system is gram. Convert one unit of mass in this system to the British system in slugs.

1.4 VISCOSITY OF WATER

Water responds to shear stress by continuously yielding an angular deformation in the direction of the shear. The rate of angular deformation is proportional to the shear stress, as shown in Figure 1.2. The schematic diagram shown in Figure 1.2 represents the physical basis of viscosity. Consider that water fills the space between two parallel plates at a distance y apart. A horizontal force T is applied to the upper plate and moves it to the right at velocity V, while the lower plate remains stationary. The shear force T is applied to overcome the water resistance R, and it must be equal to R since there is no acceleration involved in the process. The resistance per unit area of the upper plate (shear stress, $\tau = R/A = T/A$) has been found to be proportional to the rate of angular deformation in the fluid, $d\theta/dt$. The relationship may be expressed as

$$\tau \propto \frac{d\theta}{dt} = \frac{\frac{dx}{dy}}{dt} = \frac{\frac{dx}{dt}}{dy} = \frac{dv}{dy}$$

where $v = \dfrac{dx}{dt}$ is the velocity of the fluid element, or

$$\tau = \mu \frac{dv}{dy} \tag{1.2}$$

The proportionality constant, μ, is called the *absolute viscosity* of the fluid.

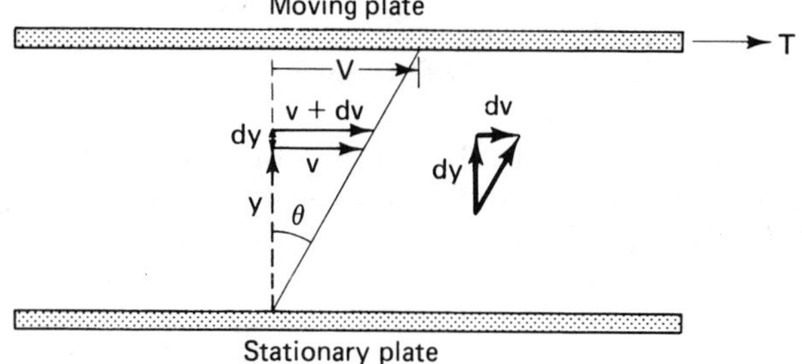

Figure 1.2 Shear stress in fluid.

The absolute viscosity has the dimension of force per unit area (stress) times the time interval considered. It is usually measured in the unit of poise (named after the French engineer-physiologist J. L. M. Poiseuille). The absolute viscosity of water at room temperature (20.2°C) is equal to 1 centipoise (cP), which is one hundredth of a poise.

$$1 \text{ poise} = 0.1 \text{ N} \cdot \text{sec/m}^2 = 100 \text{ centipoise}$$

The absolute viscosity of air is approximately 0.17 centipoise.

In engineering practice it is often convenient to introduce the term kinematic viscosity, ν, which is obtained by dividing the absolute viscosity by the mass density of the fluid at the same temperature; $\nu = \mu/\rho$. The kinematic viscosity carries the unit of cm^2/sec (with the unit of Stoke, named after the

British mathematician G. G. Stoke). The absolute viscosities and the kinematic viscosities of pure water and air are shown as functions of temperature in Table 1.3.

TABLE 1.3 Viscosities of Water and Air

	Water		Air	
Temperature °C	Viscosity, μ $N \cdot sec/m^2$	Kinematic Viscosity, ν m^2/sec	Viscosity, μ $N \cdot sec/m^2$	Kinematic Viscosity, ν m^2/sec
0	1.781×10^{-3}	1.785×10^{-6}	1.717×10^{-5}	1.329×10^{-5}
5	1.518×10^{-3}	1.519×10^{-6}	1.741×10^{-5}	1.371×10^{-5}
10	1.307×10^{-3}	1.306×10^{-6}	1.767×10^{-5}	1.417×10^{-5}
15	1.139×10^{-3}	1.139×10^{-6}	1.793×10^{-5}	1.463×10^{-5}
20	1.002×10^{-3}	1.003×10^{-6}	1.817×10^{-5}	1.509×10^{-5}
25	0.890×10^{-3}	0.893×10^{-6}	1.840×10^{-5}	1.555×10^{-5}
30	0.798×10^{-3}	0.800×10^{-6}	1.864×10^{-5}	1.601×10^{-5}
40	0.653×10^{-3}	0.658×10^{-6}	1.910×10^{-5}	1.695×10^{-5}
50	0.547×10^{-3}	0.553×10^{-6}	1.954×10^{-5}	1.794×10^{-5}
60	0.466×10^{-3}	0.474×10^{-6}	2.001×10^{-5}	1.886×10^{-5}
70	0.404×10^{-3}	0.413×10^{-6}	2.044×10^{-5}	1.986×10^{-5}
80	0.354×10^{-3}	0.364×10^{-6}	2.088×10^{-5}	2.087×10^{-5}
90	0.315×10^{-3}	0.326×10^{-6}	2.131×10^{-5}	2.193×10^{-5}
100	0.282×10^{-3}	0.294×10^{-6}	2.174×10^{-5}	2.302×10^{-5}

PROBLEMS

1.4.1 A flat plate of 100 cm^2 area is being pulled over a fixed flat surface at a constant velocity of 75 cm/sec. An oil film of unknown viscosity separates the plate and the fixed surface by a distance of 0.1 cm. If the force required to pull the plate is measured to be 51.7 N and if the viscosity of the fluid is constant, determine the viscosity.

1.4.2 A liquid flows with velocity distribution $v = y^2 - 2y$, where v is given in ft/sec and y in inches. Calculate the shear stresses at $y = 0, 1, 2, 3, 4$ if the viscosity is 380 centipoise (cP).

1.4.3 The 25-cm diameter ram of a hydraulic lift slides in a 25.015-cm diameter cylinder. The viscosity of the oil filling the gap is $0.04\ N \cdot sec/m^2$. If the speed of the ram is 15 cm/sec, determine the frictional resistance when 3 m of the ram is engaged in the cylinder. Assume concentric motion.

1.4.4 A flat circular disk of radius 1.0 m is rotated at an angular velocity of 0.65 rad/sec over a fixed flat surface. An oil film separates the disk and the surface. If the viscosity of the oil is 16 times that of water at 20°C and if the space between the disk and the fixed surface is 0.5 mm, determine the torque required to rotate the disk.

1.4.5 A rotating-cylinder viscometer consists of two concentric cylinders with a uniform gap between them. The gap of a known width is filled with the liquid to be measured. If the inner cylinder rotates at 2000 rpm and the outer cylinder remains stationary and measures a torque of 2×10^5 dyn-cm, determine the absolute viscosity of the liquid. The inner cylinder diameter is 5.0 cm, the gap width is 0.02 cm, and the liquid fills up to a height of 4.0 cm in the cylindrical gap.

1.4.6 Compare the ratios of the absolute and the kinematic viscosities of air and water at (a) 20°C, and (b) 80°C. Discuss the differences.

1.4.7 In the British system the kinematic viscosity is measured in units of ft^2/sec. Convert one unit in this system to the S.I. units in stokes.

1.4.8 Establish the equivalence between
a. Absolute viscosity units of poise and $lb\text{-}sec/ft^2$.
b. Kinematic viscosity units between stoke and the British system unit.

1.4.9 The absolute viscosity of mercury is 1.7 cP and its specific gravity is 13.6 at standard atmospheric pressure and temperature (see *Common Constants and Conversions* on the inside cover). Find the kinematic viscosity in both the British system and S.I. units.

1.5 SURFACE TENSION AND CAPILLARITY

Even at a small distance below the surface of a liquid body, liquid molecules are attracted to each other by equal forces in all directions. The molecules on the surface, however, are subjected to an inward attraction that is not balanced by the outward attraction. This causes the liquid surface to seek a minimum possible area by exerting surface tension tangent to the surface over the entire surface area. A steel needle floating on the water surface, the spherical shape of dew-drops, and the rise or fall of liquid surface in capillary tubes are the results of surface tension.

Most liquids adhere to solid surfaces. The adhesive force varies depending on the nature of the liquid and of the solid surface. If the adhesive force between the liquid and the solid surface is greater than the cohesion in the liquid molecules, the liquid tends to spread over and wet the surface, as shown in Figure 1.3(a). If the cohesion is greater, a small drop forms, as shown in Figure 1.3(b). Water wets the surface of glass but mercury does not. If we place a smallbore vertical glass tube into the free surface of water, the water surface in the tube rises. The same experiment performed with mercury will show that the mercury falls. These two typical cases are schematically presented in Figure 1.4(a) and

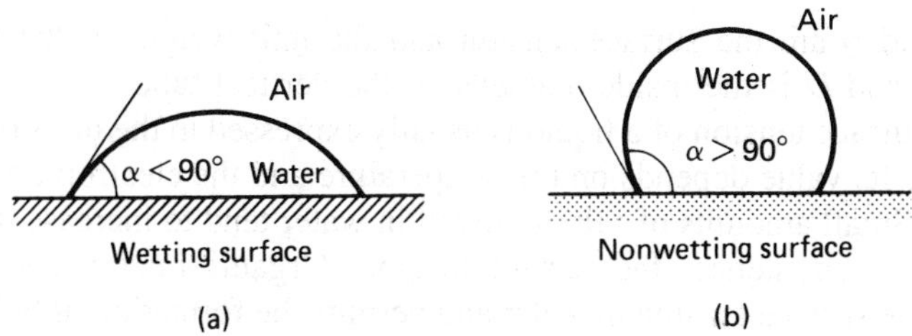

Figure 1.3 Wetting and nonwetting surfaces.

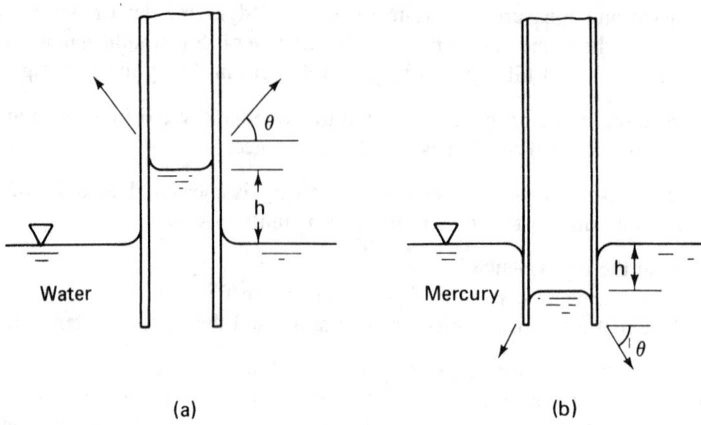

Figure 1.4 Capillary actions.

1.4(b). The phenomenon is commonly known as *capillary action*. The magnitude of the capillary rise (or depression), h, is determined by the balance of adhesive force between the liquid and solid surface and the weight of the liquid column above (or below) the liquid free surface.

The angle θ at which the liquid film meets the glass depends on the nature of the liquid and the solid surface. The upward (or downward) motion in the tube will cease when the vertical component of the surface tension force around the edge of the film equals the weight of the raised (or lowered) liquid column. When the very small volume of liquid above (or below) the base of the curved meniscus is neglected, the relationship may be expressed as

$$(\sigma \pi D) \sin \theta = \frac{\pi D^2}{4}(\gamma h)$$

Hence,

$$h = \frac{4\sigma \sin \theta}{\gamma D} \tag{1.3}$$

where σ and γ are the surface tension and the unit weight of the liquid, respectively, and D is the inside diameter of the vertical tube.

The surface tension of a liquid is usually expressed in the units of force per unit length. Its value depends on the temperature and the electrolytic content of the liquid. Small amounts of salt dissolved in water tend to increase the electrolytic content and, hence, the surface tension. Organic matter (such as soap) decreases the surface tension in water and permits the formation of bubbles. The surface tension of pure water is listed in Table 1.4.

TABLE 1.4 Surface Tension of Water

Temperature (°C)	*0°*	*10°*	*20°*	*30°*	*40°*
$\sigma(\times 10^{-2}$N/m)	7.416	7.279	7.132	6.975	6.818
σ(dyn/cm)	74.16	72.79	71.32	69.75	68.18
Temperature (°C)	*50°*	*60°*	*70°*	*80°*	*90°*
$\sigma(\times 10^{-2}$N/m)	6.786	6.611	6.436	6.260	6.071
σ(dyn/cm)	67.86	66.11	64.36	62.60	60.71

PROBLEMS

1.5.1 A small amount of solvent is added to the ground water to change its electrolytic content. As a result, the contacting angle, θ, representing the adhesion between water and soil material, is increased from 30° to 42° while the surface tension decreases by 10% (Figure P1.5.1). If the soil has a uniform pore size of 0.7 mm, determine the change of capillary rise in the soil.

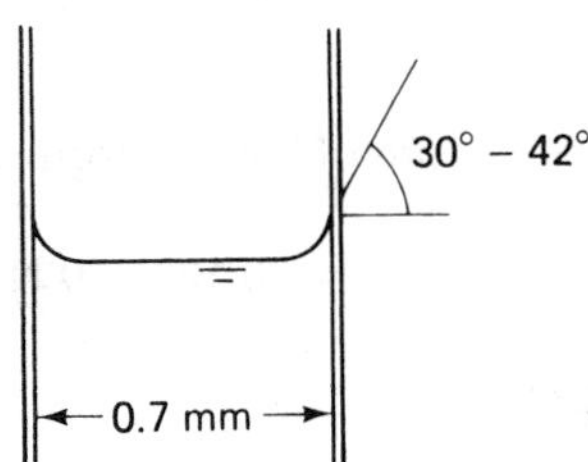

Figure P1.5.1

1.5.2 A liquid at 20°C is observed to rise to a height of 18 mm in a 0.5-mm glass tube. The contacting angle is measured to be 42.8°. Determine the surface tension of the liquid if its density is 998 kg/m^3.

1.5.3 For water in a clean glass tube, $\theta = 90°$. In an experimental measurement, water is expected to rise due to capillary action. Determine the height of the column of water if the glass tube being used is 0.2 cm in diameter and the water temperature is 20°C.

1.5.4 If mercury is used in a glass tube ($\theta = 50°$; sp.gr. $= 13.6$, $\sigma = 0.57$ N/m) to measure pressure, determine the minimum diameter of the tube so that the measurement error is less than 0.5 mm.

1.6 ELASTICITY OF WATER

Under ordinary conditions, water is commonly assumed to be incompressible. In reality, it is about 100 times as compressible as steel. It is necessary to consider the compressibility of water when water hammer problems (Chapter 4) are involved. The *compressibility of water* is inversely proportional to its *volume modulus of elasticity*, E_b, also known as the *bulk modulus of elasticity*. The pressure–volume relationship may be expressed as

$$\Delta P = -E_b\left(\frac{\Delta \text{Vol}}{\text{Vol}}\right) \tag{1.4}$$

where Vol is the initial volume. ΔP and ΔVol are the corresponding changes in pressure and volume, respectively. The negative sign means that a positive change in pressure (i.e., increment) will cause the volume to decrease (i.e., negative change). The bulk modulus of elasticity of water varies both with temperature and pressure. In the range of practical applications in typical hydraulic systems, a value of 300,000 psi (2.2×10^9 N/m^2) can be used.

Example 1.2

The specific gravity of sea water is 1.026 at sea level. Determine the density of sea water on the ocean floor 2000 m deep, where pressure is 20,231,400 N/m^2.

Solution

The change of pressure at the depth of 2000 m from that at the water surface is

$$\Delta P = P - P_{atm} = 20{,}130{,}000 \text{ N/m}^2$$

From Equation (1.4) we have

$$\Delta P = -E_b\left(\frac{\Delta \text{Vol}}{\text{Vol}_o}\right)$$

So that

$$\frac{\Delta \text{Vol}}{\text{Vol}_o} = \frac{-20{,}130{,}000}{2.2 \times 10^9} = -0.0092$$

Since

$$\rho = \frac{\text{mass}}{\text{Vol}} \quad \therefore \text{Vol} = \frac{\text{mass}}{\rho}$$

Then

$$\Delta \text{Vol} = \frac{m}{\rho} - \frac{m}{\rho_o} \quad \therefore \frac{\Delta \text{Vol}}{\text{Vol}_o} = \frac{\rho_o}{\rho} - 1$$

So that

$$\rho = \frac{\rho_0}{\left(1 + \frac{\Delta \text{Vol}}{\text{Vol}_0}\right)} = \frac{1.026}{1 - 0.0092} = 1.036$$

PROBLEMS

1.6.1 Determine the change in density of water at 20°C when pressure is suddenly changed from 25 bar to 4,500,000 dyn/cm^2.

1.6.2 The volume of a liquid decreased by 0.8% due to the increase of pressure from 5000 N/cm^2 to 10,000 N/cm^2. What is the bulk modulus of elasticity of the liquid?

1.6.3 For the tank in Problem 1.3.3 compare the weight of the same volume of water at pressure of 10 bar and at pressure of 100 bar.

1.6.4 The pressure in a 150-cm diameter pipe, 2000 m long is 30 N/cm^2. Determine the amount of water that will enter the pipe if the pressure increases to 30 bar. Assume the pipe is rigid and does not increase its volume.

1.7 FORCES IN A FLUID FIELD

Various types of forces may be exerted on a body of water at rest or in motion. In hydraulic practice, these forces usually include gravitational force, inertia force, elastic force, friction force, pressure force, and surface tension force.

These forces may be classified into three basic categories according to their physical characteristics:

1. the body forces,
2. the surface forces,
3. the line forces.

Body forces are forces that act on all particles in a body of water as a result of some external body or effect, but not due to direct contact. An example of this is the gravitational force. It acts on all particles in any body of water, as a result of the gravitational field of the earth, which may not be in direct contact with the particular water body of interest. Other body forces that are common in hydraulic practice include the inertia force and the force due to elastic effect. The body forces are usually expressed as force per unit mass (N/kg), or force per unit volume (N/m^3).

Surface forces act on the surface of the water body by directly contacting the surface. These forces may be either external or internal. The pressure force and the friction force are examples of external surface forces. The viscous force inside a fluid body may be viewed as an internal surface force. The surface forces are expressed as force per unit area (N/m^2).

The surface tension is thought of as the force in the liquid surface normal to a line drawn in the surface. Thus, it may be considered as a *line force* that has dimensions of force per unit length (N/m).

2

WATER PRESSURE AND PRESSURE FORCES

2.1 FREE SURFACE OF WATER

When water fills a containing vessel, it automatically seeks a horizontal surface upon which the pressure is constant everywhere. In practice, the free surface of water in a vessel is the surface that is not in contact with the cover of the vessel. Such a surface may be subjected to the atmospheric pressure (open vessel) or any other pressure that is exerted in the vessel (closed vessel).

2.2 ABSOLUTE AND GAGE PRESSURES

Water surface in contact with the earth's atmosphere is subjected to the atmospheric pressure, which is approximately equal to a 10.33-m-high column of water at sea level. In still water, any element located below the water surface is subjected to a pressure greater than the atmospheric pressure.

To determine the variation of pressure between any two points in water, we may consider two arbitrary points A and B along an arbitrary x-axis as shown in Figure 2.1. Consider that these points lie in the ends of a small prism of water having a cross-sectional area dA and a length L. P_A and P_B are the pressures at each end, where the cross-sectional areas are normal to the x-axis. Since the prism is at rest, all forces acting upon it must be in equilibrium in all directions. For the force components in the x-direction, we may write

$$F_x = P_A\,dA - P_B\,dA - \gamma L\,dA\sin\theta = 0$$

Here $L\sin\theta = h$ is the vertical elevation difference between the two points. The above equation reduces to

$$P_A - P_B = \gamma h \tag{2.1}$$

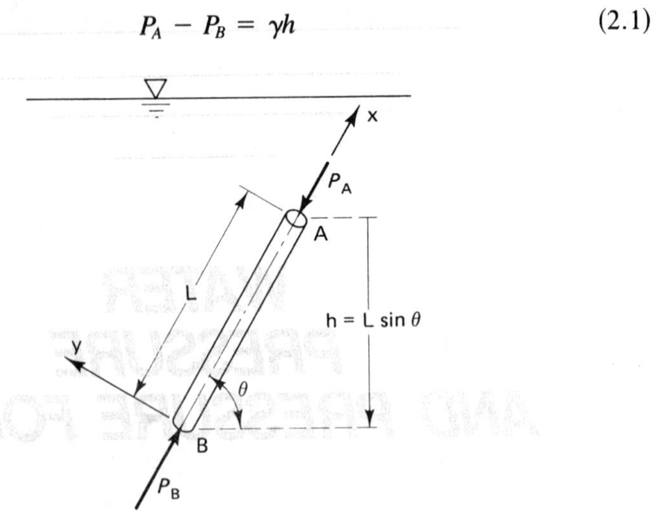

Figure 2.1 Hydrostatic pressure on a prism.

Therefore, the *difference in pressure between any two points in still water is always equal to the product of the specific weight of water and the difference in elevation between the two points.*

If the two points are on the same elevation, $h = 0$ and $P_A = P_B$. In other words, *for water at rest, the pressure at all points in a horizontal plane is the same.* If the water body has a free surface that is exposed to *atmospheric pressure*, P_{atm}, we may position point B on the free surface and write

$$(P_A)_{abs} = \gamma h + P_B = \gamma h + P_{atm} \tag{2.2}$$

This pressure, $(P_A)_{abs}$, is commonly referred to as the *absolute pressure*.

Pressure gages are usually designed to measure pressures above or below the atmospheric pressure. Pressure so measured, using atmospheric pressure as a base, is called *gage pressure*, P. Absolute pressure is always equal to gage pressure plus atmospheric pressure:

$$P = P_{abs} - P_{atm} \tag{2.3}$$

Figure 2.2 diagramatically shows the relationship between the absolute and gage pressures and two typical pressure gage dials. Comparing Equations (2.2) and (2.3), we have

$$P = \gamma h \tag{2.4}$$

or

$$h = \frac{P}{\gamma} \tag{2.5}$$

Here the pressure is expressed in terms of the height of a water column h. In hydraulics it is known as the *pressure head.*

Equation (2.1) may thus be rewritten as

$$\frac{P_A}{\gamma} - \frac{P_B}{\gamma} = \Delta h \tag{2.6}$$

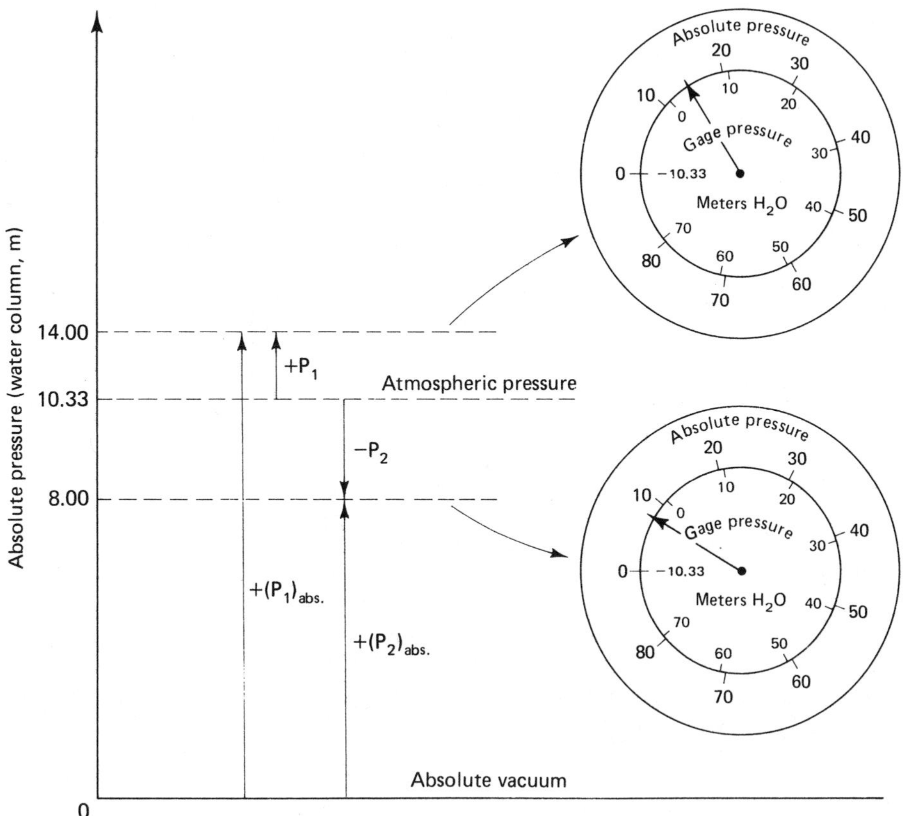

Figure 2.2 Absolute and gage pressure.

meaning that the difference in pressure heads at two points in water at rest is always equal to the difference in elevation between the two points. From this relationship we can also see that any change in pressure at point B would cause an equal change at point A, for the difference in pressure head between the two points must remain the same value h. In other words, *a pressure applied at any point in a liquid at rest is transmitted equally and undiminished in all directions to every other point in the liquid.* This principle, also known as *Pascal's law*, has been made use of in the hydraulic jacks that lift heavy weights by applying relatively small forces.

Example 2.1

The diameters of cylindrical pistons A and B are 3 cm and 20 cm, respectively. The faces of the pistons are at the same elevation and the intervening passages are filled with an incompressible hydraulic oil. A force P of 100 N is applied at the end of the lever, as shown in Figure 2.3. What weight W can the hydraulic jack support?

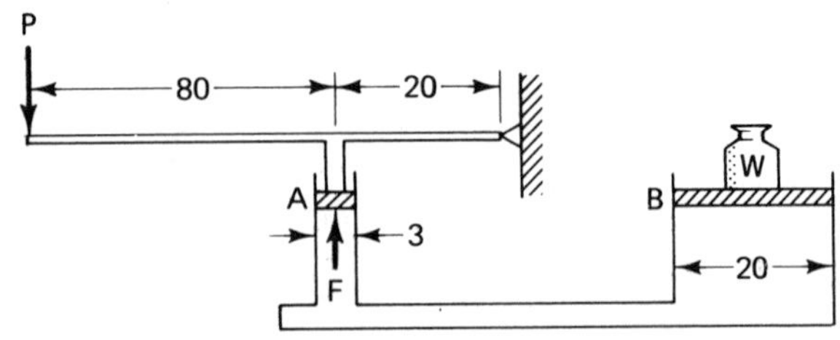

Measurements in cm

Figure 2.3 Hydraulic jack.

Solution

Balancing the moments produced by P and F, we obtain for the equilibrium condition,

$$P \cdot (80 + 20) = F \cdot (20)$$

$$100 \cdot (100) = F \cdot (20)$$

Thus

$$F = 500 \text{ N}$$

From Pascal's law, the pressure P_A applied at A should be the same as that of P_B applied at B. We may write

$$P_A = \frac{F}{\frac{1}{4}(\pi \cdot 3^2)}; \qquad P_B = \frac{W}{\frac{1}{4}(\pi \cdot 20^2)}$$

$$\frac{500}{\frac{1}{4}(\pi \cdot 3^2)} = \frac{W}{\frac{1}{4}(\pi \cdot 20^2)}$$

$$\therefore W = 500 \times \frac{20^2}{3^2} = 22{,}222 \text{ N}$$

PROBLEMS

2.2.1 The simple barometer in Figure P2.2.1 uses water at 40°C as the liquid indicator. If the liquid column rises to a height of 8.7 m in the vertical tube, compute the atmospheric pressure, neglecting surface tension effect. What is the percentage error due to vapor pressure if direct reading is used?

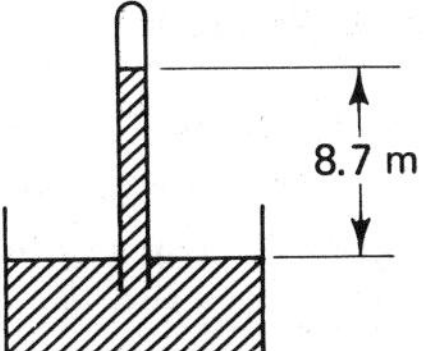

Figure P2.2.1

2.2.2 A 30-ft-high, 1-in-diameter pipe is welded to the top of a cubic container (3′ × 3′ × 3′). Both the pipe and the container are filled with water at 20°C. Determine the weight of the water and the pressure forces on the bottom and sides of the container.

2.2.3 A closed tank contains a liquid (sp. gr. = 0.80) under pressure. The pressure gage, shown in Figure P2.2.3, indicates a pressure of 3.2×10^5 dyn/cm^2. Determine the pressure at the bottom of the tank and the height of the liquid column that will rise in the vertical tube.

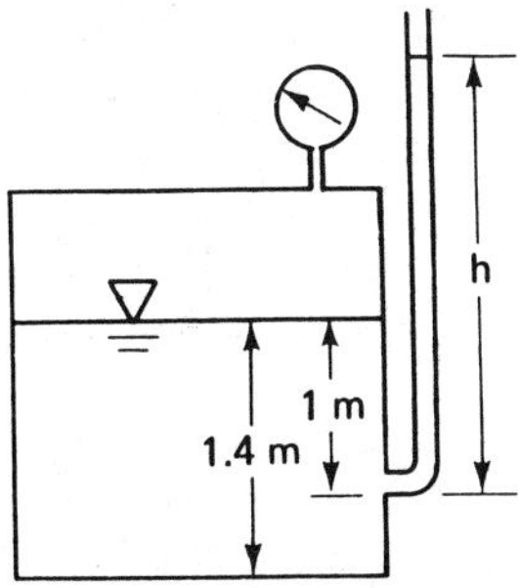

Figure P2.2.3

2.2.4 An underwater storage tank was constructed to store natural gas offshore. Determine the gas pressure in the tank (in bars) when the water elevation in the tank is 6 m below the sea level (see Figure P2.2.4). The specific gravity of sea water is 1.025.

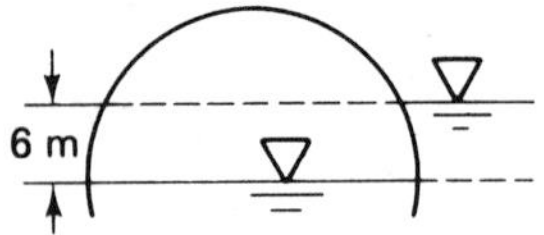

Figure P2.2.4

2.2.5 A certain freshwater fish does not survive well if it gets in water in which the absolute pressure is greater than 5 standard atmospheric pressures. How deep can the fish go before it experiences trouble?

2.2.6 A closed oil tank contains oil of specific gravity 0.85. If the pressure in the air space at the top of the oil surface measures 1.20 standard atmospheric pressures, what is the pressure at a point 5 m below the oil surface?

2.3 SURFACE OF EQUAL PRESSURE

The hydrostatic pressure in a body of water varies with the vertical distance measured from the free surface of the water body. In general, all points on a horizontal surface in the water have the same pressure (Equation 2.4). In Figure 2.4(a), points 1, 2, 3, and 4 have equal pressure, and the horizontal surface that contains these four points is a *surface of equal pressure*. In Figure 2.4(b), points 5 and 6 are on the same horizontal plane. But the pressures at 5 and 6 are not equal, because the water in the two tanks is not connected. Figure 2.4(c) shows the tanks filled with two immiscible liquids of different densities. The horizontal surface (7, 8) that passes through the interphase of the two liquids is an equal pressure surface, as the weight of the liquid columns per unit area above 7 and 8 are equal; the horizontal surface (9, 10) is *not* an equal pressure surface.

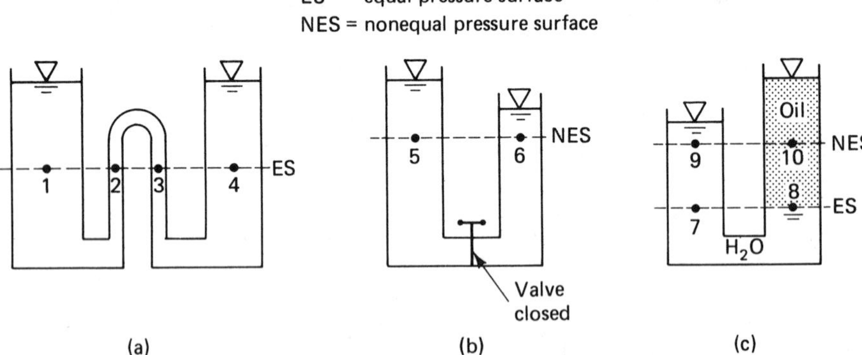

Figure 2.4 Hydraulic pressure in vessels.

The concept of equal pressure surface is a useful method in analyzing the strength or intensity of the hydrostatic pressure in a container, as demonstrated in the following section.

2.4 MANOMETERS

A manometer usually is a tube bent in the form of a U containing a fluid of known specific gravity. The difference in elevations of the liquid surfaces under pressure indicates the difference in pressure at the two ends. Basically, there are two types of manometers:

1. An *open manometer* has one end open to atmospheric pressure and is capable of measuring the gage pressure in a vessel.
2. A *differential manometer* connects each end to a different pressure vessel and is capable of measuring the pressure difference between the two vessels.

The liquid used in a manometer is usually heavier than the fluids to be measured. It must form an unblurred interface, that is, it must not mix with the adjacent liquids. The most frequently used manometer liquids are mercury (sp. gr. 13.6), water (sp. gr. 1.00), alcohol (sp. gr. 0.9), and other commercial manometer oils of various specific gravities (e.g. Meriam* Unit Oil, sp. gr. 1.00; Meriam No. 3 Oil, sp. gr. 2.95; etc).

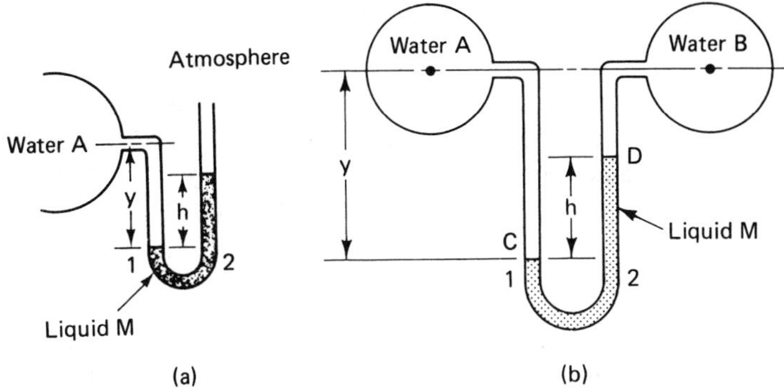

Figure 2.5 Types of manometers: (a) open manometer, (b) differential manometer.

Figure 2.5(a) shows the schematics of a typical open manometer; Figure 2.5(b) shows the schematics of a typical differential manometer. It is obvious that the higher the pressure in vessel A, the larger the difference, h, in the surface elevations in the two legs of the manometer. The accurate calculation of pressure in A, however, must involve the consideration of the densities of the fluids and the geometry involved in the entire measuring system.

* Meriam Instruments, Division of Scott & Fetzer Co., Cleveland, Ohio 44102.

A simple step-by-step procedure is suggested for manometer computation:

Step 1. Make a sketch of the manometer system, similar to that in Figure 2.5, approximately to scale.

Step 2. Draw a horizontal line at the level of the lower surface of the manometer liquid, 1. The pressures at points 1 and 2 must be the same since the system is in static equilibrium.

Step 3. (a) For open manometers, the pressure on 2 is exerted by the weight of the liquid M column above 2; and the pressure on 1 is exerted by the weight of the column of water above 1 plus the pressure in vessel A. The pressures must be equal in value. This relation may be written as follows:

$$\gamma_M \cdot h = \gamma_W \cdot y + P_A, \quad \text{or} \quad P_A = \gamma_M \cdot h - \gamma_W \cdot y$$

(b) For differential manometers, the pressure on 2 is exerted by the weight of the liquid M column above 2, the weight of the water column above D, and the pressure in vessel B, while the pressure on 1 is exerted by the weight of the water column above 1. This relationship may be expressed as:

$$\gamma_M \cdot h + \gamma_W \cdot (y - h) + P_B = \gamma_W \cdot y + P_A$$

or,

$$\Delta P = P_A - P_B = h(\gamma_M - \gamma_W)$$

Either one of these equations can be used to solve for P_A. The same procedure can be applied to any complex geometry, as demonstrated in the following example.

Example 2.2

A mercury manometer is used to measure the pressure difference in vessels A and B. Determine the pressure difference (Figure 2.6) in dyn/cm^2.

Solution

The sketch of the manometer system (Step 1) is shown in Figure 2.6. Points 3 and 4 are on a surface of equal pressure (Step 2) and so are the vessel A and points 1 and 2,

$$P_3 = P_4$$

$$P_A = P_1 = P_2$$

The pressures at points 3 and 4 are, respectively (Step 3),

$$P_3 = P_2 + \gamma_W \cdot (27 \text{ cm}) = P_A + \gamma_W \cdot (27 \text{ cm})$$

$$P_4 = P_B + \gamma_W \cdot (135 \text{ cm}) + \gamma_M \cdot (15 \text{ cm})$$

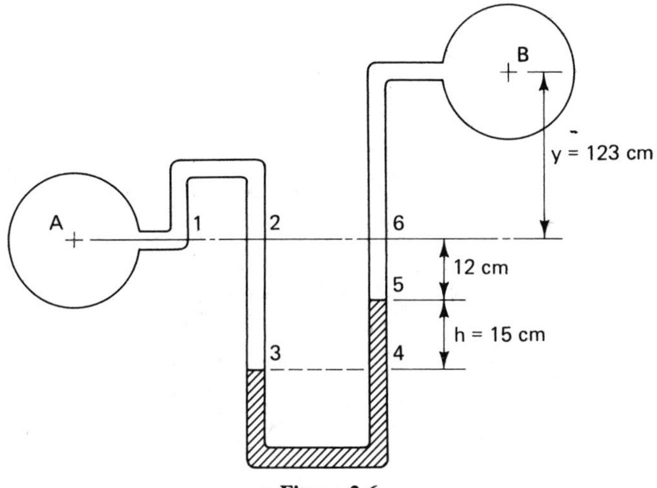

Figure 2.6

so that

$$P_3 = P_A + \gamma_W \cdot (27 \text{ cm}) = P_4 = P_B + \gamma_W \cdot (135 \text{ cm}) + \gamma_M \cdot (15 \text{ cm})$$

and

$$\begin{aligned} \Delta P = P_A - P_B &= \gamma_W \cdot (135 \text{ cm} - 27 \text{ cm}) + \gamma_M \cdot (15 \text{ cm}) \\ &= \gamma_W \cdot [108 + (13.6)(15)] \text{ cm} \\ &= 981 \frac{\text{dyn}}{\text{cm}^3} \cdot 312 \text{ cm} = 306{,}072 \frac{\text{dyn}}{\text{cm}^2} \end{aligned}$$

The open manometer in a U-tube, as described above, requires readings of liquid levels at two points. Since any change in pressure in the vessel causes a drop of liquid surface at one end and a rise in the other, a *single-reading manometer* can be made by introducing a reservoir with a large cross-sectional area, compared to that of the tube, into one leg of the manometer. A typical single-reading manometer is shown in Figure 2.7.

Due to the large area ratio between the reservoir and the tube, a small drop of surface elevation in the reservoir will cause an appreciable rise in liquid column of the other leg. If there is an increase of pressure, ΔP_A will cause the liquid surface in the reservoir to drop by a small amount Δy. Then

$$A \cdot \Delta y = a \cdot h \tag{2.7}$$

where A and a are cross-sectional areas of the reservoir and the tube, respectively.

Applying (Step 2) to points 1 and 2, we may generally write

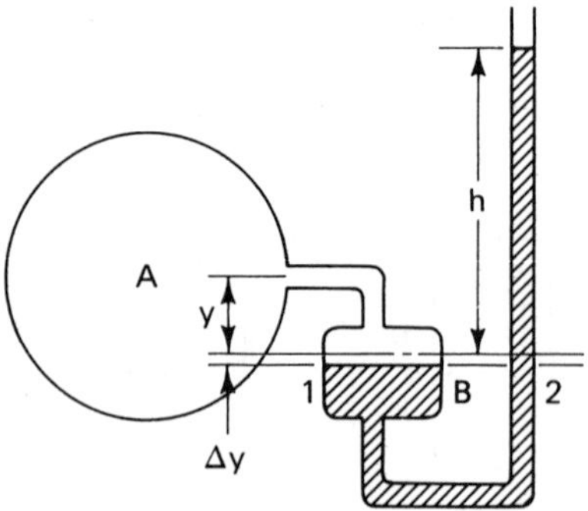

Figure 2.7 Single-reading manometer.

$$\gamma_A \cdot (y + \Delta y) + P_A = \gamma_B \cdot (h + \Delta y) \tag{2.8}$$

Simultaneous solution of Equations (2.7) and (2.8) gives the value of P_A, the pressure in the vessel, in terms of h. All other quantities in Equations (2.7) and (2.8), A, a, y, γ_A, and γ_B, are quantities predetermined in the manometer design. A single reading of h will thus determine the pressure.

Since Δy can be made negligible by introducing a very large A/a ratio, the above relationship may be further simplified to

$$\gamma_A \cdot y + P_A = \gamma_B \cdot h \tag{2.9}$$

The height reading h is the measure of the pressure in the vessel.

The solution of practical hydraulic problems frequently requires the difference in pressure between two points in a pipe or a pipe system. For this purpose, differential manometers are frequently used. A typical differential manometer is shown in Figure 2.8.

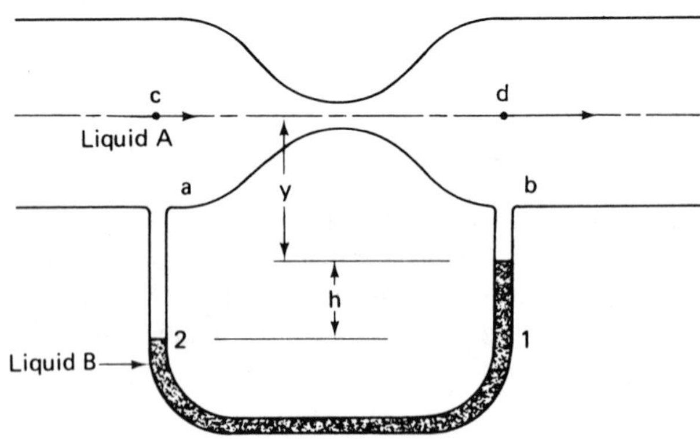

Figure 2.8 A differential manometer installed in a flow-measurement system.

The same computation steps (steps 1, 2, and 3) suggested on pages 24 and 25 can be readily applied here too: When the system is in static equilibrium, the pressure at the same elevation, points 1 and 2, must be equal. We may thus write

$$\gamma_A \cdot (y + h) + P_c = \gamma_B \cdot h + \gamma_A \cdot y + P_d$$

Hence, the pressure difference, ΔP, is expressed as

$$\Delta P = P_c - P_d = (\gamma_B - \gamma_A) \cdot h \tag{2.10}$$

PROBLEMS

2.4.1 The gage pressure on an open-top water tank reads 30 mm of mercury. Determine the water depth above the gage. Find the equivalency in bars and N/m^2 of absolute pressure at 20°C.

2.4.2 An open tank contains a layer of water of height h and a layer of oil of the same height. The oil has a specific gravity of 0.82. If the gage pressure at the bottom of the tank (referred to as ambient atmosphere) indicates 16 mm of mercury, determine the height h.

2.4.3 A U-tube with both ends open to the atmosphere contains mercury in the lower portion of the tube. If water is filled into one leg of the U-tube until the water column is 1 m above the mercury-water meniscus, what is the elevation difference between the mercury surfaces in the two legs?

2.4.4 If oil (sp. gr. 0.79) is filled into the other leg of the above manometer to a height of 60 cm, what is the elevation difference between the mercury surfaces?

2.4.5 A manometer is mounted on a city water supply main pipe to monitor the water pressure in the pipe, as shown in Figure P2.4.5. Determine the water pressure.

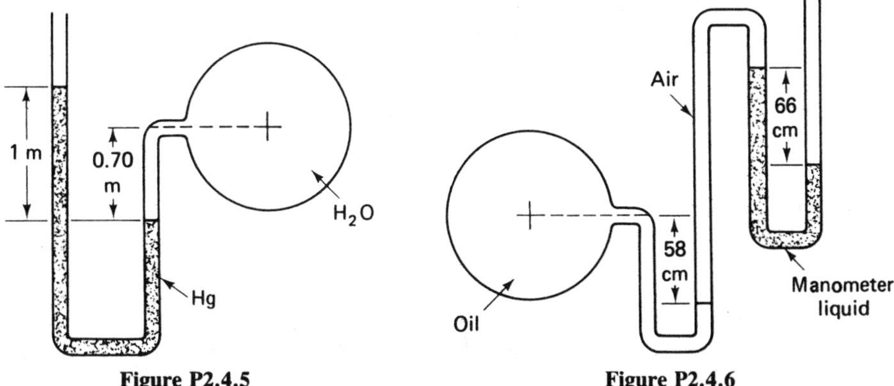

Figure P2.4.5 **Figure P2.4.6**

2.4.6 An open manometer, shown in Figure P2.4.6, is installed to measure the pressure in a pipe carrying an oil (sp. gr. 0.82). If the manometer liquid is carbon tetrachloride (sp. gr. 1.60), determine the pressure in the pipe (in meters of water column height).

2.4.7 In Figure P2.4.7, a single-reading mercury manometer is used to measure water pressure in the pipe. What is the pressure if $h_1 = 20$ cm and $h_2 = 67$ cm?

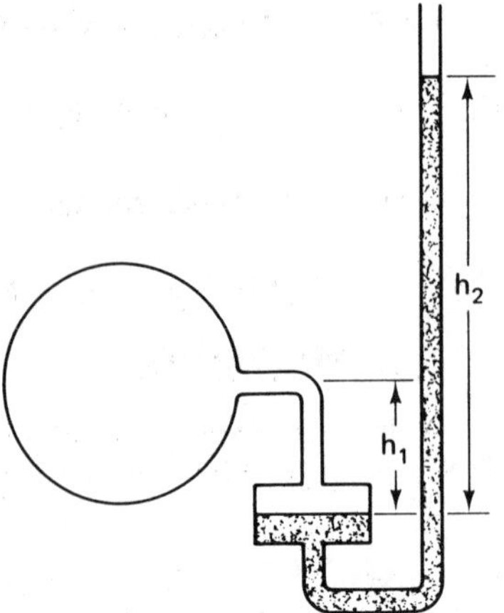

Figure P2.4.7

2.4.8 In Figure P2.4.8, fluids *A* and *B* have specific gravities of 0.75 and 1.00, respectively. If mercury is used as the manometer liquid, determine the pressure difference between *A* and *B*.

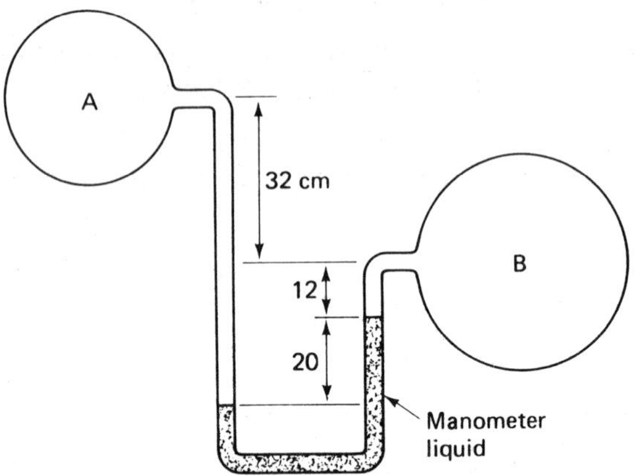

Figure P2.4.8

2.4.9 A micromanometer consists of two large reservoirs and a U-tube, as shown in Figure P2.4.9. If the density of the two liquids are, respectively, ρ_1, and ρ_2, determine an expression for the pressure difference in terms of ρ_1, ρ_2, h, d_1, and d_2.

2.4.10 For the system of manometers shown in Figure P2.4.10, determine the differential reading h.

2.4.11 The air pressure in the sealed tanks in Figure P2.4.11 are -22 cm of mercury column and 20 kN/m^2 in the left and right tanks, respectively. Determine the elevation of the manometer liquid in the right-hand column at *A*.

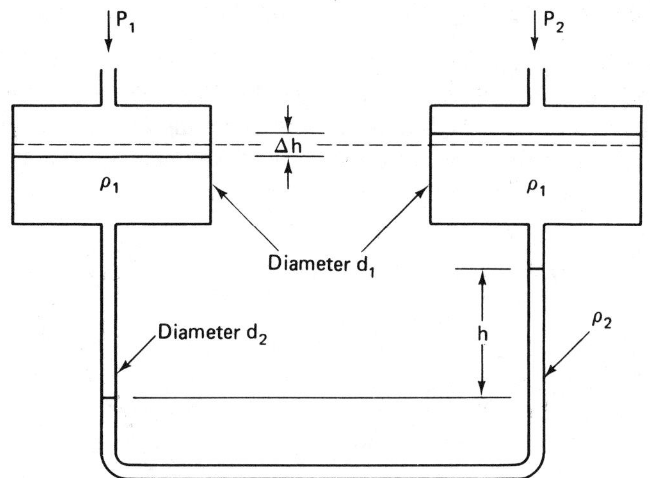

Figure P2.4.9

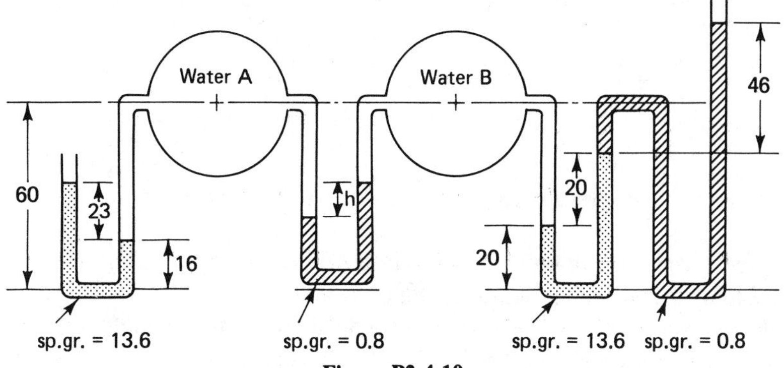

Figure P2.4.10

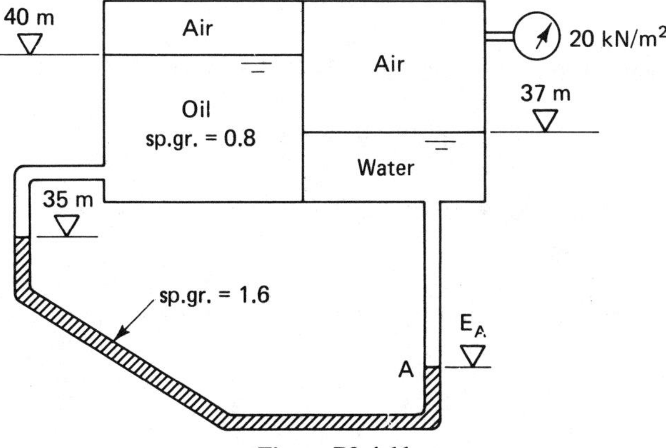

Figure P2.4.11

2.5 HYDROSTATIC FORCE ON A FLAT SURFACE

Let us take an arbitrary area *AB* on the back face of a dam that inclines at an angle θ and then place the x-axis on the line at which the water-free surface intersects with the dam surface, with the y-axis running down the direction of the dam surface. Figure 2.9(a) shows a horizontal view of the area and Figure 2.9(b) shows the projection of *AB* on the dam surface.

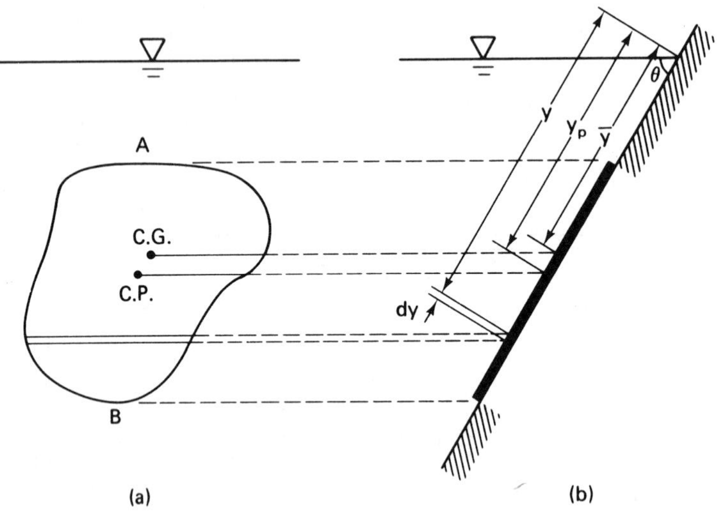

Figure 2.9 Hydrostatic pressure on a plane surface.

We may assume that the plane surface *AB* is made up of an infinite number of horizontal strips, each having a width of dy and an area of dA. The hydrostatic pressure on each strip may be considered constant because the width of each strip is very small. For a strip at depth h below the free surface, the pressure is

$$p = \gamma h = \gamma y \sin \theta$$

The total pressure force on the strip is the pressure times the area

$$dF = \gamma y \sin \theta \, dA$$

and the total pressure force over the entire *AB* plane surface is the sum of pressure on all the strips

$$\begin{aligned} F &= \int_A dF = \int_A \gamma \cdot y \cdot \sin \theta \cdot dA = \gamma \cdot \sin \theta \cdot \int_A y dA \\ &= \gamma \cdot \sin \theta \cdot A \cdot \bar{y} \end{aligned} \tag{2.11}$$

where $\bar{y} = \int_A y \, dA/A$ is the distance measured from the x-axis to the centroid (C.G.), of the *AB* plane.

Substituting $\bar{h}$, the vertical distance of the centroid below the water surface, for $\bar{y} \sin \theta$, we have

$$F = \gamma \cdot \bar{h} \cdot A \tag{2.12}$$

This equation states the fact that *the total hydrostatic pressure force on any submerged plane surface is equal to the product of the surface area and the pressure acting at the centroid of the plane surface.*

Pressure forces acting on a plane surface are parallel and distributed over every part of the surface. These parallel forces can be analytically replaced by a single resultant force P of the same magnitude and direction as shown in Equation (2.12). The point on the plane surface at which this resultant force acts is known as the *center of pressure*. Considering the plane surface as a free body, we see that the distributed forces can be replaced by the single resultant force at the pressure center without altering any reactions or moments in the system. Designating Y_P as the distance measured from the x-axis to the center of pressure, we may thus write

$$F \cdot Y_p = \int_A y \cdot dF$$

Hence

$$Y_P = \frac{\int_A y dF}{F} \tag{2.13}$$

Substituting the relationships $dF = \gamma y \sin \theta \cdot dA$ and $F = \gamma \sin \theta \cdot A \cdot \bar{y}$, we may write Equation (2.13) as

$$Y_p = \frac{\int_A y^2 \, dA}{A\bar{y}} \tag{2.14}$$

in which $\int_A y^2 dA = I_x$ and $A \cdot \bar{y} = M_x$ are, respectively, the moment of inertia and the static moment of the plane surface AB with respect to the x-axis. Therefore,

$$Y_P = \frac{I_x}{M_x} \tag{2.15}$$

With respect to the centroid of the plane, they may be written as

$$Y_p = \frac{I_0 + A\bar{y}^2}{A\bar{y}} = \frac{I_0}{A\bar{y}} + \bar{y} \tag{2.16}$$

where I_0 is the moment of inertia of the plane with respect to its own centroid, A is the plane surface area, and $\bar{y}$ is the distance between the centroid and the x-axis.

The center of pressure of any submerged plane surface is always below the centroid of the surface. All three quantities in the first term on the right-hand side of Equation (2.16) are positive quantities.

The centroid, area, and moment of inertia with respect to the centroid of certain common geometrical plane surfaces are given in Table 2.1.

TABLE 2.1 Surface Area, Center of Gravity, and Moment of Inertia of Certain Simple Geometrical Plates

Shape	Area	Centroid	Moment of inertia about the neutral axis
Rectangle 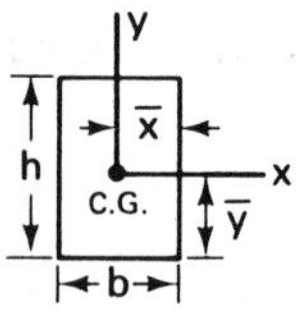	$b \cdot h$	$\bar{x} = \frac{1}{2} b$ $\bar{y} = \frac{1}{2} h$	$I_0 = \frac{1}{12} b \cdot h^3$
Triangle 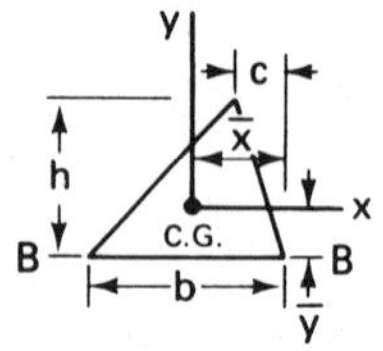	$\frac{1}{2} b \cdot h$	$\bar{x} = \frac{b + c}{3}$ $\bar{y} = h/3$	$I_0 = \frac{1}{36} b \cdot h^3$
Circle 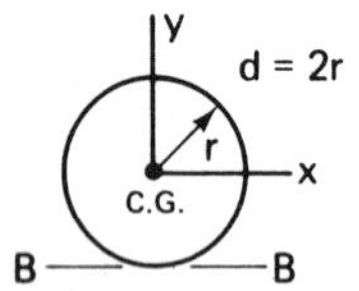	$\frac{1}{4} \pi d^2$	$\bar{x} = \frac{1}{2} d$ $\bar{y} = \frac{1}{2} d$	$I_0 = \frac{1}{64} \pi d^4$
Trapezoid 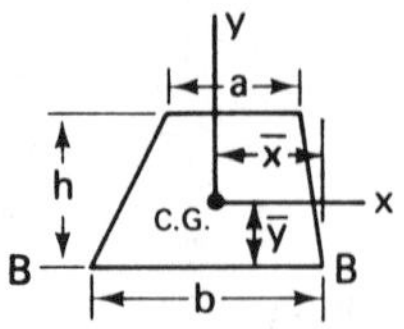	$\frac{h\,(a + b)}{2}$	$\bar{x} =$ $\bar{y} = \frac{h\,(2a + b)}{3\,(a + b)}$	$I_0 = \frac{h^3\,(a^2 + 4ab + b^2)}{36\,(a + b)}$

TABLE 2.1 Cont.

Shape	Area	Centroid	Moment of inertia about the neutral axis
Ellipse	πbh	$\bar{x} = b$ $\bar{y} = h$	$I_0 = \frac{\pi}{4} b \cdot h^3$
Semi-ellipse 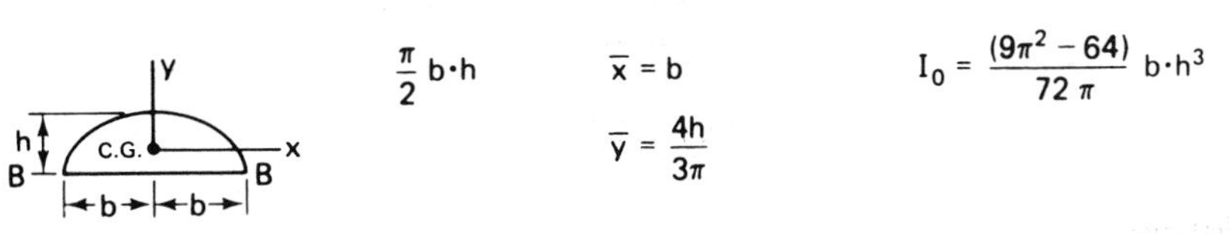	$\frac{\pi}{2} b \cdot h$	$\bar{x} = b$ $\bar{y} = \frac{4h}{3\pi}$	$I_0 = \frac{(9\pi^2 - 64)}{72\pi} b \cdot h^3$
Parabolic section 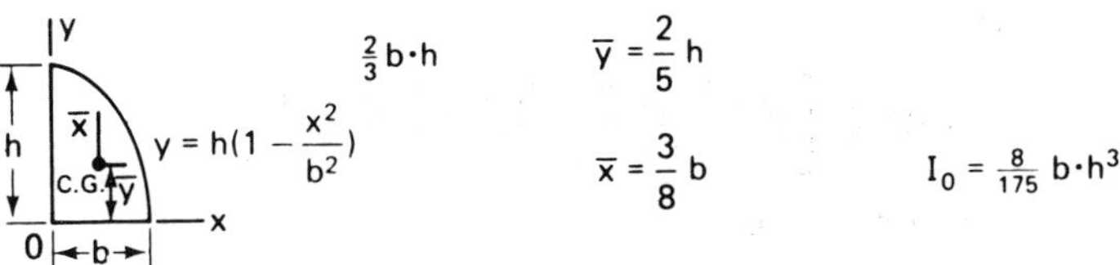	$\frac{2}{3} b \cdot h$	$\bar{y} = \frac{2}{5} h$ $\bar{x} = \frac{3}{8} b$	$I_0 = \frac{8}{175} b \cdot h^3$
Semicircle 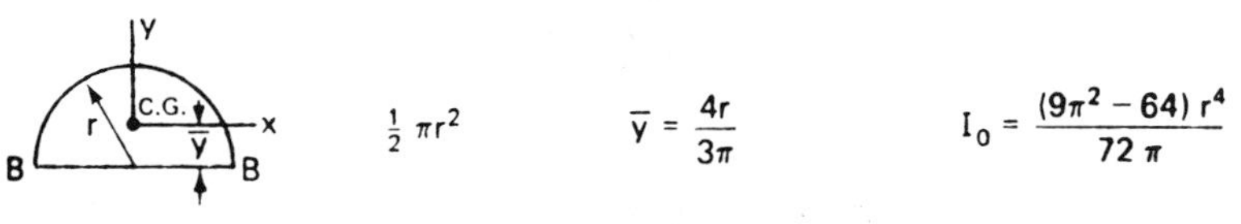	$\frac{1}{2} \pi r^2$	$\bar{y} = \frac{4r}{3\pi}$	$I_0 = \frac{(9\pi^2 - 64) r^4}{72\pi}$

Example 2.3

A vertical trapezoidal gate with its upper edge located 5 m below the free surface of water is shown in Figure 2.10. Determine the total pressure force and the center of pressure on the gate.

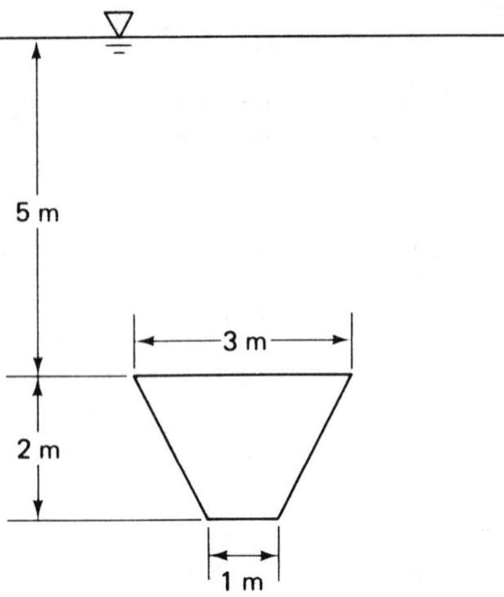

Figure 2.10

Solution

The total pressure force is

$$F = \gamma \cdot \bar{h} \cdot A$$

$$= 9810 \cdot \left[5 + \frac{2 \cdot (2 \cdot 1 + 3)}{3 \cdot (1 + 3)}\right] \cdot \left[\frac{2 \cdot (3 + 1)}{2}\right]$$

$$= 228{,}900 \text{ N}$$

The location of the center of pressure is, (Equation 2.16),

$$Y_P = \frac{I_o}{A\bar{y}} + \bar{y}$$

where

$$I_o = \frac{2^3 \cdot [1^2 + 4(1)(3) + 3^2]}{36 \cdot (1 + 3)} = 1.22 \text{ m}^4$$

$$\bar{y} = 5.83 \text{ m}$$

$$A = 4 \text{ m}^2$$

Thus

$$Y_p = \frac{1.22}{4 \cdot (5.83)} + 5.83 = 5.88 \text{ m}$$

below the water surface.

Example 2.4

An inverted semicircular gate (see Figure 2.11) is installed at 45° with respect to the free water surface. The top of the gate is 5 m below the water surface. Determine the total pressure and the center of pressure on the gate.

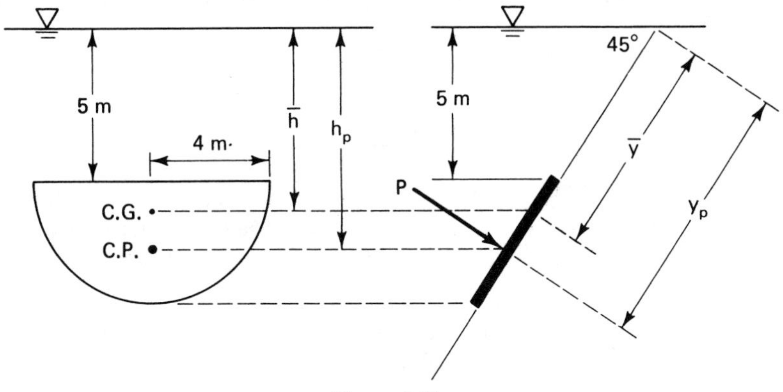

Figure 2.11

Solution

From Equation (2.11)

$$P = \gamma \cdot \sin\theta \cdot A \cdot \bar{y}$$

where

$$A = \frac{1}{2}\left(\pi \cdot 4^2\right) = 25.13 \text{ m}^2$$

and

$$\bar{y} = 5 \sec 45° + \frac{4 \cdot 4}{3\pi} = 8.77 \text{ m}$$

Therefore, $P = 9810 \cdot (\sin 45°)(25.13)(8.77) = 1{,}528{,}728$ N. This is the total hydrostatic force active on the gate. From Equation (2.16)

$$Y_p = \frac{I_o}{A\bar{y}} + \bar{y}$$

From Table 2.1

$$I_o = \frac{9\pi^2 - 64}{72\pi} r^4$$

$$= 28.10 \text{ m}^4$$

Therefore,

$$Y_p = \frac{28.10}{25.13(8.77)} + 8.77 = 8.90 \text{ m}$$

This is the distance measured from the top of the water surface to the pressure center of the gate.

PROBLEMS

2.5.1 A square gate 3 m by 3 m lies in a vertical plane. Determine the total pressure force on the gate and the distance between the center of pressure and the centroid when the upper edge of the gate is at the water surface. Compare with values when the upper edge is 15 m below the water surface.

2.5.2 Calculate the magnitude and the location of the resultant pressure force on the annnular gate shown in Figure P2.5.2.

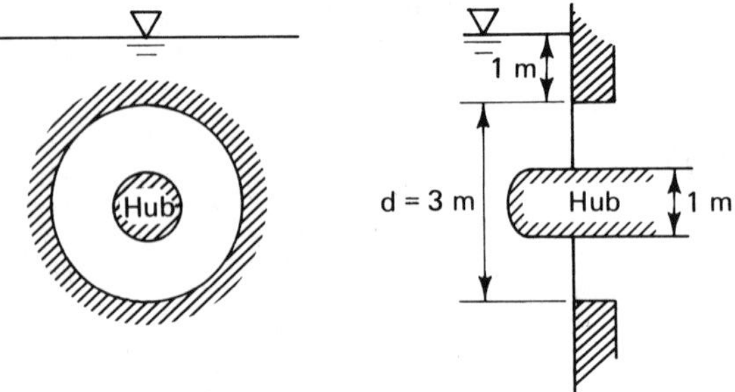

Figure P2.5.2

2.5.3 A sliding gate 10 ft wide and 5 ft high is installed in a vertical plane and has a coefficient of friction against the guides of 0.2. If the gate weighs 2 tons and its upper edge is 30 ft below water surface, calculate the vertical force required to lift the gate.

2.5.4 The rectangular gate in Figure P2.5.4 is hinged at A and separates water in the reservoir from the tail water tunnel. If the uniform thickness gate has a dimension of 2 m $\times$ 3 m and weighs 2 tons, determine the maximum height h for which the gate will stay closed.

2.5.5 A circular gate is installed on a vertical wall as shown in Figure P2.5.5. Determine the horizontal force, F, necessary in order to hold the gate at the edge (in terms of diameter, D, and height, h). Neglect friction at the pivot.

2.5.6 Figure P2.5.6 shows a vertical, rectangular gate, 3 m in height. The gate is designed to open automatically when h increases to 1 m. Determine the location of the horizontal axis of rotation 0–0′.

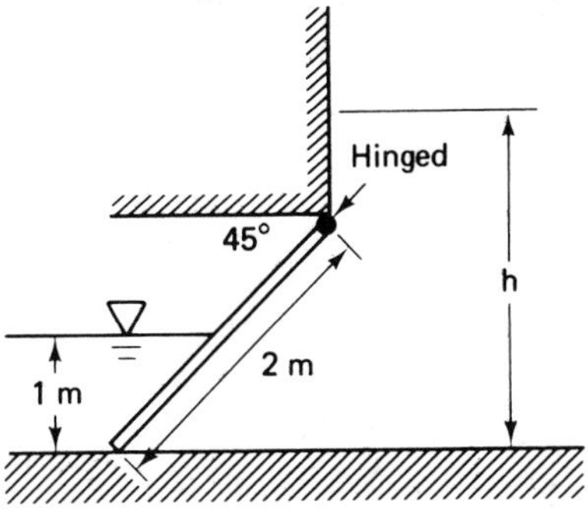

Figure P2.5.4

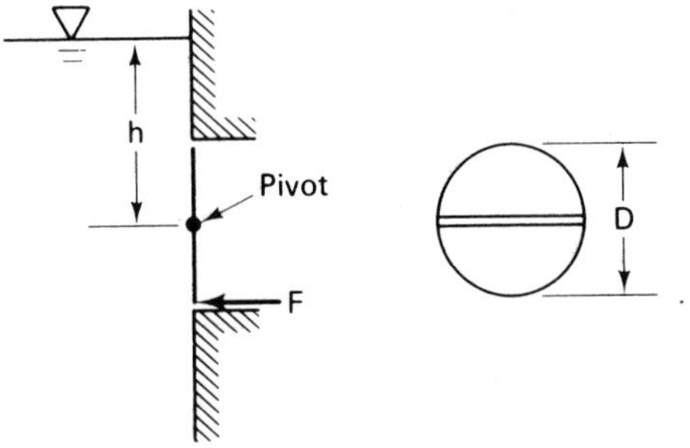

Figure P2.5.5

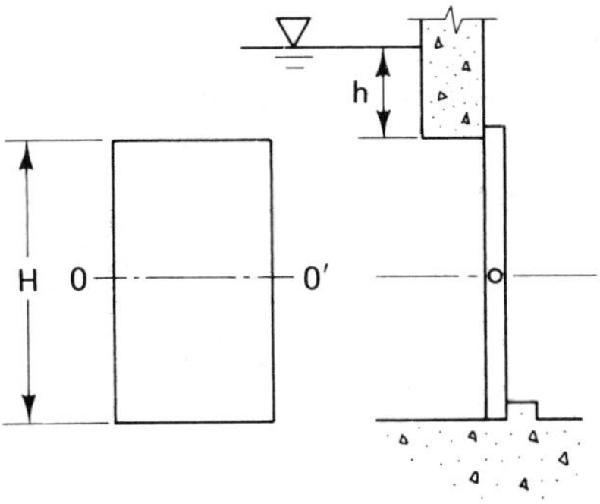

Figure P2.5.6

2.5.7 A vertical, rectangular gate 3 m high and 2 m wide is filled with water to a depth of 5 m above its upper edge. Locate the horizontal line that divides this area so that a) the forces on the upper and lower portions are the same and b) the moments of the forces about the line are the same.

2.5.8 A concrete dam with a triangular cross section as shown in Figure P2.5.8 is built to hold 10 m of water. The specific gravity of concrete is 1.67. Determine the moment generated with respect to the toe of the dam, A.

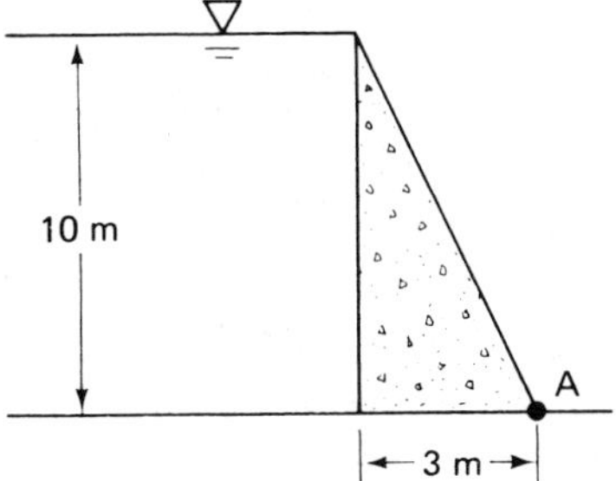

Figure P2.5.8

2.5.9 Calculate the minimum weight of the cover necessary to keep the 10-m wide cover of the box in Figure P2.5.9 closed.

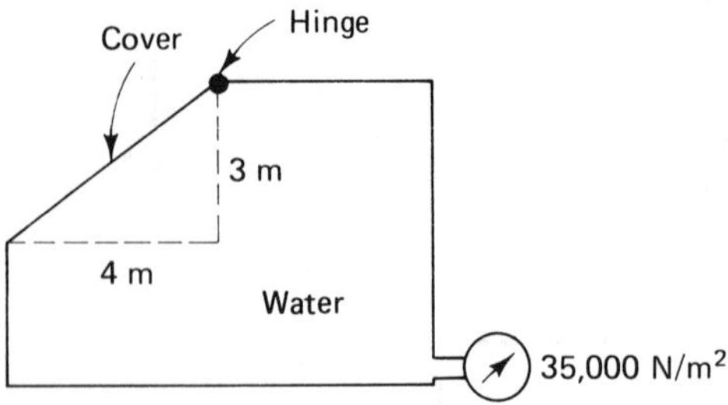

Figure P2.5.9

2.5.10 In Figure P2.5.10 the wicket dam is 5 m high and 2 m wide and is pivoted at its center. Determine the reaction force in the supporting member *AB*.

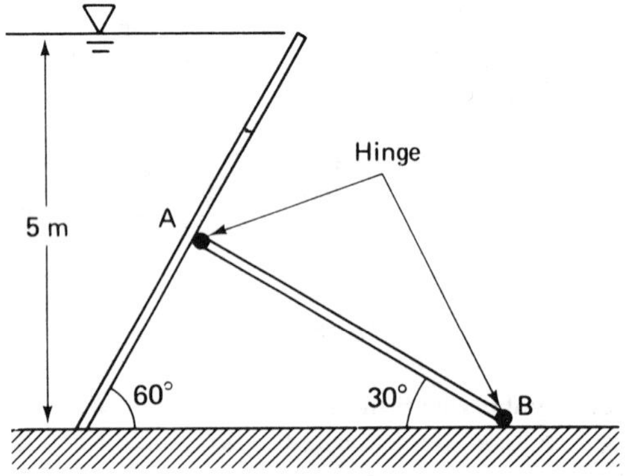

Figure P2.5.10

2.5.11 Determine the depth of water that will force the 8-ft-wide rectangular gate in Figure P2.5.11 to open. Neglect its own weight.

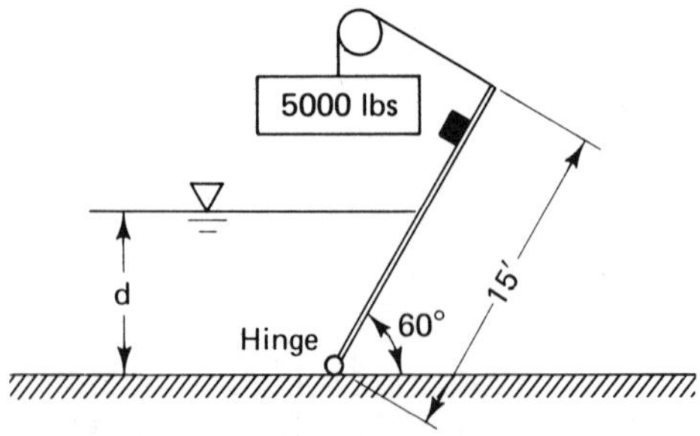

Figure P2.5.11

2.5.12 Neglecting the weight of the hinged gate, determine the depth *h* at which the gate will open in Figure P2.5.12.

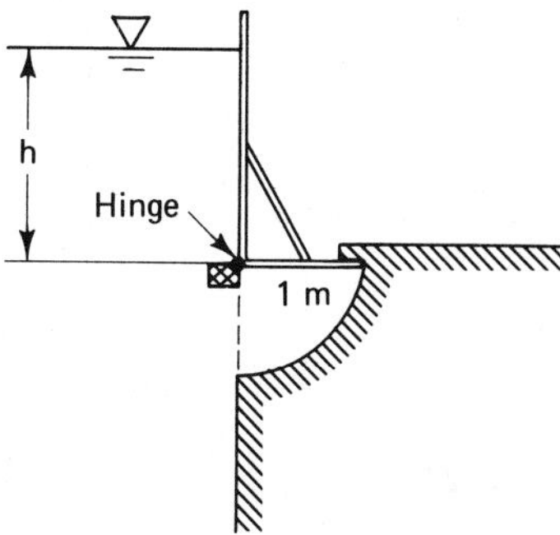

Figure P2.5.12

2.5.13 The circular gate shown in Figure P2.5.13 is hinged at the horizontal diameter. If it is in equilibrium, determine the relationship between h_A and h_B as a function of γ_A, γ_B, and d.

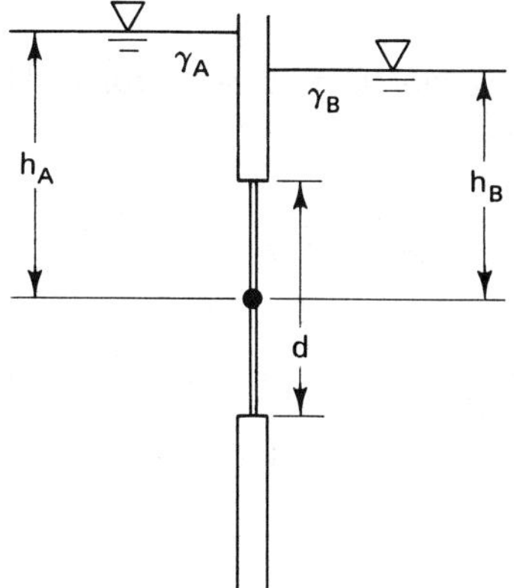

Figure P2.5.13

2.6 HYDROSTATIC FORCES ON CURVED SURFACES

Pressure force on a curved surface can be best analyzed by resolving the total pressure force on the surface's horizontal and vertical components. Figure 2.12 shows a curved wall of a container gate having a unit width normal to the plane of the paper.

Since the water body in the container is stationary, every part of the water body must be in equilibrium, or, each of the force components must satisfy the equilibrium conditions, i.e., $\Sigma F_x = 0$ and $\Sigma F_y = 0$.

In the free body diagram of water ABA', where $A'B$ is the vertical projection of AB, we can clearly see that the horizontal pressure component F_H must equal the horizontal pressure exerted on plane surface $A'B$. The vertical pressure component, F_V, must equal the total weight of the water body above gate AB, since the vertical component of the pressure force exerted on plane surface

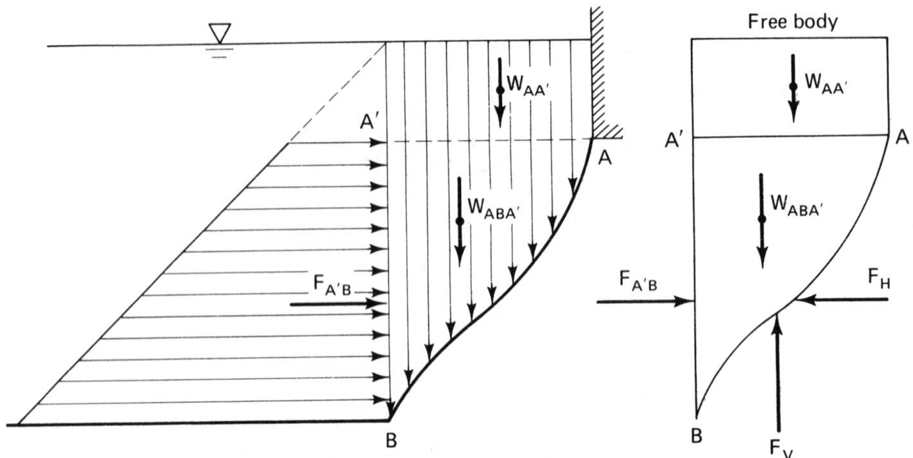

Figure 2.12 Hydrostatic pressure on a curved surface.

$A'B$ is zero and the plane cannot take any shear force. The horizontal and vertical pressure force on the gate may be expressed as

$$\Sigma F_x = F_{A'B} - F_H = 0$$

$$\therefore F_H = F_{A'B}$$

$$\Sigma F_y = F_V - (W_{AA'} + W_{ABA'}) = 0$$

$$\therefore F_V = W_{AA'} + W_{ABA'}$$

We may thus state

1. *The horizontal component of the total hydrostatic pressure force on any surface is always equal to the total pressure on the vertical projection of the surface. The resultant force of the horizontal component can be located through the center of pressure of this projection.*
2. *The vertical component of the total hydrostatic pressure force on any surface is always equal to the weight of the entire water column above the surface extending vertically to the free surface. The resultant force of the vertical component can be located through the centroid of this column.*

Example 2.5

Determine the total hydrostatic pressure and the center of pressure on the 3-m-long, 2-m-high quadrant gate in Figure 2.13.

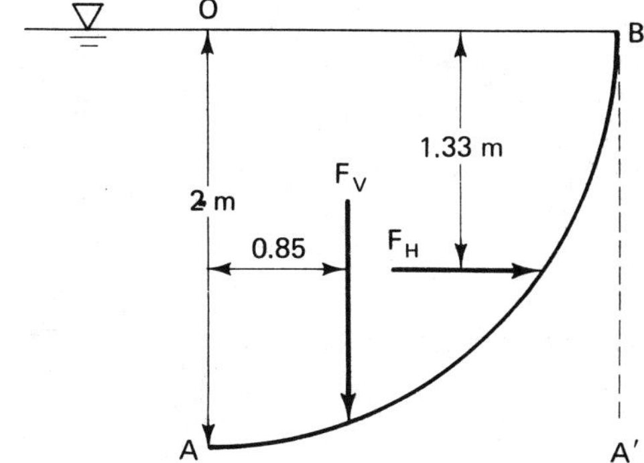

Figure 2.13

Solution

The horizontal component is equal to the hydrostatic pressure force on the projection plane $A'B$.

$$F_H = \gamma \cdot \bar{h} \cdot A = 9810\left(\frac{1}{2} \cdot 2 \cdot 2 \cdot 3\right) = 58{,}860 \text{ N}$$

The pressure center of the horizontal component is located at $2/3 \times 2 = 1.33$ m below the free surface. The vertical component is equal to the weight of water in the volume AOB. The direction of this pressure component is downward.

$$F_v = \gamma V = 9810\left[\frac{1}{4}\,\pi \cdot (2^2)\right](3) = 92{,}460 \text{ N}$$

The pressure center is located at $4(2)/3\pi = 0.85$ m. The resultant force of water on AB is

$$F = \sqrt{(58{,}860)^2 + (92{,}460)^2} = 109{,}600 \text{ N}$$

$$\theta = \tan^{-1}\left(\frac{F_v}{F_H}\right) = \tan^{-1}\frac{92{,}460}{58{,}860} = 57.5^\circ$$

Example 2.6

Determine the total hydrostatic pressure and the center of pressure on the semicylindrical gate shown in Figure 2.14.

Solution

The horizontal component of the pressure force per unit width is

$$F_H = \gamma \cdot \bar{h} \cdot \frac{A_v}{L} = \gamma \cdot \frac{H}{2} \cdot H = \frac{1}{2}\gamma H^2$$

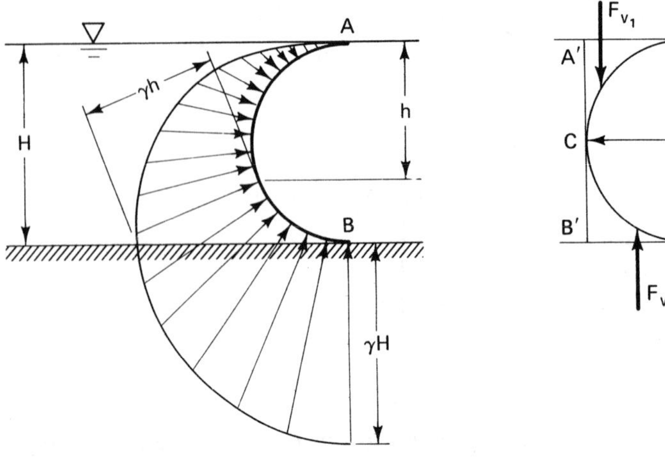

Figure 2.14

The pressure center of this component is located at a distance of H/3 from the bottom.

The vertical component can be determined as follows. The volume *AA'C* over the upper half of the gate, *AC*, produces a downward vertical pressure force component

$$F_{v_1} = -\gamma \cdot \left(\frac{H^2}{4} - \frac{\pi H^2}{16}\right)$$

The vertical pressure force component exerted by the water on the lower half of the gate, *CB*, is upward and equivalent to the weight of water replaced by the volume *AA'CB*

$$F_{v_2} = \gamma \cdot \left(\frac{H^2}{4} + \frac{\pi H^2}{16}\right)$$

The direction of the vertical force is upward and proportional to the weight of the water replaced by the volume *ACB*.

$$F_v = F_{v_1} + F_{v_2} = \gamma \cdot \left[-\left(\frac{H^2}{4} - \frac{\pi H^2}{16}\right) + \left(\frac{H^2}{4} + \frac{\pi H^2}{16}\right)\right]$$

$$= \gamma \cdot \frac{\pi}{8} H^2$$

The resultant force is, then

$$F = \sqrt{\frac{1}{4} + \frac{\pi^2}{64}} \cdot \gamma H^2$$

$$\theta = \tan^{-1}\frac{F_v}{F_H} = \tan^{-1}\left(\frac{\pi}{4}\right) = 38.15°$$

Since all pressure forces are concurrent at the center of the gate, 0, the resultant force must also act through 0.

PROBLEMS

2.6.1 The tainter gate section shown in Figure P2.6.1 has a cylindrical surface with a 10-m radius and is supported by a structural frame hinged at *O*. The gate is 12 m long (in the direction perpendicular to the paper). Determine the magnitude and location of the total hydrostatic force on the gate (*Hint*: Horizontal and vertical force components).

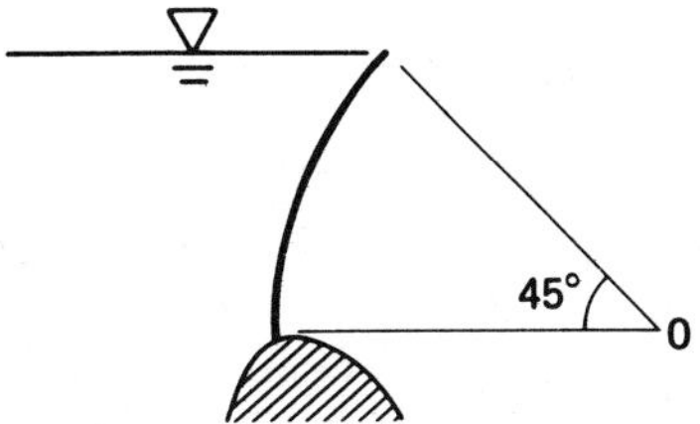

Figure P2.6.1

2.6.2 Calculate the magnitude, direction, and location of the pressure–force components on the gate shown in Figure P2.6.2.

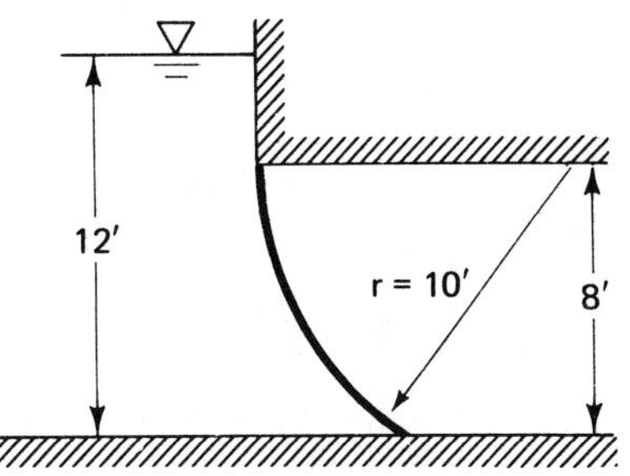

Figure P2.6.2

2.6.3 Determine the relationship between γ_1, and γ_2 in Figure P2.6.3 if the weightless triangular gate is in equilibrium in the position shown.

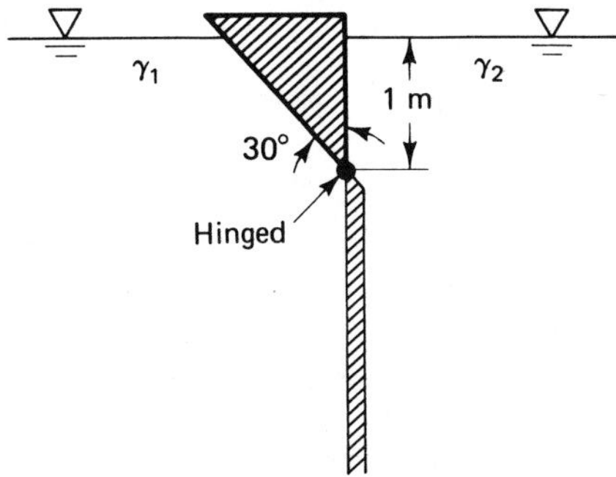

Figure P2.6.3

2.6.4 In Figure P2.6.4 the corner plate of a large hull is curved with a radius of 1.75 m. When the barge is submerged in sea water (sp. gr. 1.025), determine the magnitude and location of pressure on the curved plate *AB* per unit length of the hull.

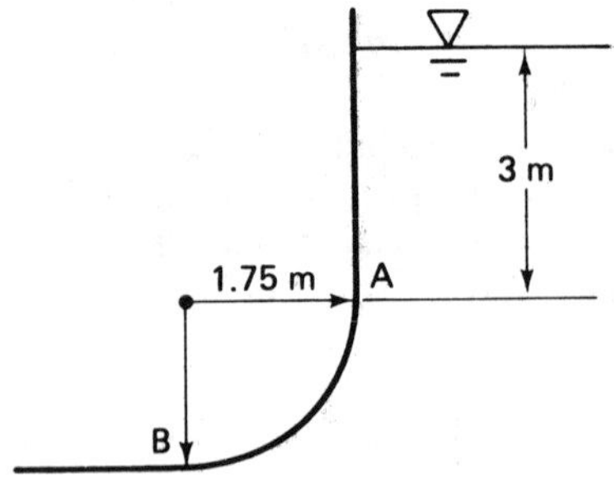

Figure P2.6.4

2.6.5 The cylindrical dome in Figure P2.6.5 is 10 m long and is secured to the top of an oil tank by bolts. If the oil has a specific gravity of 0.86 and the pressure gage reads 2.42 bar, determine the total tension force in the bolts. Neglect the weight of the cover.

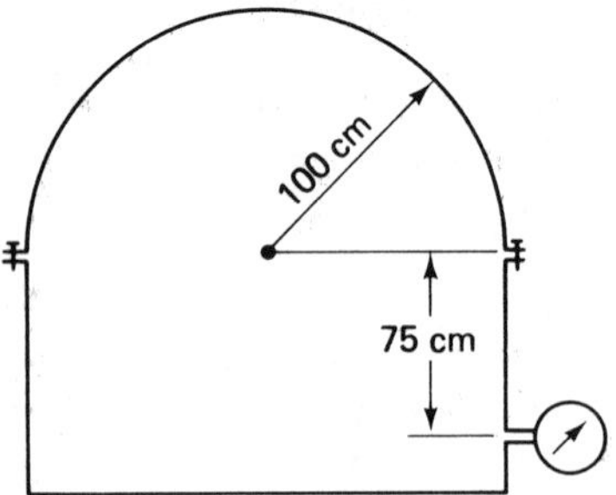

Figure P2.6.5

2.6.6 An inverted spherical shell of diameter *d* as shown in Figure P2.6.6 is used to cover a tank filled with water at 20°C. Determine the minimum weight the shell needs to be to hold itself in place.

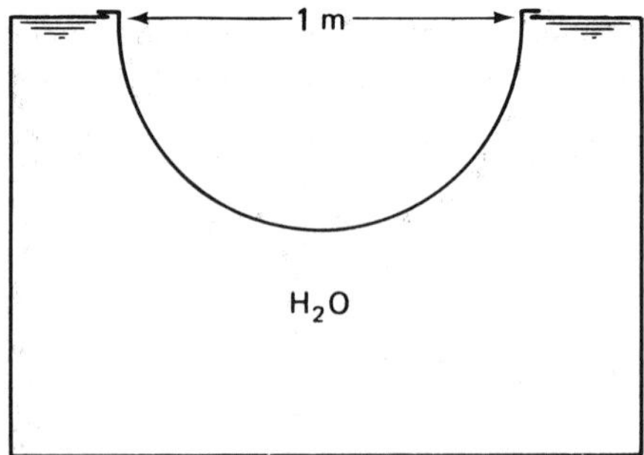

Figure P2.6.6

2.6.7 Calculate the vertical and horizontal hydrostatic forces acting on the curved surface *ABC* in Figure P2.6.7.

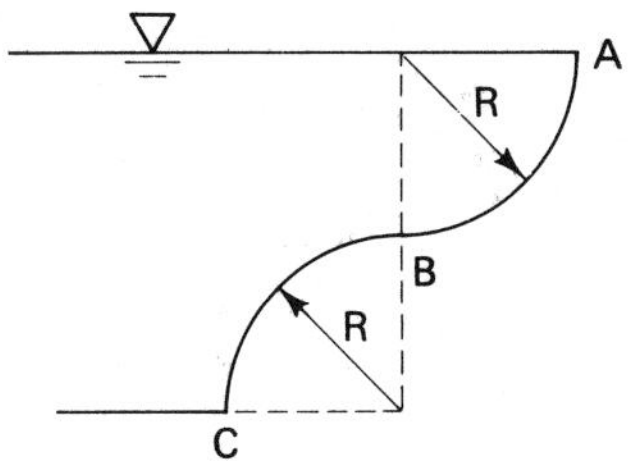

Figure P2.6.7

2.6.8 The homogeneous cylinder (sp. gr. 2.0) in Figure P2.6.8 is 1 m long and $\sqrt{2}$ m in diameter and blocks a 1 m square opening between reservoirs *A* and *B* (sp. gr. A = 0.8, sp. gr. B = 1.5). Determine the magnitude and direction of the reactions at the upper and lower edges (u,ℓ) of the opening.

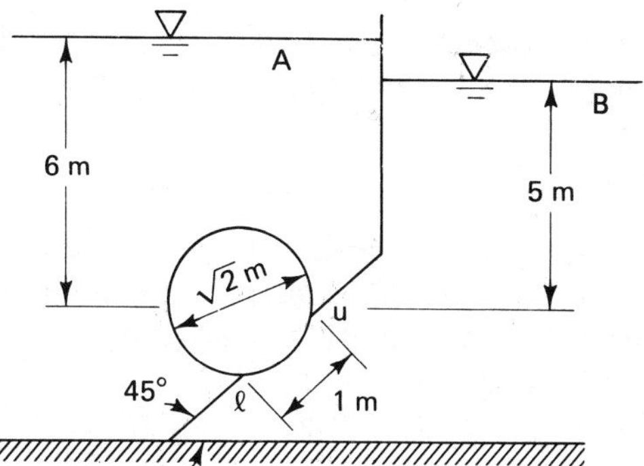

Figure P2.6.8

2.6.9 Calculate the magnitude and location of the vertical and horizontal components of the hydrostatic force on the surface shown in Figure P2.6.9 (spherical in top of an inverted conical shape). The liquid is water and the radius $R = 1.5$ m.

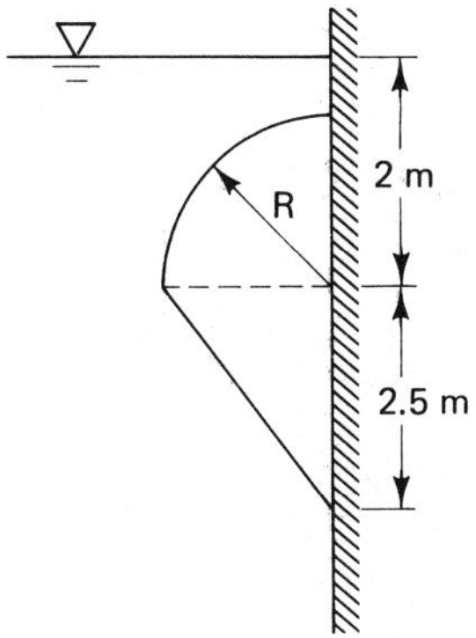

Figure P2.6.9

2.6.10 In Figure P2.6.10 a homogeneous cone plugs the 0.1 m diameter orifice between reservoir *A* that contains water, and reservoir *B* that contains oil (sp. gr. 0.8). Determine the specific weight of the cone if it unplugs when h_0 reaches 1.5 m.

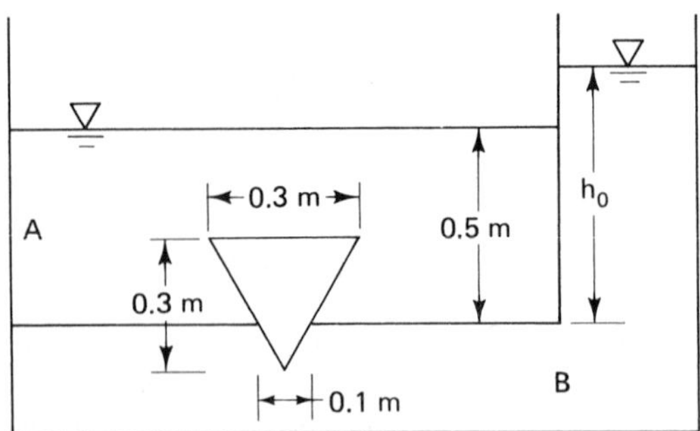

Figure P2.6.10

2.6.11 What would be the specific weight of the cone if reservoir *B* in Figure P2.6.10 contains air at the pressure of 12,000 N/m^2 instead of oil?

2.7 BUOYANCY

Archimedes (about 250 B.C.) discovered that *the weight of a submerged body is reduced by an amount equal to the weight of the liquid displaced by the body. Archimedes' principle,* as we now call it, can be easily proven by using the formula described in Section 2.5.

Assume that a solid body of arbitrary shape, *AB*, is submerged in water as shown in Figure 2.15. A vertical plane *MN* may then be drawn through the body in the direction normal to the paper. One observes that the horizontal pressure force component in the direction of the paper, F_H and F'_H, must be equal because they both are measured by the same vertical projection area *MN*. The horizontal pressure force component in the direction normal to the paper must also be equal for the same reason because they share the projection in the plane of the paper.

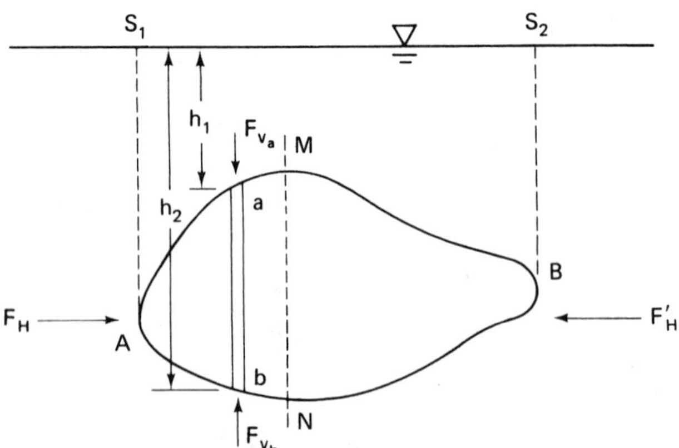

Figure 2.15 Bouyancy of a submerged body.

The vertical pressure force component can be analyzed by taking a small vertical prism *ab* with a cross-sectional area ΔA. The vertical pressure force on top of the prism is $(\gamma h_1 \cdot dA)$ acting downward. The vertical force on the bottom of the prism is $(\gamma h_2 \cdot dA)$ acting upward. The difference gives the resultant vertical force component on the prism

$$F_V = \gamma h_2 \cdot dA - \gamma h_1 \cdot dA = \gamma(h_2 - h_1)\, dA \qquad \text{(upward)}$$

which is exactly equal to the weight of water column *ab* replaced by the prism. In other words, the weight of the submerged prism is reduced by an amount equal to the weight of the liquid replaced by the prism. A summation of the vertical forces on all the prisms that make up the entire submerged body *AB* gives the proof of Archimedes' principle.

Archimedes' principle may also be looked on as an action of the difference of vertical pressure forces on the two surfaces *ANB* and *AMB*. The vertical pressure force on surface *ANB* is equal to the weight of the hypothetical water column (volume of S_1ANBS_2) acting upward; and the vertical pressure force on surface *AMB* is equal to the weight of the water column S_1AMBS_2 acting downward. Since the volume S_1ANBS_2 is larger than the volume S_1AMBS_2 by an amount exactly equal to the volume of the submerged body *AMBN*, the net difference is a force equal to the weight of water that would be contained in the volume *AMBN* acting upward. This force is the buoyancy on the body.

A floating body is a body partially submerged due to the balance of the body weight and the buoyancy force.

2.8 FLOTATION STABILITY

The stability of a floating body is determined by the relative positions of the center of gravity of body *G* and the *center of buoyancy B*, which is the center of gravity of the liquid volume replaced by the body, as shown in Figure 2.16.

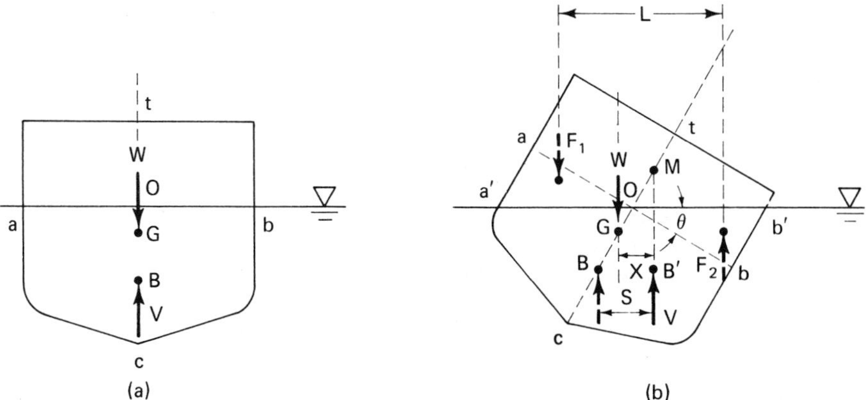

Figure 2.16 Center of buoyancy and metacenter of a floating body.

The body is in equilibrium if its center of gravity and its center of buoyancy lie on the same vertical line, as in Figure 2.16(a). This equilibrium may be disturbed by a variety of causes, for example, wind or wave action. The floating body is made to heel through an angle θ as shown in Figure 2.16(b). When the floating body is in the heeled position, the center of gravity of the body remains unchanged but the center of buoyancy, which is now the center of gravity of area $a'cb'$, has been changed from B to B'. The buoyance force V, acting upward through B', and the weight of the body W, acting downward through G, constitute a couple, $W \cdot X$, which resists further overturning and tends to restore the body to its original equilibrium position.

By extending the line of vertical buoyancy force action through the center of buoyancy B', we see that the vertical line intersects the axis of original symmetry $c = t$ at a point M. The point M is known as the *metacenter* of the floating body, and the distance between the center of gravity and the metacenter is known as the *metacentric height*. The metacentric height is a measure of the flotation stability of the body. When the angle of inclination is small, the position of M does not change materially with the tilting position. The metacentric height and the righting moment can be determined in the following way.

Since the tilting of a floating body does not change the total weight of the body, the total displacement volume is not changed. The roll through an angle θ only changes the shape of the displaced volume due to the adding of the immersion wedge bob' and the subtracting of the emersion wedge aoa'. In this new position, the total buoyancy force γVol is shifted through a horizontal distance S to B', and it may be considered as the result of compounding the force γVol in its original position at B and a couple F_1 and F_2, which caused γVol to change position. The moment of the resultant force about point B must equal the algebraic sum of the moments of the components

$$\begin{aligned}\gamma \cdot \text{Vol} \cdot S &= \gamma \cdot \text{Vol} \cdot (\text{zero}) + \text{moment of the force couple} \\ &= 0 + \gamma \cdot \text{vol} \cdot L\end{aligned}$$

or

$$\gamma \cdot \text{Vol} \cdot S = \gamma \cdot \text{vol} \cdot L$$

$$S = \frac{\text{vol}}{\text{Vol}} \cdot L \tag{a}$$

where vol is the volume of wedge bob' (or aoa') and L is the horizontal distance between the centers of gravity of the two wedges.

But, according to the geometric relation, we have

$$S = \overline{MB} \cdot \sin\theta \qquad \text{or} \qquad \overline{MB} = \frac{S}{\sin\theta} \tag{b}$$

Comparing Equations (a) and (b), we get

$$\overline{MB} = \frac{\text{vol} \cdot L}{\text{Vol} \cdot \sin\theta}$$

For a small angle, $\sin\theta \simeq \theta$, the above relationship may be simplified

$$\overline{MB} = \frac{\text{vol}\cdot L}{\text{Vol}\cdot\theta}$$

The buoyancy force produced by wedge bob' (see Figure 2.17) can be estimated by considering a small prism of the wedge. Assume that the prism has a horizontal area, dA, and is located at a distance x from the axis of rotation, O. The height of the prism is $x\cdot(\tan\theta)$. For small angle θ, it may be approximated by $x\cdot\theta$. Thus, the buoyancy force produced by this small prism is $\gamma x\theta\, dA$. The moment of this force about the axis of rotation O is $\gamma x^2\theta\, dA$. The sum of the moment produced by each of the prisms in the wedge gives the moment of the immersed wedge. The moment produced by the force couple is, therefore,

$$\gamma\cdot\text{vol}\cdot L = F\cdot L = \int_A \gamma x^2\theta\, dA = \gamma\theta\int_A x^2\, dA$$

But $\int_A x^2\, dA$ is the moment of interia of the waterline cross-sectional area of the floating body about the axis of rotation O.

$$I_O = \int_A x^2\, dA$$

Hence, we have

$$\text{vol}\cdot L = I_O\theta$$

For small angles of tilting, the moment of inertia for upright cross section aob and the inclined cross section $a'ob'$ may be approximated by a constant value. Therefore,

$$\overline{MB} = \frac{I_O}{\text{Vol}} \tag{2.17}$$

The metacentric height, defined as the distance between the metacenter M and the center of gravity G can be estimated

$$\overline{GM} = \overline{MB} \pm \overline{GB} = \frac{I_0}{\text{Vol}} \pm \overline{GB} \tag{2.18}$$

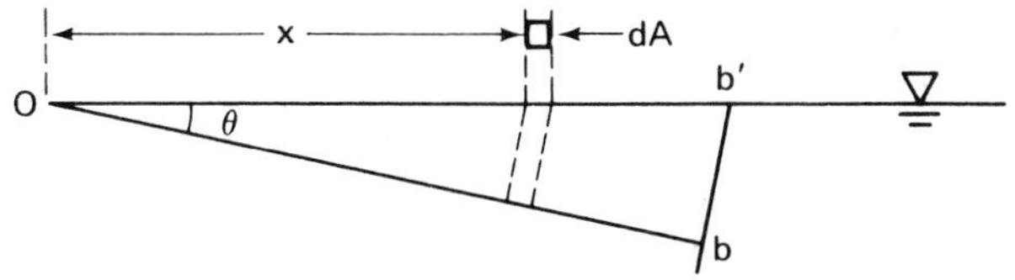

x
dA
O
θ
b'
b

Figure 2.17

The distance between the center of gravity and the center of buoyancy $\overline{GB}$ in the upright position, shown in Figure 2.16, can be determined by the sectional geometry or the design data of the vessel.

The (±) sign, taken as the center of gravity of a floating body, may be located above or below the center of buoyancy. For greater stability of flotation, it is obviously advantageous to have a negative value for $\overline{GB}$ or to make the center of gravity as low as possible.

The *righting moment,* when tilted, is

$$M = W \cdot \overline{GM} \cdot \sin \theta \tag{2.19}$$

The stability of buoyant bodies under various conditions may be summarized as follows:

1. A floating body is stable if the center of gravity is below the metacenter. Otherwise, it is unstable.
2. A submerged body is stable if the center of gravity is below the center of buoyancy.

Example 2.7

A 3 m × 4 m rectangular box caisson is 2 m deep (see Figure 2.18). It has a draft of 1.2 m when it floats in an upright position. Compute (a) the metacentric height and (b) the righting moment in sea water (sp. gr. 1.03) when the angle of heel is 8°.

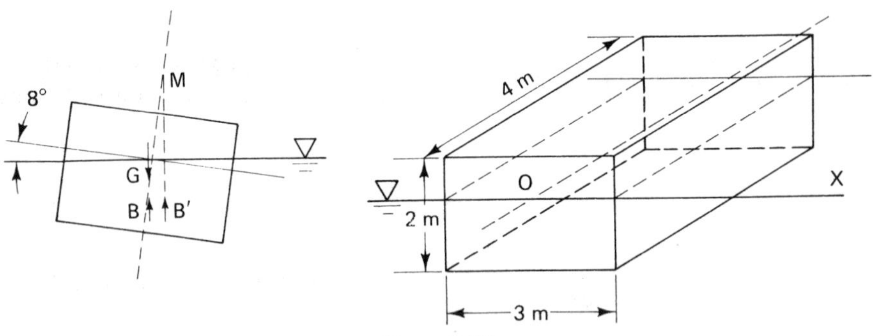

Figure 2.18

Solution

From Equation 2.18

$$\overline{GB} = \overline{MB} - \overline{GB}$$

where

$$\overline{MB} = \frac{I_o}{\text{Vol}}$$

and I_o is the waterline area moment of inertia of the box about its longitudinal axis through O.
Then

$$\overline{GM} = \frac{\frac{1}{12} b^3 \ell}{b \cdot \ell \cdot 1.2} - \left(\frac{h}{2} - \frac{1.2}{2}\right)$$

$$= 0.225 \text{ m}$$

The specific gravity of sea water is 1.03, and from Equation 2.19, the right moment is

$$M = W \cdot \overline{GM} \sin \theta$$

$$= 9810 \cdot 1.03 \cdot (4 \cdot 3 \cdot 1.2) \cdot (0.225)(\sin 8^\circ)$$

$$= 4560 \text{ N} \cdot \text{m}$$

PROBLEMS

2.8.1 A solid brass sphere (see Figure P2.8.1) of 30 cm diameter is used to weight down a cylindrical buoy in sea water (sp. gr. 1.03). The buoy has a height of 2 m and a specific gravity of 0.5, and is tied to the sphere at one end. What rise in tide, h, will be required to lift the anchor off the bottom?

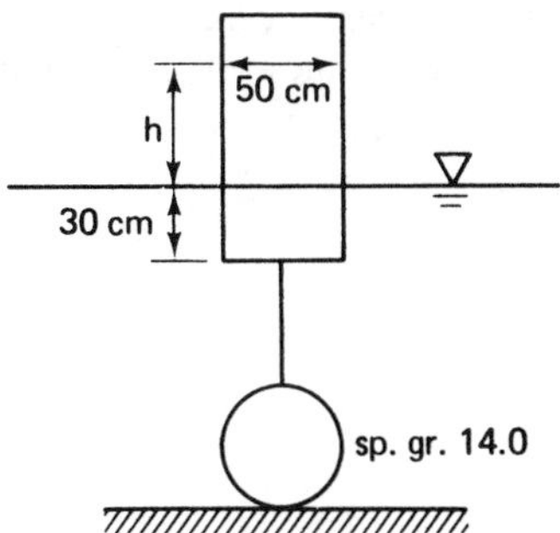

Figure P2.8.1

2.8.2 A concrete block that has a total volume of 1 m^3 and a specific gravity of 1.67 is tied to one end of a long, hollow cylinder (similar arrangement as in Problem 2.8.1). The cylinder is 3 m long and is 72 cm in diameter. In deep water, 12 cm of the cylinder remain above the water surface. Determine the weight of the cylinder.

2.8.3 The solid floating prism shown in Figure P2.8.3 has two components. Determine γ_A and γ_B if $\gamma_B = 2\gamma_A$.

2.8.4 In Figure P2.8.4 the buoy of radius R opens the square gate AB when water rises to the half-buoy height. Determine R if the weight of the buoy and the gate are negligible.

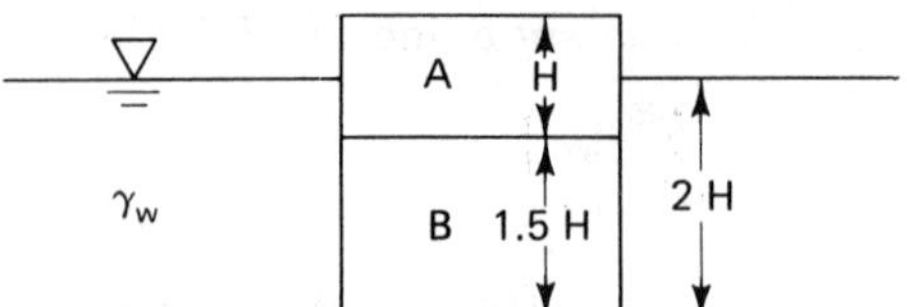

Figure P2.8.3

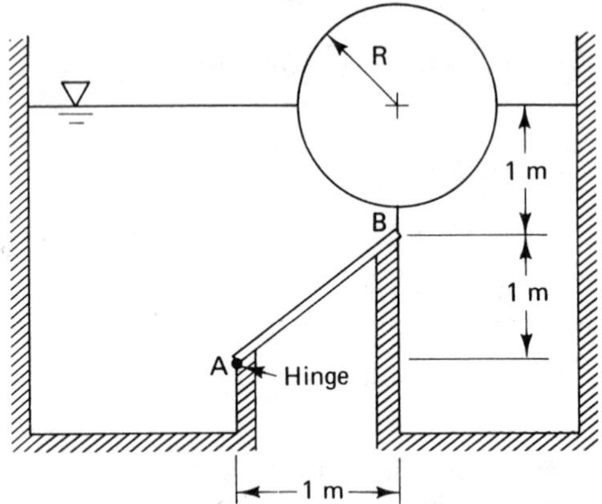

Figure P2.8.4

2.8.5 The floating rod shown in Figure P2.8.5 weighs 150 lbs and the water surface is 7 ft above the hinge. Calculate the angle θ assuming line force distribution, and determine the error incurred.

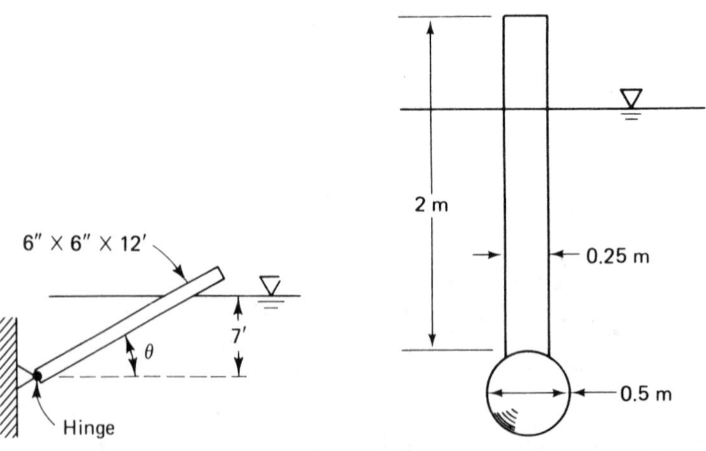

Figure P2.8.5 **Figure P2.8.6**

2.8.6 Figure P2.8.6 shows a buoy that consists of a wooden pole 25 cm in diameter and 2 m long, with a spherical weight at the bottom. The specific weight of the wood is 0.62 and that of the bottom weight is 1.62. Determine the position of the center of buoyancy from the top of the buoy, and the metacenter height.

2.8.7 A rectangular barge is 12 m long, 5 m wide, and 2 m deep. The center of gravity is 1.0 m from the bottom and the barge draws 1.5 m of water. Find the metacenter height and the distance measured from the bottom to the metacenter for the following angles of heel: 4°, 8°, and 12°. What are the respective righting moments?

2.8.8 A 10-m-diameter cylindrical caisson floats upright in the ocean (sp. gr. 1.03). The center of gravity measures 2 m from the bottom, which is 8 m below the water surface. Determine the metacenter height and the righting moment when the caisson is tipped through an angle of 10°.

2.8.9 A wooden block is 2 m long, 1 m wide, and 1 m deep. What is the specific gravity of the block if the metacenter is at the same point as the center of gravity? Is the floating block stable? Explain.

2.8.10 A 12-m-long, 4.8-m-wide, and 4.2-m-deep rectangular pontoon has a draft of 2 m in sea water (sp. gr. 1.03). Assuming the load is uniformly distributed on the bottom of the pontoon, and the maximum design angle of heel is 15°, determine the distance that the center of gravity can be moved from the center line towards the edge of the pontoon.

3

WATER FLOW IN PIPES

3.1 DESCRIPTION OF A PIPE FLOW

In hydraulics the term *pipe flow* usually refers to a full water flow in closed conduits of circular cross sections under a certain pressure gradient. For a given discharge Q, the pipe flow at any section of the pipe can be described by the pipe cross section, the pipe elevation, the pressure, and the flow velocity in the pipe.

The elevation h of a particular section in the pipe is usually measured with respect to a horizontal reference datum, such as the *mean sea level* (MSL) at a certain estuary. The pressure in a pipe generally varies from one point to another, but a mean value is normally used over a particular section of interest. In other words, the regional pressure variation in a given cross section is commonly neglected unless otherwise specified.

In most engineering computations the section mean velocity V, defined as the discharge, Q, divided by the cross-sectional area, A, is used.

$$V = \frac{Q}{A} \tag{3.1}$$

The velocity distribution within a cross section in a pipe, however, has special meaning in hydraulics. Its significance and importance in hydraulics are discussed next.

3.2 THE REYNOLDS NUMBER

Near the end of the nineteenth century the French engineer Osborne Reynolds performed a very carefully prepared pipe flow experiment. Figure 3.1 shows the schematics of a typical setup for the Reynolds experiment. A long, straight, glass pipe of small bore was installed in a large tank with glass sides. A control valve, C, was installed at the outlet end of the glass pipe to regulate the outflow. A small bottle, B, filled with colored water and a regulating cock at the bottle's neck were used to introduce a fine stream of colored water into the entrance of the glass pipe when the flow was initiated. Water in the large tank was allowed to settle very quietly in a room for several hours so that water in every part of the tank became totally stationary. Valve C was then partially opened to allow a very slow flow in the pipe. At this time the colored water appeared as a straight line extending to the downstream end, indicating *laminar flow* in the pipe. The valve was opened up slowly to allow the pipe flow rate to increase gradually until a certain velocity was reached at which time the thread of color suddenly broke up and mixed with the surrounding water, which showed that the pipe flow became turbulent at this point.

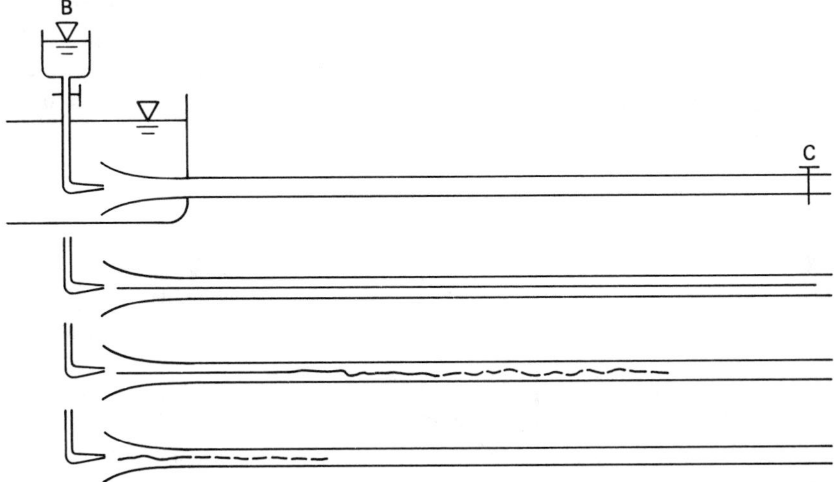

Figure 3.1 Reynolds' apparatus.

Reynolds found that the transition from laminar to turbulent flow in a pipe actually depends not only on the velocity, but also on the pipe diameter and the viscosity of the fluid. The relationship can be described by the ratio of the inertia force to the viscous force in the pipe. This ratio is commonly known as the

Reynolds number, N_R, and it is shown (see also Chapter 10) in the following expression:

$$N_R = \frac{DV}{\nu} \tag{3.2}$$

In the case of pipe flow, D is the pipe diameter, V is the mean velocity, and ν is the kinematic viscosity of the fluid, defined by the ratio of absolute viscosity μ and the fluid density ρ,

$$\nu = \frac{\mu}{\rho} \tag{3.3}$$

It has been found and verified by many carefully prepared experiments that for flows in circular pipes the *critical Reynolds number* is about 2000. At this point the laminar pipe flow changes to a turbulent one. The transition from laminar to turbulent flow does not always happen at exactly $N_R = 2000$ but varies somewhat from one experiment to another due to differences in experimental conditions. This range of Reynolds number between laminar and turbulent flow is commonly known as the *transitional range* (see Figure 3.8).

Laminar flow occurs in a circular pipe when fluid flows in orderly laminae; this is analogous to the telescoping of a large number of thin-walled concentric tubes. The outer tube adheres to the pipe wall while the tube next to it moves with slightly higher velocity. The velocity of each successive tube increases gradually and reaches a maximum velocity near the center of the pipe. In this case, the velocity distribution takes the form of a paraboloid of revolution with the mean velocity V equal to one half of the maximum center line velocity, as shown in Figure 3.2

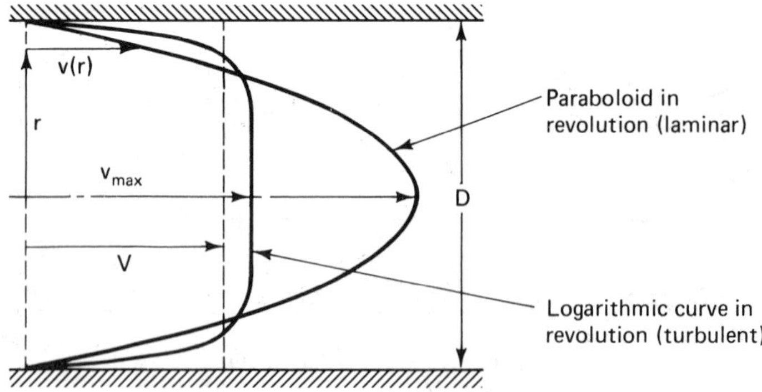

Figure 3.2 Velocity profiles of laminar and turbulent flows in circular pipes.

In turbulent flow, the turbulent motion causes the slower water particles adjacent to the pipe wall to mix continuously with the high-speed particles in the midstream; as a result, the high-speed particles in the midstream are slowed down due to the mixing. For this reason, the velocity distribution in turbulent flow is more uniform (flatter) than that in laminar flow. The velocity profiles in turbulent pipe flows have been shown to take the general form of a logarithmic curve in revolution. Turbulent mixing activities increase with the Reynolds number; hence, the velocity distribution becomes flatter as the Reynolds number increases.

Under ordinary circumstances, water loses energy as it flows through a pipe. A major portion of the energy loss is caused by

1. friction against pipe walls
2. viscous dissipation due to internal actions of the flow.

Wall friction on a moving column of water depends on the roughness of the wall material, e, and the velocity gradient $(dV/dr)|_{r=R_0}$ at the wall. For the same flow rate, it is seen (Figure 3.2) that turbulent flow has a higher wall velocity gradient than that of laminar flows; hence, a higher friction loss may be expected as the Reynolds number increases. At the same time, the internal mixing activities are also intensified as the flow becomes more turbulent, which indicates an increasing rate of viscous dissipation in the flows. As a consequence, the rate of energy loss in pipe flow varies as a function of the Reynolds number and the roughness of the pipe wall. The engineering application of this consequence will be discussed later in the chapter.

Example 3.1

A 40-mm-diameter circular pipe carries water at 20°C. Calculate the largest flowrate for which laminar flow can be expected.

Solution

The kinematic viscosity of water at 20°C is $\nu = 1 \times 10^{-6}$ m²/sec (Table 1.3). Taking $N_R = 2000$ as the conservative upper limit for laminar flow, $D = 0.04$ m.

$$N_R = \frac{DV}{\nu} = \frac{0.04\ V}{1 \times 10^{-6}} = 2000$$

$$V = 2000 \cdot 1 \times 10^{-6}/0.04 = 0.05 \text{ m/sec}$$

The flowrate

$$Q = AV = \frac{\pi}{4}(0.04)^2 \times 0.05 = 6.28 \times 10^{-5} \text{ m}^3\text{/sec}$$

3.3 FORCES IN PIPE FLOWS

Figure 3.3 shows a section of water flowing in a circular pipe. For a general description of the flow, the cross-sectional area and the elevation of the pipe are allowed to vary along the longitudinal distance of the pipe.

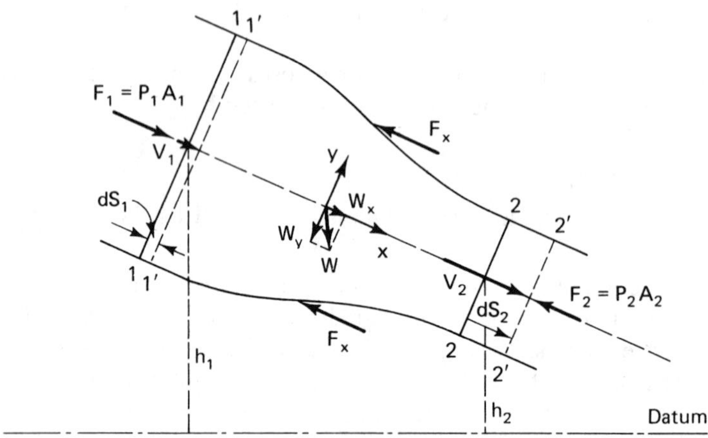

Figure 3.3 General description of flow in pipes.

A *control volume* is considered between sections 1–1 and 2–2. After a short time interval, *dt*, the mass originally occupying the control volume, has advanced to a new position, between sections 1′–1′, and 2′–2′.

For incompressible and steady flows, the mass flux that enters the control volume, $\rho \cdot d\mathrm{Vol}_{1\text{-}1'}$, should equal the mass flux that leaves the control volume, $\rho \cdot d\mathrm{Vol}_{2\text{-}2'}$. This is the principle of conservation of mass.

$$\rho \frac{d\mathrm{Vol}_{1-1'}}{dt} = \rho A_1 \frac{dS_1}{dt} = \rho A_1 V_1 = \rho A_2 V_2 = \rho Q$$

or

$$A_1 V_1 = A_2 V_2 = Q \tag{3.4}$$

Equation (3.4) is known to hydraulic engineers as the *continuity equation* for incompressible, steady flows.

Applying Newton's second law to the moving mass in the control volume,

$$\Sigma \vec{F} = M\vec{a} = M\frac{d\vec{V}}{dt} = \frac{M\vec{V}_2 - M\vec{V}_1}{\Delta t} \tag{3.5}$$

In this equation, both the forces and velocities are vector quantities. They must be balanced in every direction considered. Along the axial direction of the flow, the external forces exerted on the control volume may be expressed as

$$\Sigma F_x = P_1 A_1 - P_2 A_2 - F_x + W_x \tag{3.6}$$

where V_1, V_2 and P_1, P_2 are the velocities and pressures at sections 1–1 and 2–2, respectively. F_x is the axial direction force exerted on the control volume by the wall of the pipe. W_x is the axial component of the weight of the liquid in the control volume.

Substituting Equations (3.4) and (3.6) into Equation (3.5), the *principle of conservation of momentum* may be expressed, for the axial direction, as

$$\Sigma F_x = \rho Q (V_{x_2} - V_{x_1}) \tag{3.7}$$

Similarly, for the other directions,

$$\Sigma F_y = \rho Q (V_{y_2} - V_{y_1}) \tag{3.7a}$$

$$\Sigma F_z = \rho Q (V_{z_2} - V_{z_1}) \tag{3.7b}$$

In general, we may write in vector quantities

$$\Sigma \vec{F} = \rho Q (\vec{V}_2 - \vec{V}_1) \tag{3.7c}$$

Example 3.2

A horizontal nozzle, $d_B = 20$ mm, discharges 0.01 m^3 water into the air. The circular supply pipe's diameter $d_A = 40$ mm. The nozzle is held in place by a hinge mechanism as shown in Figure 3.4. Determine the reaction force at the hinge, if the pressure at A is 500,000 N/m^2.

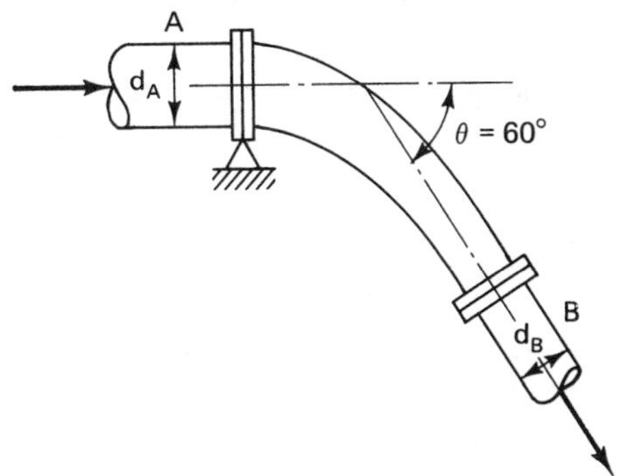

Figure 3.4

Solution

The hinge provides force to support the weight and the hydrodynamic force of the system. This force can be computed from the momentum equation (Equation 3.7),

$$\Sigma \vec{F} = \rho Q (\vec{V}_B - \vec{V}_A) \qquad \text{(Eq. 3.7c)}$$

the hydrostatic forces are

$$\vec{F}_A = P_A A_A \vec{i} = (500{,}000)\frac{\pi(0.04)^2}{4}\vec{i} = 628.32\ \vec{i}$$

$$\vec{F}_B = P_B A_B(-\cos 60\ \vec{i} + \sin 60\ \vec{j}) = \vec{0}$$

and the velocities are

$$\vec{V}_B = \frac{Q}{A_B}(\cos 60\ \vec{i} - \sin 60\ \vec{j}) = 31.83\left(\frac{1}{2}\vec{i} - \frac{3}{2}\vec{j}\right)$$

$$\vec{V}_A = \frac{Q}{A_A}\vec{i} = 7.96\ \vec{i}$$

so that

$$628.32\ \vec{i} + \vec{F} = (1000)(0.01)(15.92\ \vec{i} - 27.57\ \vec{j} - 7.96\ \vec{i})$$

$$\vec{F} = -548.7\ N\vec{i} - 275.7\ N\vec{j}$$

PROBLEMS

3.3.1 A 1-m pipe is carrying 1 m^3/sec water. The pipe has a 90° bend in the horizontal plane. The entrance and exit pressures to the bend are measured in height of water columns of 42 m and 41 m, respectively. Determine the force exerted by the water in the bend.

3.3.2 What is the force exerted on a fixed blade that deflects a 1-in. diameter jet with a velocity of 105 ft/sec by 180°?

3.3.3 At the end of a 0.6-m-diameter pipe, the pressure is 270,000 N/m^2. The pipe is connected to a 0.3-m-diameter nozzle. If the flowrate is 1.1 m^3/sec, determine the total force on the connection.

3.4 ENERGY HEAD IN PIPE FLOWS

Water flow in pipes may contain energy in various forms. The major portion of the energy is contained in three basic forms

1. kinetic energy,
2. potential energy,
3. pressure energy.

The three forms of energy may be demonstrated by using the general section of a pipe flow, as shown in Figure 3.3. The section of pipe flow approximately represents the concept of a stream tube that is a cylindrical passage with its surface everywhere parallel to the flow velocity; therefore, the flow cannot cross its surface.

Consider a control volume similar to that described in Figure 3.3, in time interval dt, the water particles at section 1–1 move to 1′–1′ with the velocity V_1. In the same time interval, the particles at section 2–2 move to 2′–2′ with the velocity V_2. To satisfy the continuity condition,

$$A_1 V_1 \, dt = A_2 V_2 \, dt$$

The work done by the pressure force acting on section 1–1 in the time dt is the product of the total pressure force and the distance through which it acts, or

$$P_1 A_1 \, ds_1 = P_1 A_1 V_1 \, dt \tag{3.8}$$

Similarly, the work done by the pressure force on section 2–2 is

$$-P_2 A_2 \, ds_2 = -P_2 A_2 V_2 \, dt \tag{3.9}$$

being negative because P_2 is in the opposite direction to the distance ds_2 traveled.

The work done by gravity on the entire mass of water in moving from 1122 to 1′1′2′2′ is the same as the work done if 111′1′ were moved to the position 222′2′ and the mass 1′1′22 was left undisturbed. The gravity force acting on the mass 111′1′ is equal to the volume $A_1 V_1 \, dt$ times the specific weight $\gamma = \rho g$. If h_1 and h_2 represent the elevations of the center of mass of 111′1′ and 222′2′ above a certain horizontal datum line, respectively, the work done by gravity force to move the mass from h_1 to h_2 is

$$\rho g A_1 V_1 \, dt \cdot (h_1 - h_2) \tag{3.10}$$

The net gain in kinetic energy in the entire mass is

$$\frac{1}{2} M V_2^2 - \frac{1}{2} M V_1^2 = \frac{1}{2} \rho A_1 V_1 \, dt \, (V_2^2 - V_1^2) \tag{3.11}$$

since the total work done on a given mass by all the forces is equal to the change in kinetic energy. The above equations (3.8, 3.9, 3.10, and 3.11) may be combined to give

$$P_1 Q \, dt - P_2 Q \, dt + \rho g Q \, dt \, (h_1 - h_2) = \frac{1}{2} \rho Q \, dt \, (V_2^2 - V_1^2)$$

Dividing both sides by $\rho g Q \, dt$, results in the *Bernoulli Equation* expressed in terms of energy per unit weight of water, or the energy head,

$$\frac{V_1^2}{2g} + \frac{P_1}{\gamma} + h_1 = \frac{V_2^2}{2g} + \frac{P_2}{\gamma} + h_2 \tag{3.12}$$

In reality, however, certain amount of energy loss occurs when the water mass flows from one section to another. The engineering application of the energy equation is discussed as follows.

The algebraic sum of the kinetic head, the pressure head, and the elevation head accounts for nearly all the energy contained in the unit weight of water flowing through a particular section of a pipe.

Figure 3.5 shows, schematically, the energy heads at two sections along a pipeline. At section 1, the upstream section, the three energy heads are $V_1^2/2g$, P_1/γ, and h_1, respectively. The algebraic sum of these three heads gives the point a above the energy datum. The distance measured between points a and b on the energy datum represents the total energy contained in each unit weight of water that passes through section 1.

$$H_1 = \frac{V_1^2}{2g} + \frac{P_1}{\gamma} + h_1 \tag{3.13}$$

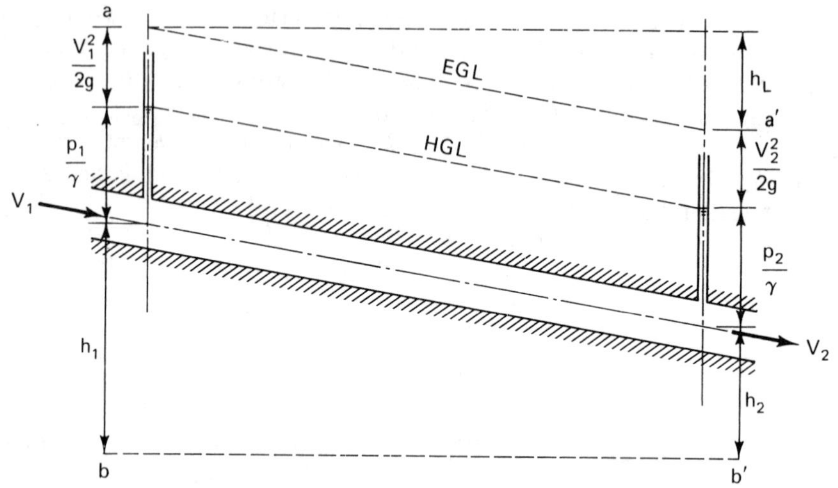

Figure 3.5 Energy head and head loss in pipe flow.

During the journey between the two sections, a certain amount of the energy is lost. The remaining total energy in each unit weight of water at section 2 can be measured by the distance between points a' and b' on the energy datum. This distance, as shown in Figure 3.5, is the sum of the kinetic head, the pressure head, and the potential head at section 2.

$$H_2 = \frac{V_2^2}{2g} + \frac{P_2}{\gamma} + h_2 \tag{3.14}$$

The difference in elevations between points a and a' represents the head loss, h_L, between sections 1 and 2. The energy relationship between the two sections can be written in the following form:

$$\frac{V_1^2}{2g} + \frac{P_1}{\gamma} + h_1 = \frac{V_2^2}{2g} + \frac{P_2}{\gamma} + h_2 + h_L \tag{3.15}$$

This relationship is commonly referred to as *Bernoulli's equation* by hydraulic engineers.

Example 3.3

A 25-cm circular pipe carries 0.16 m^3/sec of water under the pressure of 2000 dyn/cm^2. The pipe is laid at an elevation of 10.71 m above the mean sea level (MSL) of Freeport, Texas. What is the energy head measured with respect to the MSL?

Solution

The continuity condition, Equation (3.4), requires that

$$Q = AV$$

Hence,

$$V = \frac{Q}{A} = \frac{0.16}{\frac{\pi}{4} \cdot \left(\frac{25}{100}\right)^2} = 3.26 \text{ m/sec}$$

The total energy head measured with respect to the MSL is

$$\frac{V^2}{2g} + \frac{P}{\gamma} + h = \frac{(3.26)^2}{2(9.81)} + \frac{2000\left(\frac{10^4}{10^5}\right)}{9810} + 10.71$$

$$= 11.27 \text{ m}$$

Example 3.4

The elevated water tank shown in Figure 3.6 is being drained to an underground storage through a 12 in. diameter pipe. The flowrate is 3200 gallons per minute (GPM) and the total head loss is 11.53 ft. Determine the water surface elevation in the tank.

Solution

Energy relationship (Equation 3.13) can be established between section 1 at the reservoir, and section 2 at the end of the pipe.

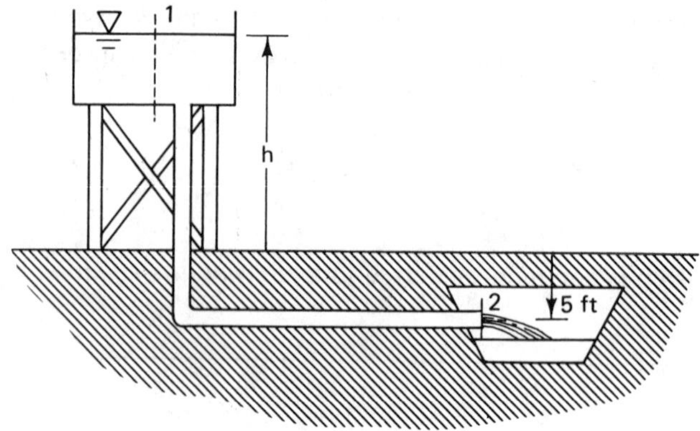

Figure 3.6

$$\frac{V_1^2}{2g} + \frac{P_1}{\gamma} + h_1 = \frac{V_2^2}{2g} + \frac{P_2}{\gamma} + h_2 + h_L$$

The water velocity in the reservoir, being small compared to that in the pipe, can be neglected. Furthermore, both sections are exposed to atmospheric pressure, so that

$$P_1 = P_2 = 0$$

The mean velocity is,

$$V = \frac{Q}{A} = \frac{3200 \text{ GPM}}{\pi r^2} = \frac{3200 \text{ GPM}}{(0.5 \text{ ft})^2 \pi} \cdot \frac{1 \text{ ft}^3/\text{s}}{448.83 \text{ GPM}} = 9.08 \text{ ft/sec}$$

and

$$h = h_1 = \frac{V_2^2}{2g} + h_2 + h_L = \frac{(9.08)^2}{2(32.2)} - 5 + 11.53 = 7.81 \text{ ft}$$

3.5 LOSS OF HEAD DUE TO FRICTION

Energy loss through friction in the length of a pipeline is commonly termed as the *major loss*, h_f. This is the loss of head due to pipe friction and to the viscous dissipation in flowing water. Several studies have been performed during the past century on laws that govern the loss of head by pipe friction. It has been learned from these studies that resistance to flow in a pipe is

1. Independent of the pressure under which the water flows,
2. Linearly proportional to the pipe length, L,

3. Inversely proportional to some power of the pipe diameter, D,
4. Proportional to some power of the mean velocity, V,
5. Related to the roughness of the pipe, if the flow is turbulent.

Several experimental formulas have been developed in the past. Some of these formulas have faithfully been used in various hydraulic engineering practices.

The most popular pipe flow formula was derived by Henri Darcy (1803–1858), Julius Weisbach (1806–1871), and others about the middle of the nineteenth century. The formula takes the following form:

$$h_f = f\left(\frac{L}{D}\right)\frac{V^2}{2g} \tag{3.16}$$

This formula is commonly known as the Darcy-Weisbach formula. It is conveniently expressed in terms of the velocity head in the pipe. Moreover, it is dimensionally uniform since in engineering practice the *friction factor, f,* is treated as a dimensionless numerical factor; h_f and $V^2/2g$ are both in units of length.

In laminar flow the friction factor f can be determined by balancing the viscous force and the pressure force at the two end sections of a horizontal pipe separated by a distance L. In a cylindrical portion of radius r (see Figure 3.7) the difference in pressure force between the two ends of the cylinder is $(P_1 - P_2) \cdot \pi r^2$ and the viscous force on the cylinder is equal to $2\pi rL \cdot \tau$. The values of shear stress τ have been previously shown, Equation (1.2), to be $\mu(dv/dr)$. Under equilibrium condition, when the pressure force and the viscous force on the cylinder of water are balanced, we may write

$$-2\pi rL \cdot \left(\mu \frac{dv}{dr}\right) = (P_1 - P_2)\pi r^2$$

The minus sign is used because the velocity decreases as the radial position, r, increases; dv/dr is always negative in a pipe flow. This equation can be inte-

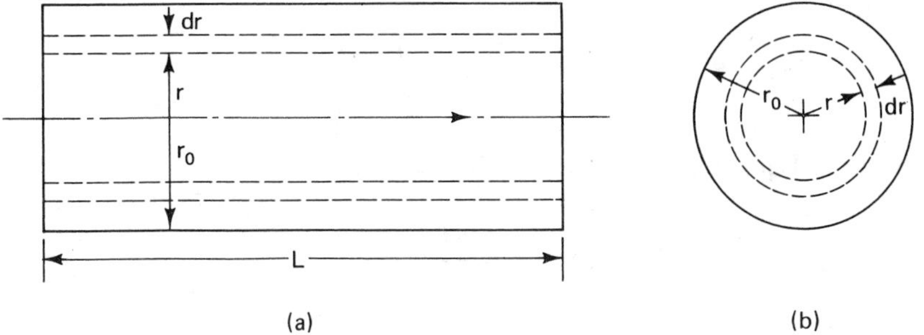

Figure 3.7 Geometry of a circular pipe.

grated to give the general expression of flow velocity in terms of radial position r.

$$v = \frac{P_1 - P_2}{4\mu L}(r_0^2 - r^2) \tag{3.17}$$

where r_0 is the inner radius of the pipe, and the equation shows that the velocity distribution in a laminar pipe flow is a parabolic function of radius, r. The total discharge through the pipe can be obtained by integrating the discharge through the elemental area, $2\pi r \cdot dr$

$$\begin{aligned} Q &= \int dQ = \int v \cdot dA = \int_{r=0}^{r=r_0} \frac{P_1 - P_2}{4\mu L}(r_0^2 - r^2) \cdot 2\pi r \cdot dr \\ &= \frac{\pi r_0^4(P_1 - P_2)}{8\mu L} = \frac{\pi D^4(P_1 - P_2)}{128\mu L} \end{aligned} \tag{3.18}$$

This relationship is also known as the Hagen-Poiseuille law* of laminar flow. The mean velocity is

$$\begin{aligned} V &= \frac{Q}{A} = \frac{\pi D^4(P_1 - P_2)}{128\mu L}\left(\frac{1}{\frac{1}{4}\pi D^2}\right) \\ V &= \frac{(P_1 - P_2)D^2}{32\mu L} \end{aligned} \tag{3.19}$$

For a horizontal uniform pipe, Bernoulli's equation leads to

$$h_f = \frac{P_1 - P_2}{\gamma}$$

The Darcy-Weisbach formula may be written as

$$\frac{P_1 - P_2}{\gamma} = f\left(\frac{L}{D}\right)\frac{V^2}{2g} \tag{3.20}$$

Comparing Equations (3.19) and (3.20), we have

$$f = \frac{64}{\gamma}\frac{\mu g}{VD}$$

* Experimentally derived by G.W. Hagen (1839) and later independently obtained by J.L.M. Poiseuille (1840).

Since $\gamma = \rho g$,

$$f = \frac{64\mu}{\rho VD} = \frac{64}{N_R} \tag{3.21}$$

which indicates a direct relationship between the friction factor, f, and the Reynolds number, N_R, for laminar pipe flow. It is independent of the surface roughness of the pipe.

When the Reynolds number approaches a higher value, say, $N_R >> 2000$, the flow in the pipe becomes practically turbulent and the value of f then becomes less dependent on the Reynolds number but more dependent on the *relative roughness height*, e/D, of the pipe. The quantity e is the measure of the average roughness height of the pipe wall, as previously discussed, and D is the pipe diameter. The roughness height of commercial pipes is commonly described by providing a value of e to the pipe. It means that the pipe has the same value of f at high Reynolds numbers that would be obtained if the pipe were coated with sand grains of a uniform size e. The roughness height for certain common commercial pipe materials is provided in Table 3.1.

TABLE 3.1 Roughness Height, *e*, for Certain Common Pipe Materials

Pipe Material	e(mm)	e(ft)
Glass, drawn brass, copper (new)	0.0015	0.000005
Seamless commercial steel (new)	0.004	0.000013
Commercial steel (enamel coated)	0.0048	0.000016
Commercial steel (new)	0.045	0.00015
Wrought iron (new)	0.045	0.00015
Asphalted cast iron (new)	0.12	0.0004
Galvanized iron	0.15	0.0005
Cast iron (new)	0.26	0.00085
Wood Stave (new)	0.18 ~ 0.9	0.0006 ~ 0.003
Concrete (steel forms, smooth)	0.18	0.0006
Concrete (good joints, average)	0.36	0.0012
Concrete (rough, visible, form marks)	0.60	0.002
Riveted steel (new)	0.9 ~ 9.0	0.003-0.03
Corrugated metal	45	0.15

It has been determined that immediately next to the pipe wall there exists a very thin layer of flow commonly referred to as the *laminar sublayer* even though the pipe flow is turbulent. The thickness of the laminar sublayer δ' decreases with an increase in the pipe's Reynolds number. A pipe is said to be *hydraulically smooth* if the average roughness height is less than the thickness of the laminar sublayer. In a hydraulically smooth pipe flow the friction factor f is not affected by the surface roughness of the pipe.

Based on laboratory experimental data, it has been found that if $\delta' > 1.7\, e$, the effect of surface roughness is completely submerged by the laminar sublayer and the pipe flow is hydraulically smooth. In this case, Th. von Karman* developed an equation for the friction factor f

$$\frac{1}{\sqrt{f}} = 2 \log \left(\frac{N_R \sqrt{f}}{2.51} \right) \tag{3.22}$$

At high Reynolds numbers, δ' becomes very small. If $\delta' < 0.08\, e$, it has been found that the friction factor, f, becomes independent of the Reynolds number and depends only on the relative roughness height. In this case, the pipe behaves as a *hydraulically rough pipe*, and von Karman found that the friction factor f can be expressed as

$$\frac{1}{\sqrt{f}} = 2 \log \left(3.7 \frac{D}{e} \right) \tag{3.23}$$

In between these two extreme cases, if $0.08\, e < \delta' < 1.7\, e$, the pipe behaves neither smoothly nor completely roughly. C.F. Colebrook† devised an approximate relationship for this intermediate range

$$\frac{1}{\sqrt{f}} = -2 \log \left(\frac{\frac{e}{D}}{3.7} + \frac{2.51}{N_R \sqrt{f}} \right) \tag{3.24}$$

Obviously, it is cumbersome to use any of these three equations in engineering practice. A convenient chart was prepared by Lewis F. Moody‡ (see Figure 3.8) and is commonly called the *Moody diagram of friction factors for pipe flow*.

The chart clearly shows the four zones of pipe flow

1. a laminar flow zone where the friction factor f is a simple linear function of the Reynolds number;
2. a critical zone (shaded) where values are uncertain because the flow might be neither laminar nor truly turbulent;
3. a transitional zone where f is a function of both the Reynolds number and the relative roughness height of the pipe wall;
4. a zone of fully developed turbulence where the value of f depends solely on the relative roughness height and is independent of the Reynolds number.

* Th. von Karman, "Mechanische Ähnlichkeit und Turbulenz" (Mechanical similitude and turbulence), Proc. 3rd International Congress for Applied Mechanics, Stockholm, Vol. I (1930).

† C.F. Colebrook, "Turbulent flow in pipes, with particular reference to the transition region between smooth and rough pipe laws," *Jour. Ist. Civil Engrs.*, London (Feb. 1939).

‡ L.F. Moody, "Friction factors for pipe flow," *Trans. ASME*, 66 (1944).

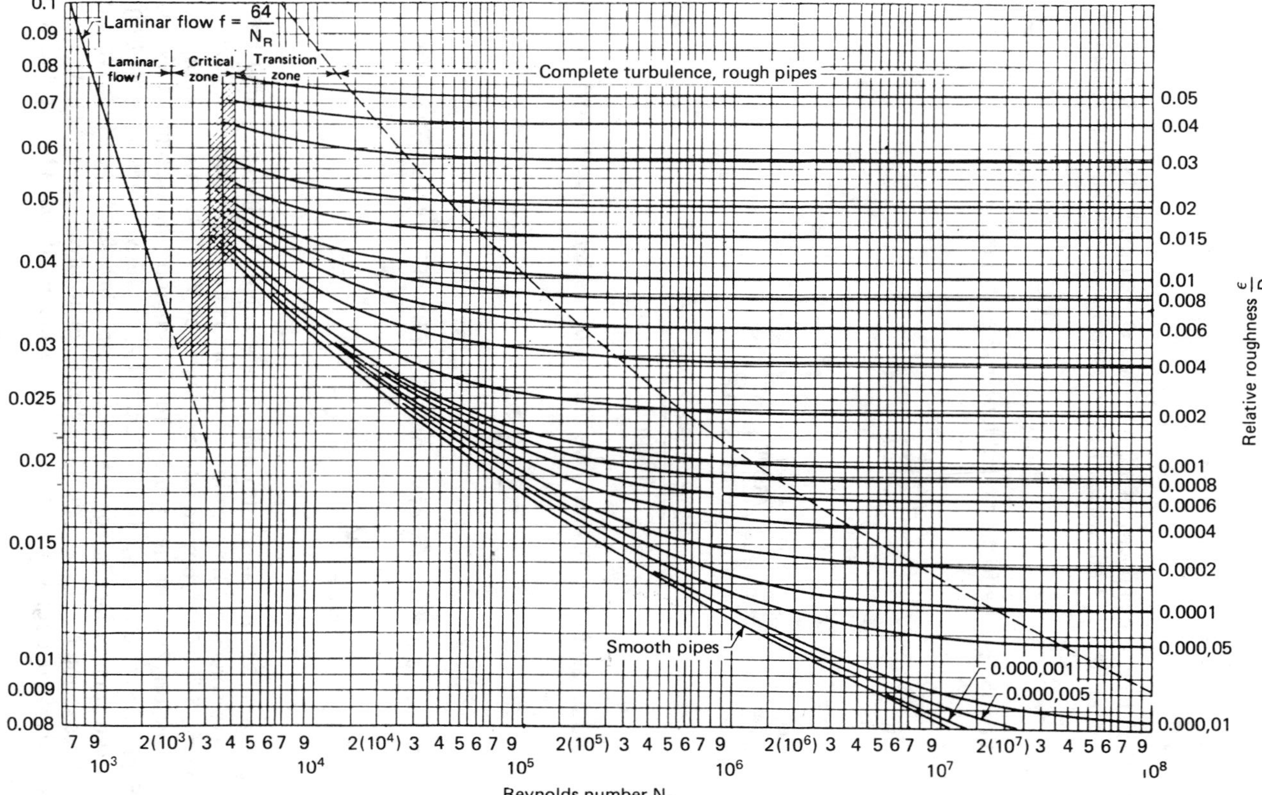

Figure 3.8 Friction factors for flow in pipes, the Moody diagram (From L. F. Moody, "Friction factors for pipe flow," *Trans. ASME*, vol. 66, 1944.)

Figure 3.8 may be used together with Table 3.1 to obtain the friction factor f for circular pipes.

Example 3.5

Compute the discharge capacity of a 3-m, wood-stave pipe in its best condition carrying water at 10°C. It is allowed to have a head loss of 2 m/km of pipe length.

Solution

From Equation (3.16), the friction head loss in the pipe is,

$$h_f = f\left(\frac{L}{D}\right)\frac{V^2}{2g}$$

Hence,

$$2 = f\left(\frac{1000}{3}\right)\frac{V^2}{2(9.81)}$$

$$V^2 = \frac{2 \cdot 3 \cdot 2 \cdot 9.81}{1000f} = 0.12/f \tag{1}$$

From Table 3.1, taking $e = 0.3$ mm, we obtain for the 3-m, wood-stave pipe:

$$\frac{e}{D} = 0.0001$$

At 10°C the kinematic viscosity of water is $\nu = 1.31 \cdot 10^{-6}$ m^2/sec. Therefore,

$$N_R = \frac{DV}{\nu} = \frac{3V}{1.31 \cdot 10^{-6}} = 2.29 \cdot 10^6 V \tag{2}$$

By using Figure 3.8, Equations (1) and (2) are solved by iteration until both conditions are satisfied. The iteration procedure is demonstrated as follows:

Assume $f = 0.02$. From Equation (1) we get $V = 2.45$ m/sec, using this value in Equation (2) we get $N_R = 5.6 \times 10^6$. This number is taken to the Moody chart (Figure 3.8) to obtain the friction factor $f = 0.0122$, which is different from the assumed friction factor. For the second iteration, let $f = 0.0122$; we get $V = 3.14$ m/sec and $N_R = 7.2 \times 10^6$. The Moody chart gives $f = 0.0121$, which is considered close enough to the assumed value and will be used to calculate the flow velocity,

$$V = 3.15 \text{ m/sec}$$

This value and the cross-sectional area give the discharge.

$$Q = AV = \pi\left(\frac{3^2}{4}\right)(3.15)$$

$$= 22.27\ \text{m}^3/\text{sec}$$

Example 3.6

Estimate the size of a uniform, horizontal welded-steel pipe installed to carry 500 ℓ/sec of water at 10°C. The allowable pressure loss is 5 m/km of pipe length.

Solution

Bernoulli's equation can be applied to two pipe sections 1 km apart,

$$\frac{V_1^2}{2g} + \frac{P_1}{\gamma} + h_1 = \frac{V_2^2}{2g} + \frac{P_2}{\gamma} + h_2 + h_L$$

For a uniform, horizontal pipe with no localized head losses

$$V_1 = V_2$$

$$h_1 = h_2$$

$$h_L = h_f$$

and the energy equation reduces to

$$\frac{P_1}{\gamma} - \frac{P_2}{\gamma} = h_L = 5\ \text{m}$$

From Equation 3.16

$$h_f = f\frac{L}{D}\frac{V^2}{2g} = f\frac{L}{D}\frac{Q^2}{2g(\pi D^2/4)^2} = \frac{8fLQ^2}{g\pi^2 D^5}$$

Therefore,

$$D^5 = \frac{8fLQ^2}{g\pi^2 h_f} = 4.13\ f \qquad \text{(a)}$$

where $L = 1000$ m, and $h_f = 5$ m. At 10°C, $\nu = 1.31 \times 10^{-6}$ m²/sec. Assuming welded-steel roughness to be in the lower range of riveted steel, $e = 0.9$ mm, the diameter can then be found using the Moody chart (Figure 3.8) by means of an iteration procedure as follows.

Let $D = 0.8$ m, then

$$V = \frac{Q}{A} = \frac{0.5\ \text{m}^3/\text{sec}}{\pi(0.4\ \text{m})^2} = 0.995\ \text{m/sec}$$

and

$$N_R = \frac{VD}{\gamma} = \frac{(0.995)(0.8)}{1.31 \times 10^{-6}} = 6.1 \times 10^5$$

$$e/D = \frac{0.9 \text{ mm}}{800 \text{ mm}} = 0.0011$$

Entering these values into the Moody chart, we get $f = 0.021$. A better estimate of D can be obtained by substituting the latter values into Equation (a), which gives:

$$D = (0.021 \times 4.13)^{1/5} = 0.61 \text{ m}$$

A new iteration provides $V = 1.71$ m/sec, $N_R = 8.0 \times 10^5$, $e/D = 0.0015$, $f = 0.022$, and $D = 0.62$ m. More iterations will produce the same result.

PROBLEMS

3.5.1 A 30-cm circular cast-iron pipeline, 2 km long, carries water at 10°C. What is the maximum discharge if a 4.6-m head loss is allowed?

3.5.2 A new, asphalted cast-iron pipe, 45 cm in diameter, carries water from a reservoir to a location 1.37 km away and discharges the water into the air. If the entrance to the pipe is 3 m below the water surface in the reservoir and the pipe is laid on a 1/100 slope, determine the discharge at 10°C.

3.5.3 A horizontal, commercial steel pipe, 0.5 m in diameter, carries 0.4 m^3/sec of water at 20°C. Calculate the pressure change in the pipe per kilometer length.

3.5.4 A 6-cm galvanized iron pipe is installed on a 1/50 slope and carries water at 85°C. What is the pressure drop in the 20-m-long pipe when the discharge is 8.2 ℓ/sec?

3.5.5 Two sections, A and B, are 5 km apart along a 4-m riveted-steel pipe in its best condition. A is 100 m higher than B. If the water temperature is 15°C and the pressure heads measured at A and B are 8.3 m and 76.7 m, respectively, what is the flow rate?

3.5.6 The city water company wants to transport 1800 m^3 of water per day to a plant from a reservoir 8 km away. The water surface elevation at the reservoir is 5 m above the entrance of the pipe, and the pipe may be laid on a 1/500 slope. What is the minimum diameter of a concrete pipe that may be used if the water temperature varies between 5°C and 20°C? Assume pipe discharges into open air and neglect minor losses.

3.5.7 Water flows through a hydraulically smooth pipe and the friction factor is 0.03. Determine the friction factor if the flow increases ten times.

3.5.8 Water flowrate in a 6-in. galvanized iron pipe is 1.34 ft^3/sec. Determine the pressure drop in 1000 ft of horizontal pipe if the water temperature is 68°F.

3.5.9 Water flows through a 20-cm commercial steel pipe at 10°C. The flowrate is 80 ℓ/sec and pressure is constant throughout the length of the pipe. Determine the slope on which it is laid.

3.5.10 Determine the pressure change in 1 km of pipe laid on a 1/200 slope. The new cast-iron pipe is 15 cm in diameter and carries a discharge of 0.070 m^3/sec at 20°C.

3.5.11 Determine the flowrate of water at 10°C that will cause a pressure drop of 17,250 N/m^2 in 350 m of horizontal, cast-iron pipe, 60 cm in diameter.

3.5.12 Water at 10°C is transported through a 200-m-long uniform cast-iron pipe with head loss of 7.8 m. Determine the diameter of the wrought iron pipe that will convey 10 ℓ/sec.

3.5.13 Water flows in a new, horizonal cast-iron pipe (30 cm in diameter). A leakage in the pipeline is noted. A pair of pressure gages located upstream of the leak indicate a pressure drop of 23,000 N/m^2. Another pair of pressure gages located downstream of the leak indicate a pressure drop of 21,000 N/m^2. The distance between the gages in each pair is 100 m. Determine the magnitude of the leak.

3.6 EMPIRICAL FORMULAS FOR FRICTION HEAD LOSS

Throughout the history of civilization hydraulic engineers have designed and built systems to deliver water for people to use. In the earlier days, many of these designs were based on empirical formulas. Generally speaking, empirical design formulas were developed from the experiences of dealing with fluid flow under certain conditions in a specific range. Normally, they do not have a sound analytical basis. For this reason, the empirical formulas may not be dimensionally correct, and, if so, can only be applicable to the conditions and ranges specified.

One of the best examples is the Hazen-Williams formula, which was developed for water flow in larger pipes ($D \geq 5$ cm) within a moderate range of water velocity ($V \leq 3$ m/s). This formula has been used extensively for the designing of water-supply systems in the United States. The Hazen-Williams formula, originally developed for the British measurement system, has been written in the form

$$V = 1.318 C_{HW} R_h^{0.63} S^{0.54} \tag{3.25}$$

where S is the slope of the *energy gradient line* (EGL), or head loss per unit length of the pipe ($S = h_f/L$), and R_h is the *hydraulic radius*, defined as the water cross-sectional area, A, divided by the *wetted perimeter*, P. For a circular pipe, $A = \pi D^2/4$ and $P = \pi D$, the hydraulic radius is

$$R_h = \frac{A}{P} = \frac{\frac{\pi D^2}{4}}{\pi D} = \frac{D}{4} \tag{3.26}$$

The Hazen-Williams coefficient, C_{HW}, is not a function of the flow conditions (i.e., Reynolds number). Its values range from 140 for very smooth, straight pipe down to 90 or 80 for old, unlined tuberculated pipe. Generally, the value of 100 is taken for average conditions. The values of C_{HW} for commonly used water-carrying conduits are listed in Table 3.2.

TABLE 3.2 Hazen-Williams Coefficient, C_{HW}, for Different Types of Pipe

Pipe Materials	C_{HW}
Asbestos Cement	140
Brass	130–140
Brick sewer	100
Cast-iron	
New, unlined	130
10 yr. old	107–113
20 yr. old	89–100
30 yr. old	75–90
40 yr. old	64–83
Concrete or concrete lined	
Steel forms	140
Wooden forms	120
Centrifugally spun	135
Copper	130–140
Galvanized iron	120
Glass	140
Lead	130–140
Plastic	140–150
Steel	
Coal-tar enamel lined	145–150
New unlined	140–150
Riveted	110
Tin	130
Vitrified clay (good condition)	110–140
Wood stave (average condition)	120

Note that the Hazen-Williams formula as shown in Equation (3.25) is applicable *only* for the British units in which the velocity is measured in feet per second and the hydraulic radius R_h is measured in feet. When used in S.I. units, the Hazen-Williams formula may be written in the following form:

$$V = 0.85 C_{HW} R_h^{0.63} S^{0.54} \tag{3.27}$$

Solution for the Hazen-Williams formula, Equation (3.27), may be obtained by direct computation or by the use of a solution chart, as demonstrated in the following example.

Example 3.7

A 100-m-long pipe with $D = 20$ cm and $C_{HW} = 120$ carries a discharge of 30 ℓ/sec. Determine the head loss in the pipe.

Solution

a. **Direct Computation**

$$\text{Area:}\quad A = \frac{\pi D^2}{4} = \frac{\pi}{4}(0.2)^2 = 0.0314 \text{ m}^2,$$

$$\text{Wetted Perimeter:}\quad P = \pi D = 0.2\pi = 0.628 \text{ m}$$

$$\text{Hydraulic Radius:}\quad R_h = A/P = \frac{0.0314}{0.628} = 0.05$$

Applying Equation 3.27,

$$V = \frac{Q}{A} = 0.85\, C_{HW} R_h^{0.63} S^{0.54}$$

$$\frac{0.03}{0.0314} = 0.85 \times 120(0.05)^{0.63}\left(\frac{h_f}{100}\right)^{0.54}$$

$$h_f = 0.58 \text{ m}$$

b. **Solution Chart**

Solution charts, such as Figure 3.9, are useful to provide a quick, qualitative estimation of the problem. In engineering practice, such an estimation is frequently found helpful in understanding the magnitude of the problem. Solutions obtained from the chart do not have the resolution needed for engineering design purposes.

Applying the conditions given to the chart, the following steps are suggested:

1. Locate the discharge, $Q = 0.03$ m^3/sec, and the diameter of the pipe, $D = 0.2$ m, in the appropriate columns.
2. Connect these two points with a straight line and extend the line to the right until it meets the "Turning Point" column. Extending the line further to the far right column gives the mean velocity, V, in the pipe.

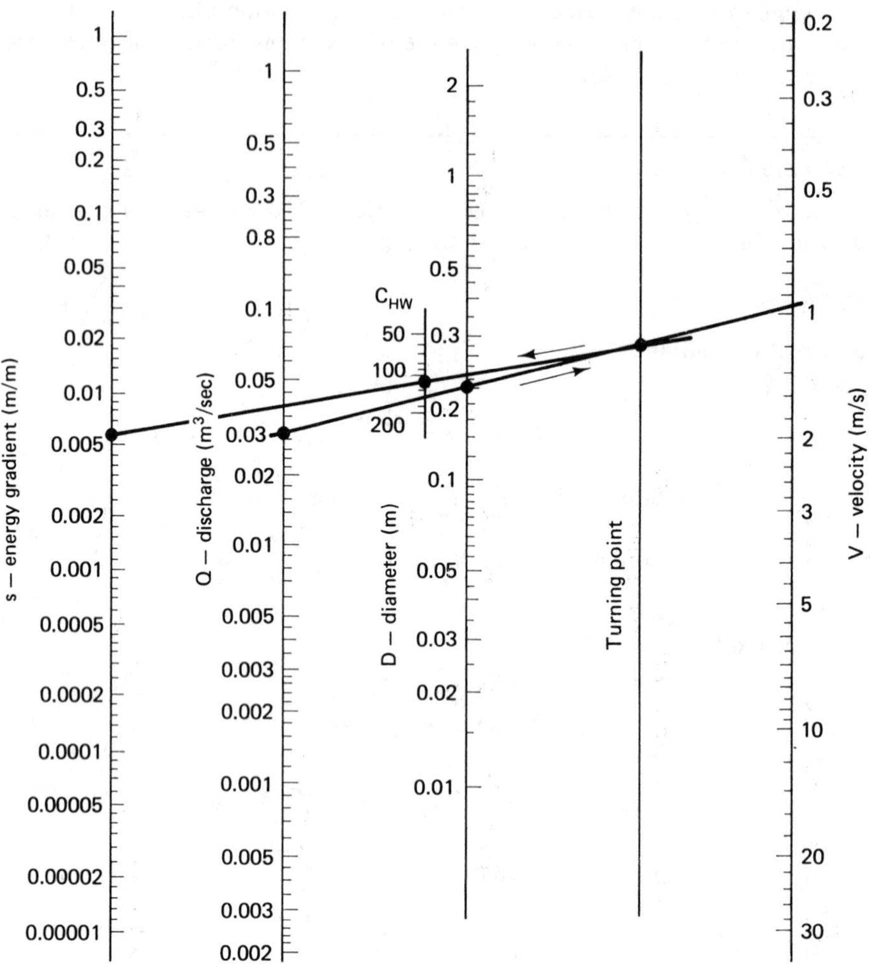

Figure 3.9 Solution chart for the Hazen-Williams formula.

3. Locate the value $C_{HW} = 120$ on the short column. Draw a straight line to connect this point and the point on the "Turning point" column. Extend the line further to meet the far left column and the value of the energy slope, $S = 0.006$ m/m, is obtained.

The energy slope is defined as the energy loss per unit pipe length, or

$$S = \frac{h_f}{L} = \frac{h_f}{100} = 0.006 \text{ m/m}$$

Hence,

$$h_f = 0.006 \text{ m/m} \cdot 100 \text{ m} = 0.60 \text{ m}$$

Another popular empirical formula is the *Manning formula* originally developed in the metric units. The Manning formula has extensively been used for open channel designs (discussed in detail in Chapter 6). It is also quite commonly used for pipe flows. The Manning formula may be expressed in the following form:

$$V = \frac{1}{n} R_h^{2/3} S^{1/2} \tag{3.28}$$

where n is Manning's coefficient of roughness, specifically known to hydraulic engineers as Manning's n.

In English units, the Manning formula is written as

$$V = \frac{1.49}{n} R_h^{2/3} S^{1/2} \tag{3.29}$$

where the hydraulic radius, R_h, is measured in feet, and the velocity is measured in units of feet per second. Table 3.3 contains typical values of n for water flow in common pipes.

TABLE 3.3 Manning's Roughness Coefficient, *n*, for Pipe Flows

	Manning's n	
Type of Pipe	*Min.*	*Max.*
Glass, brass, or copper	0.009	0.013
Smooth cement surface	0.010	0.013
Wood-stave	0.010	0.013
Vitrified sewer pipe	0.010	0.017
Cast-iron	0.011	0.015
Concrete, precast	0.011	0.015
Cement mortar surfaces	0.011	0.015
Common-clay drainage tile	0.011	0.017
Wrought iron	0.012	0.017
Brick with cement mortar	0.012	0.017
Riveted-steel	0.017	0.020
Cement rubble surfaces	0.017	0.030
Corrugated metal storm drain	0.020	0.024

Solution to the Manning formula may also be obtained by direct computation or from a solution chart as provided in Figure 3.10. The following example demonstrates the use of (a) direct computation, and (b) the solution chart, for solution of the Manning formula.

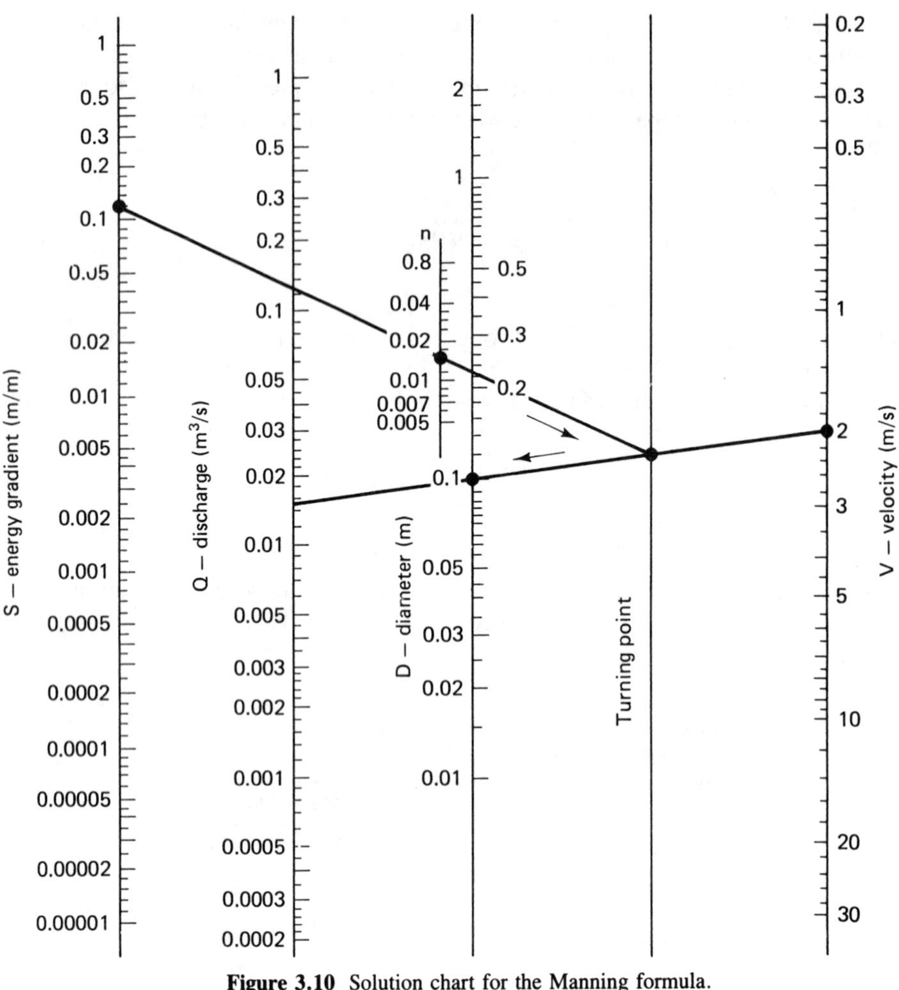

Figure 3.10 Solution chart for the Manning formula.

Example 3.8

A horizontal pipe with 10 cm uniform diameter is 200 m long. Manning's roughness coefficient is $n = 0.015$ and the measured pressure drop is 24.6 m in water column. Determine the discharge.

Solutions

a. **Direct Computation:**

$$\text{Area: } A = \frac{\pi}{4}D^2 = \frac{\pi}{4}(0.1)^2 = 0.00785 \text{ m}^2$$

$$\text{Wetted Perimeter: } P = \pi D = 0.1\ \pi = 0.314 \text{ m}$$

$$\text{Hydraulic Radius: } R_h = A/P = \frac{0.00785}{0.314} = 0.025 \text{ m}$$

$$\text{Energy Slope: } S = h_f/L = \frac{24.6 \text{ m}}{200 \text{ m}} = 0.123$$

Manning's Roughness Coefficient: $n = 0.015$

Substituting the above quantities into the Manning equation, Equation (3.28), we have

$$V = \frac{Q}{A} = \frac{1}{n} R_h^{2/3} S^{1/2}$$

$$Q = \frac{1}{0.015}(0.00785)(0.025)^{2/3}(0.123)^{1/2} = 0.0157 \text{ m}^3/\text{sec}$$

b. **Solution Chart:**
Applying the given conditions to the chart, Figure 3.10, the following steps are suggested:

1. Locate the values of the energy slope, $S = 0.123$ m/m, and the Manning's roughness coefficient, $n = 0.015$, in the appropriate columns.
2. Connect the two points with a straight line and extend the line to the right to meet the "Turning Point" column.
3. Locate the pipe diameter, $D = 0.1$ m, on the middle column. Connect this point and the point on the "Turning Point" column with a straight line. Extrapolate the line to the left to obtain the discharge, $Q = 0.015$ m^3/sec; and extrapolate the line to the right to obtain the mean velocity, $V = 2$ m/sec.

PROBLEMS

3.6.1 A 6-km-long, new cast-iron pipeline carries 320 ℓ/sec of water at 30°C. The pipe diameter is 30 cm. Compare the head loss calculated from (a) the Hazen-Williams formula, (b) the Manning formula, and (c) the Darcy-Weisbach formula.

3.6.2 An 80-cm riveted-steel pipe carries water at 10°C from reservoir A to reservoir B. The length of the pipe is 3.2 km. The elevation difference between the two reservoirs is 102 m. Calculate the discharge by (a) the Hazen-Williams formula, (b) the Manning formula, and (c) the Darcy-Weisbach formula.

3.6.3 Repeat the calculation in Problem 3.4.4 but use (a) the Manning formula, and (b) the Hazen-Williams formula, and compare the results with those obtained from the Darcy-Weisbach formula. Discuss the differences.

3.6.4 Two reservoirs 1200 m apart are connected by a 50-cm smooth concrete pipe. If the two reservoirs have an elevation difference of 5 m, determine the discharge in the pipe by (a) the Hazen-Williams formula, (b) the Manning formula, and (c) the Darcy-Weisbach formula.

3.6.5 Research in the library and find, in addition to the three introduced in this section, six empirical formulas involving head loss in pipe lines. List the author(s) and limitations of each formula.

3.6.6 A lined concrete tunnel with semicircular cross section (radius 1.0 m) has a Hazen-Williams coefficient of 120. What is the head loss if the tunnel carries a discharge of 3 m^3/sec over a 1.2-km distance?

3.6.7 The elevation difference between two reservoirs 2000 m apart is 20 m.

(a) Compute the flowrate if a 30-cm commercial steel (C_{HW} = 140) pipeline connects the reservoirs.

(b) Compute the flowrate if two 20-cm commercial steel pipelines are used instead.

3.6.8 In Problem 3.5.7, use Manning's equation (commercial steel, $n = 0.013$, $e = 0.045$ *mm*) to compute the flowrate. Compare the results from these two empirical formulas (Mannings, Equation 3.29 and Hazen-Williams, Equation 3.26) to that of the Darcy-Weisbach equation (Equation 3.16).

3.6.9 Based on Problems 3.5.7. and 3.5.8, determine the percent error in flowrate that will occur if,

(a) the estimate of C_{HW} is off by 15%,

(b) the estimate of n is off by 15%.

3.7 LOSS OF HEAD DUE TO CONTRACTION

Sudden contraction in a pipe usually causes a marked drop in pressure in the pipe due to both the increase in velocity and the loss of energy to turbulence. The phenomenon of a sudden contraction is schematically represented in Figure 3.11.

The vertical distance measured between the energy gradient line (EGL) and the pipe centerline represents the total energy head at any particular location along the pipe. The vertical distance measured between the hydraulic gradient line (HGL) and the pipe centerline represents the pressure head, P/γ, and the distance between the EGL and HGL is the velocity head, $V^2/2g$, at the location. After point B the HGL begins to drop as the stream picks up speed and a region of stagnant water appears at the corner of contraction C. Immediately downstream from the contraction the streamlines separate from the pipe wall and form a high-speed jet that reattaches to the wall at point D. The phenomenon that takes place between C and D is known to hydraulic engineers as the *vena contracta*, which will be discussed in detail in Chapter 9. Most of the energy loss in a pipe

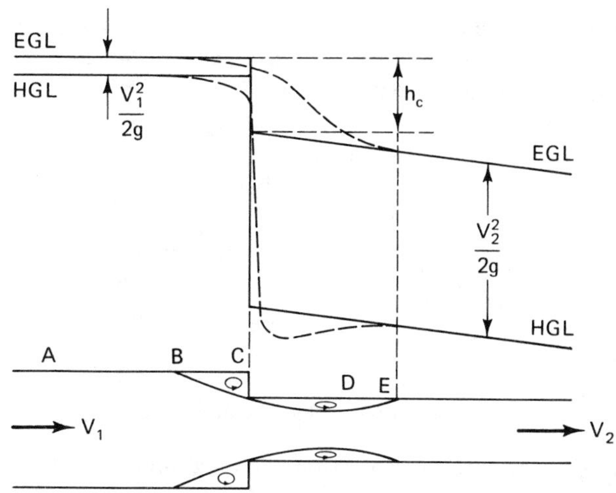

Figure 3.11 Head loss and pressure variation due to sudden contraction.

contraction takes place between C and D where the jet stream velocity is high and the pressure is low. A certain amount of pressure recovers between D and E as the jet stream gradually dissipates and normal pipe flow resumes. Downstream from point E the EGL and HGL lines become again parallel to each other, but they take a steeper slope where a higher rate of energy dissipation in the smaller pipe is expected.

The loss of head in a sudden contraction may be represented in terms of velocity head in the smaller pipe as

$$h_c = K_c\left(\frac{V_2^2}{2g}\right) \tag{3.30}$$

Here, K_c is the coefficient of contraction. Its value varies with the ratio of contraction, D_2/D_1, as shown in Table 3.4.

TABLE 3.4 Values of the Coefficient K_c for Sudden Contraction

Velocity in Smaller Pipe V(m/sec)	Ratio of Smaller to Larger Pipe Diameters, D_2/D_1									
	0.0	0.1	0.2	0.3	0.4	0.5	0.6	0.7	0.8	0.9
1	0.49	0.49	0.48	0.45	0.42	0.38	0.28	0.18	0.07	0.03
2	0.48	0.48	0.47	0.44	0.41	0.37	0.28	0.18	0.09	0.04
3	0.47	0.46	0.45	0.43	0.40	0.36	0.28	0.18	0.10	0.04
6	0.44	0.43	0.42	0.40	0.37	0.33	0.27	0.19	0.11	0.05
12	0.38	0.36	0.35	0.33	0.31	0.29	0.25	0.20	0.13	0.06

Head loss due to pipe contraction may be greatly reduced by introducing a gradual pipe transition known as a *confusor* and shown in Figure 3.12. The head loss in this case may be expressed as

$$h'_c = K'_c\left(\frac{V_2^2}{2g}\right) \tag{3.31}$$

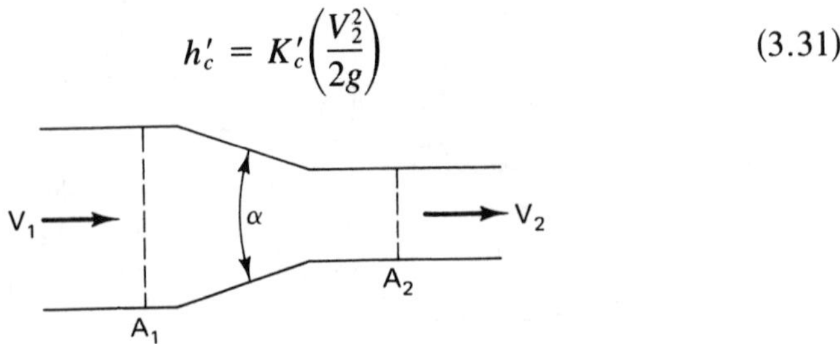

Figure 3.12 Pipe confusor.

The values of K'_c vary with the transition angle, α, and the area ratio A_2/A_1, as shown in Figure 3.13.

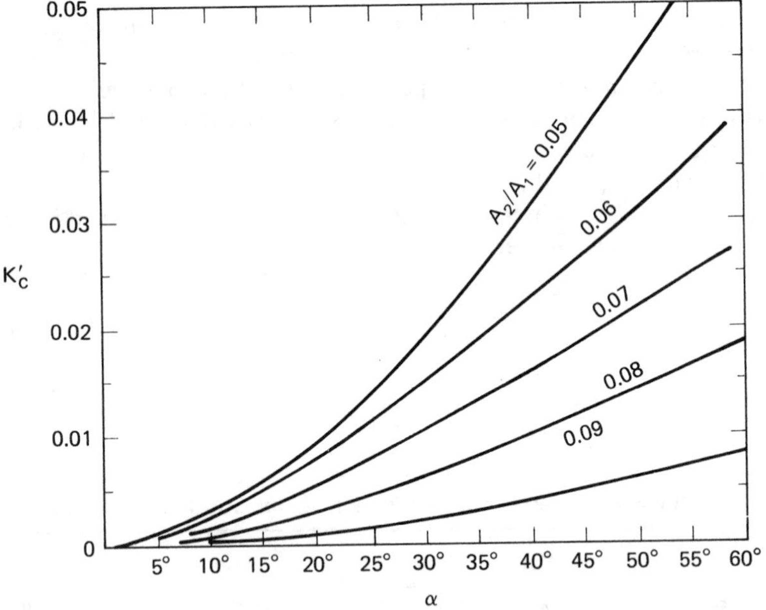

Figure 3.13 Coefficient K'_c for pipe confusors.

The loss of head at the entrance to a pipe from a large reservoir is a special case of loss of head due to contraction. Since the water cross-sectional area in the reservoir is very large compared to that of the pipe, a ratio of contraction of zero may be taken. For a square-cornered entrance, where the entrance of the pipe is flush with the reservoir wall, as shown in Figure 3.14 (a), the K_c values shown for $D_2/D_1 = 0.0$ in Table 3.4 are used.

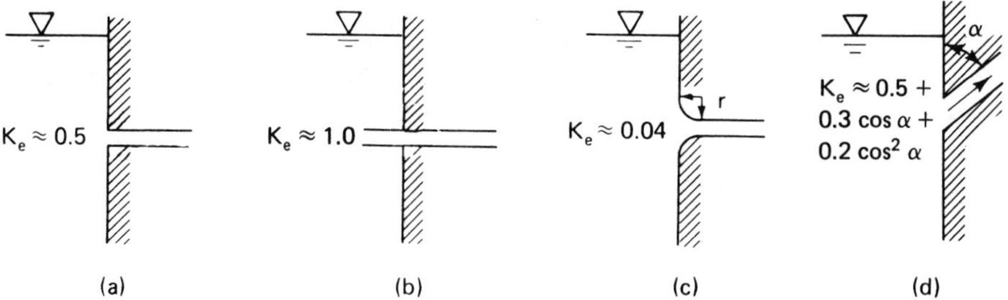

Figure 3.14 Coefficient K_e for pipe entrances.

The general formula for head loss at the entrance to a pipe is also expressed in terms of the velocity head of the pipe.

$$h_e = K_e\left(\frac{V^2}{2g}\right) \tag{3.32}$$

The approximate values for the coefficient of entrance loss K_e for different entrance conditions are shown in Figure 3.14 (a), (b), (c), and (d).

3.8 LOSS OF HEAD DUE TO ENLARGEMENT

The behavior of the energy gradient line and the hydraulic gradient line in the vicinity of a sudden pipe enlargement is schematically demonstrated in Figure 3.15. At the corner of the sudden enlargement A, the stream line separates from the wall of the large pipe and leaves an area of relative stagnancy between A and B, in which a large recirculating eddy is formed to fill the space. Most of the energy loss in a sudden enlargement takes place between A and B where the stream lines reattach to the wall. A recovery of pressure may take place here as a result of the decrease of velocity in the pipe. The high-speed jet stream

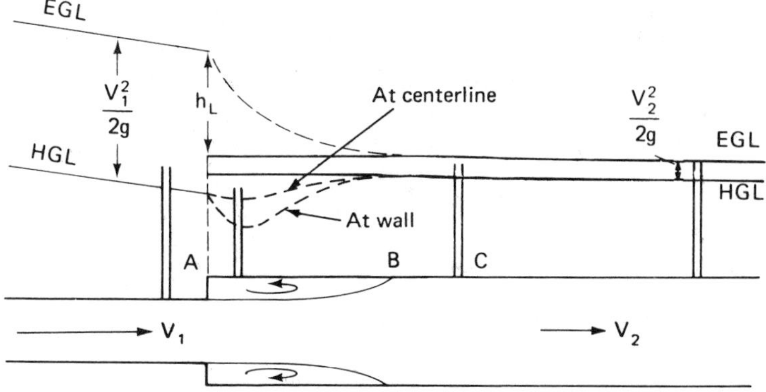

Figure 3.15 Head loss due to sudden enlargement.

gradually slows down and reaches an equilibrium at point C. From this point downstream the normal pipe flow conditions resume and the energy gradient line takes on a smaller slope than that of the approaching pipe, as expected.

The loss of head due to sudden enlargement in a pipe can be derived from the momentum considerations (e.g., Daugherty and Franzini, 1976*). The magnitude of the head loss may be expressed as

$$h_E = \frac{(V_1 - V_2)^2}{2g} \tag{3.33}$$

It means that, physically, the change in velocity heads between the two sections of pipe is the head loss in the sudden expansion.

The head loss due to pipe enlargement may be greatly reduced by introducing a gradual pipe transition known as a *diffusor* and shown in Figure 3.16. The head loss in this pipe transition case may be expressed as

$$h'_E = K'_E \frac{V_1^2 - V_2^2}{2g} \tag{3.34}$$

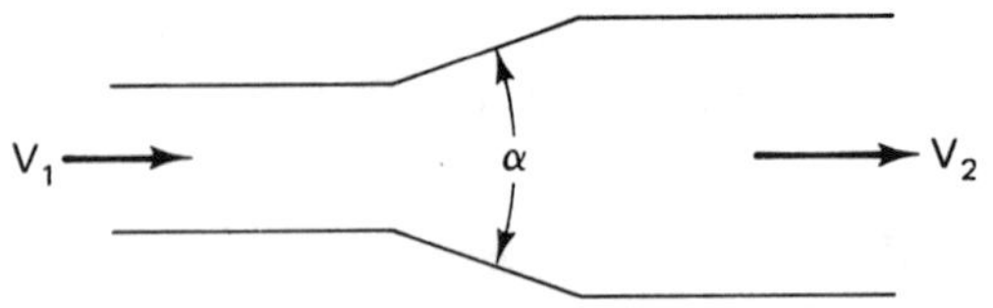

V1 α V2

Figure 3.16

The values of K'_E vary with the diffusor angle α:

α	10°	20°	30°	40°	50°	60°	75°
K'_E	.078	.31	.49	.60	.67	.72	.72

A submerged pipe discharging into a large reservoir is a special case of loss of head due to enlargement. The flow velocity V in the pipe is discharged from the end of a pipe into a reservoir that is so large that the velocity within it is almost zero. From Equation (3.33) we see that the entire velocity head of the pipe flow is dissipated and that the *discharge head loss* is

$$h_d = \frac{V^2}{2g} \tag{3.35}$$

The phenomenon of discharge loss is shown in Figure 3.17.

* R.L. Daugherty, and J.B. Franzini, *Fluid Mechanics with Engineering Applications* (New York: McGraw-Hill Book Company, 1977).

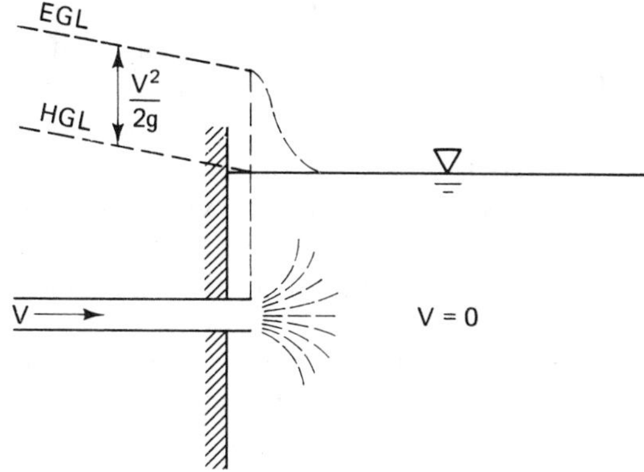

Figure 3.17 Head loss of discharge.

3.9 LOSS OF HEAD IN BENDS

Pipe flow around a bend experiences an increase of pressure along the outer wall and a decrease of pressure along the inner wall. A certain distance downstream from the bend the velocity and pressure resume normal distributions. In order to achieve this, the inner wall pressure must rise to bring the velocity back to the normal value. The velocity near the inner wall of the pipe is lower than that at the outer wall and it must also increase to the normal value. The simultaneous demand of energy may cause separation of the stream from the inner wall, as shown in Figure 3.18(a). In addition, the unbalanced pressure at the bend causes secondary current, as shown in Figure 3.18(b). This transverse current and the axial velocity form a pair of spiral flows that persist as far as 100 diameters downstream from the bend. Thus, the head loss at the bend is combined with the distorted flow conditions downstream from the bend until the spiral flows are dissipated by viscous friction.

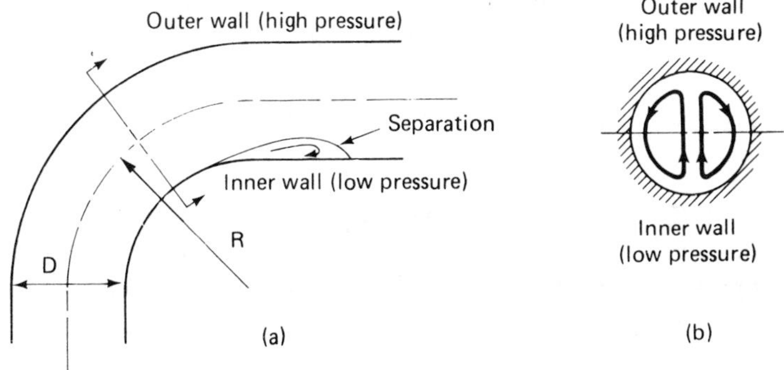

Figure 3.18 Head loss at a bend: (a) flow separation in a bend; (b) secondary flow at a bend.

The head loss produced at a bend was found to be dependent on the ratio of the radius of curvature of the bend, R, to the diameter of the pipe, D. Since the spiral flow produced by a bend extends some distance downstream from the bend, the head loss produced by different pipe bends placed close together cannot be treated by simply adding the losses of each one separately. The total loss of a series of bends placed close together depends not only on the spacing between the bends but also on the direction of the bends. Detailed analysis of head loss produced by a series of bends is a rather complex matter and it can only be analyzed on an individual case-by-case basis.

In hydraulic design the loss of head due to a bend, in excess of that which would occur in a straight pipe of equal length, may be expressed in terms of the velocity head as

$$h_b = K_b \frac{V^2}{2g} \tag{3.36}$$

For a smooth pipe bend of 90°, the values of K_b for various values of R/D as determined by Beij* are listed in the following table. The bend loss has also been found nearly proportional to the angle of bend, α, for pipe bends other than 90° in steel pipes and drawn tubings.

R/D	1	2	4	6	10	16	20
K_b	0.35	0.19	0.17	0.22	0.32	0.38	0.42

3.10 LOSS OF HEAD THROUGH VALVES

Valves are installed in pipelines to control flow by imposing high head losses through the valves. Depending on how the particular valve is designed, a certain amount of energy loss usually takes place even when the valve is fully open. Similar to all other losses in pipes, the head loss through valves may also be expressed in terms of velocity head in the pipe

$$h_v = K_v \frac{V^2}{2g} \tag{3.37}$$

The values of K_v vary with the design of the valves. From the viewpoint of designing a hydraulic system, it is necessary to determine the required loss of head through a valve in the system. The values of K_v for common valves are listed in Table 3.5.

* K.H. Beij, "Pressure Losses for Fluid Flow in 90° Pipe Bends" *Jour. Research Natl. Bur. Standards*, 21 (1938).

TABLE 3.5 Values of K_v for Certain Common Hydraulic Valves

A. Gate valves

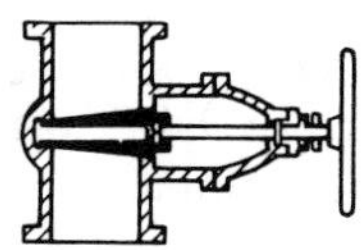

Closed

$K_V = 0.15$ (fully open)

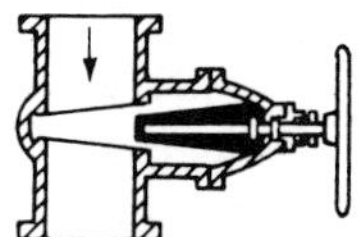

Open

B. Globe valves:

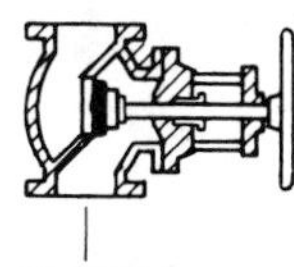

Closed

$K_V = 10.0$ (fully open)

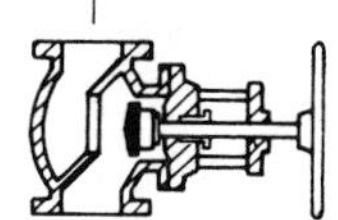

Open

C. Check valves:

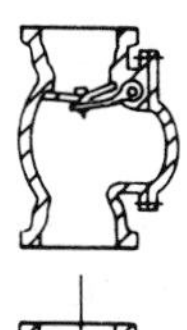

Closed

Hinge (Swing type)

Swing type	$K_V = 2.5$	(fully open)
Ball type	$K_V = 70.0$	(fully open)
Lift type	$K_V = 12.0$	(fully open)

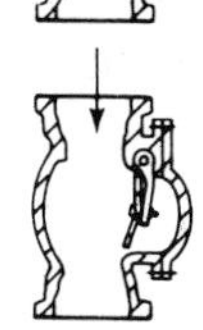

Open

D. Rotary valves:

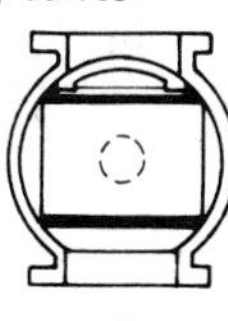

Closed

$K_V = 10.0$ (fully open)

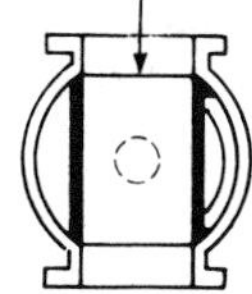

Open

Example 3.9

Figure 3.19 shows two sections of cast-iron pipe connected in series that bring water from a reservoir and discharge it into air at a location 100 m below the water surface elevation in the reservoir through a rotary valve. If the water temperature is 10°C, and square connections are used, determine the discharge.

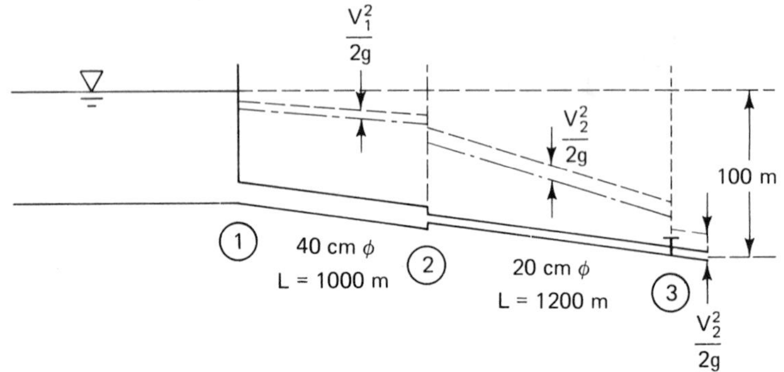

Figure 3.19

Solution

Bernoulli's equation can be written for section 1 at the reservoir and section 3 at the discharge end

$$\frac{V_1^2}{2g} + \frac{P_1}{\gamma} + h_1 = \frac{V_3^2}{2g} + \frac{P_3}{\gamma} + h_3 + h_L$$

Selecting the reference datum at section 3, $h_3 = 0$. Since the reservoir and the discharge end are both exposed to atmospheric pressure and the velocity head at the reservoir can be neglected, we have,

$$h_1 = 100 = \frac{V_3^2}{2g} + h_L$$

The total available energy, 100 m of water column, is equal to the velocity head at the discharge end plus all the head losses incurred in the pipeline system. This relationship, as shown in Figure 3.19, may be expressed as

$$h_e + h_{f_1} + h_c + h_{f_2} + h_v + \frac{V_2^2}{2g} = 100$$

where h_e is the head loss at entrance. For square-cornered entrance, Equation (3.32) gives:

$$h_e = (0.5)\frac{V_1^2}{2g}$$

h_{f_1} is the head loss due to friction in pipe section 1–2. From Equation (3.16),

$$h_{f_1} = f_1 \frac{1000}{0.40} \frac{V_1^2}{2g}$$

h_c is the head loss due to sudden contraction at section 2. From Table 3.4 and Equation (3.30), (assume $K_c = 0.33$ for a first trial).

$$h_c = K_c \frac{V_2^2}{2g} = 0.33 \frac{V_2^2}{2g}$$

h_{f_2} is the head loss due to friction in pipe section 2–3:

$$h_{f_2} = f_2 \frac{1200}{0.20} \frac{V_2^2}{2g}$$

h_v is the head loss at the valve. From Table 3.5 and Equation (3.37)

$$h_v = K_v \frac{V_2^2}{2g} = (10) \frac{V_2^2}{2g}$$

Therefore,

$$100 = \left(1 + 10 + f_2 \frac{1200}{0.20} + 0.33\right) \frac{V_2^2}{2g} + \left(f_1 \frac{1000}{0.40} + 0.5\right) \frac{V_1^2}{2g}$$

From the continuity equation (3.4), we have

$$A_1 V_1 = A_2 V_2$$

$$\frac{\pi}{4}(0.4)^2 V_1 = \frac{\pi}{4}(0.2)^2 V_2$$

The above relations give

$$V_2^2 = \frac{1962}{11.36 + 156.25 f_1 + 6000 f_2}$$

To evaluate f_1 and f_2, we have:

$$N_{R_1} = \frac{D_1 V_1}{\nu} = \frac{0.4}{1.31 \cdot 10^{-6}} V_1 = 7.63 \cdot 10^4 \cdot V_2$$

$$N_{R_2} = \frac{D_2 V_2}{\nu} = \frac{0.2}{1.31 \cdot 10^{-6}} V_2 = 1.53 \cdot 10^5 \cdot V_2$$

where $\nu = 1.31 \cdot 10^{-6}$ at 10°C. For the 40-cm pipe, $e/D = 0.00065$, which yields $f_1 \approx 0.0178$. For the 20-cm pipe, $e/D = 0.0013$; so $f_2 \approx 0.0205$. Solving the above equation for V_2 provides the following:

$$V_2^2 = \frac{1962}{11.36 + 156.25(0.0178) + 6000(0.0205)}$$

$$V_2 = 3.78 \text{ m/s}$$

$$V_1 = 0.25(3.78) = 0.94 \text{ m/s}$$

Hence,

$$N_{R_1} = 7.63 \cdot 10^4(3.78) = 2.88 \cdot 10^5; \qquad f_1 = 0.019$$

$$N_{R_2} = 1.53 \cdot 10^5(3.78) = 5.78 \cdot 10^5; \qquad f_2 = 0.021$$

These f values do not agree with those stated previously, a second trial must be made. For the second trial, we assume that $K_c = 0.35$, $f_1 = 0.019$ and $f_2 = 0.021$. Repeating the above calculation, we get $V_2 = 3.74$, $N_{R_1} = 2.9 \cdot 10^5$, $N_{R_2} = 5.72 \cdot 10^5$. From Figure 3.7, we have,

$$f_1 \cong 0.019; \qquad f_2 \cong 0.021.$$

Therefore, the discharge is

$$Q = A_2 V_2 = \frac{\pi}{4}(0.2)^2(3.74) = 0.12 \text{ m}^3/\text{sec}$$

PROBLEMS

3.10.1 A 75-m-long cast-iron pipe, 15 cm in diameter, takes water from a lake at 2 m below the lake surface. The pipe is laid on a 1/25 slope and has a 90° bend ($R/D = 1.0$) at midlength. What is the flow rate if the pipe discharges freely into the air, if the water temperature in the pipe is 20°C?

3.10.2 A globe valve is installed in a 1-m steel penstock that brings water from a high dam to a hydroelectric power plant. The entrance is 30 m below the reservoir water surface elevation, and the penstock, 440 m long, is laid on a 1/2 slope. Determine the maximum discharge.

3.10.3 A 30-m-high water tower supplies water to a residential area by means of a 20-cm-diameter, 800-m-long steel pipe. In order to preserve maximum pressure head at the delivery, it is considered to replace 94% of the pipe length with a larger (30-cm-diameter) steel pipe connected by a 30° confusor. If the peak discharge demand is 0.25 m³/sec, find out how much pressure head it may gain at 20°C.

3.10.4 A cylindrical tank, 5 m in diameter, is filled with water to a 3 m depth. A short horizontal pipe with a 20-cm diameter and a lift-type check valve is used to drain the tank from the bottom. How long does it take to drain 50% of the tank?

3.10.5 Water flows through an abrupt expansion from a 20-cm pipe to a 40-cm pipe. Determine the flowrate if the pressure drop across the horizontal expansion is 17,000 N/m^2.

3.10.6 Water flows at a rate of 120 ℓ/sec in a pipe with a 30 cm diameter. Determine the contraction loss if the diameter is suddenly reduced to 15 cm. Compare with the head loss incurred when the 15-cm pipe suddenly expands to 30 cm.

3.10.7 In the previous problem, determine the corresponding head losses if a 15° confusor and a 15° diffuser are used instead of the abrupt reduction and expansion.

3.10.8 A swing-type check valve and a globe valve are installed in series in a 20-cm pipe. Determine the flowrate if the total head loss is 3.6 m. Neglect the friction and other losses.

3.10.9 A 60-cm pipe has a horizontal bend with a radius of curvature of 3.6 m. Determine the flowrate if the pressure drop through the bend is 980 N/m^2.

PIPELINES AND PIPE NETWORKS

In general, when a number of pipes are connected together to transport water in a given project, they perform as a system that may include series pipes, parallel pipes, branching pipes, elbows, valves, meters, and other devices. The arrangement is known as a *pipeline* if all elements are connected in series. Otherwise, it is known as a *pipe network*.

While the basic knowledge of pipe flows discussed in Chapter 3 is applicable to each individual pipe in the system, the design and analysis of a pipeline or pipe network does create certain complex problems unique to the system. This is particularly true if the system consists of a large number of pipes like those that frequently occur in the water distribution network of large metropolitan areas.

The physical phenomena and problems that are pertinent to pipelines or pipe networks, and the special techniques developed for the analysis and design of such systems are discussed in the following sections.

4.1 PIPELINE CONNECTING TWO RESERVOIRS

A *pipeline* is a connection of one or more pipes in series and is designed to transport water from one reservoir to another. There are three principal types of pipeline problem:

1. Given the flow rate and the pipe combinations, determine the total head loss.
2. Given the allowable total head loss and the combination of the pipes, determine the flow rate.
3. Given the flow rate and the allowable total head loss, determine the pipe diameter.

The first type of problem can be solved by the direct approach, but the second and third types involve trial-and-error procedures as shown in the following examples.

Example 4.1

Two new cast-iron pipes in series connect two reservoirs (see Figure 4.1). Both pipes are 300 m long and have diameters of 0.6 m and 0.4 m, respectively. The elevation of water surface in reservoir *A* is 80 m. The discharge of 10°C water from reservoir *A* to reservoir *B* is 0.5 m^3/sec. Find the elevation of the surface of reservoir *B*. Assume a sudden contraction at the junction and a square-edge entrance.

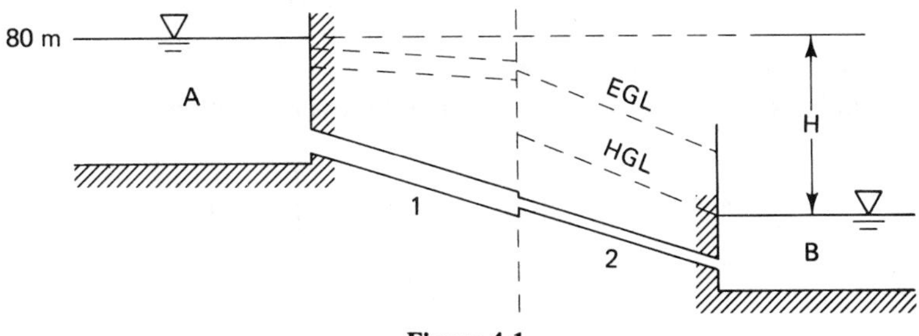

Figure 4.1

Solution

Applying the energy equation (Equation 3.15) between the reservoirs, *A* and *B*,

$$\frac{V_A^2}{2g} + \frac{P_A}{\gamma} + h_A = \frac{V_B^2}{2g} + \frac{P_B}{\gamma} + h_B + h_L$$

Since $P_A = P_B = 0$, and the velocity heads can be neglected in a reservoir

$$h_B = h_A - h_L$$

Designating subscript 1 for the upstream pipe and subscript 2 for the downstream pipe to compute the head loss, we may write

$$V_1 = \frac{Q}{A_1} = \frac{0.5}{\frac{1}{4}\cdot\pi\cdot(0.6)^2} = 1.77 \text{ m/sec}$$

$$V_2 = \frac{Q}{A_2} = \frac{0.5}{\frac{1}{4}\cdot\pi\cdot(0.4)^2} = 3.98 \text{ m/sec}$$

$$N_{R_1} = \frac{V_1 D_1}{\nu} = \frac{1.77(0.6)}{1.31\cdot 10^{-6}} = 8.08\cdot 10^5$$

$$N_{R_2} = \frac{V_2 D_2}{\nu} = \frac{3.98(0.4)}{1.31\cdot 10^{-6}} = 1.22\cdot 10^6$$

From Table 3.1 we have

$$\frac{e}{D_1} = \frac{0.26}{600} = 0.00043$$

$$\frac{e}{D_2} = \frac{0.26}{400} = 0.00065$$

From the Moody chart (Figure 3.8) we have

$$f_1 = 0.017 \quad \text{and} \quad f_2 = 0.018$$

The total energy head $= h_e + h_{f_1} + h_c + h_{f_2} + h_d$. From Equations (3.16), (3.30), (3.32), and (3.35) we may write

$$h_L = \left(0.5 + f_1\frac{L_1}{D_1}\right)\frac{V_1^2}{2g} + \left(0.24 + f_2\frac{L_2}{D_2} + 1\right)\frac{V_2}{2g}$$

$$= 13.34 \text{ m}$$

The elevation of the surface of reservoir B is

$$80 - 13.34 = 66.66 \text{ m}$$

Example 4.2

Pipeline AB connects two reservoirs. The difference in elevation between the two reservoirs is 10 m. The pipeline consists of an upstream section, $D_1 = 0.75$ m and $L_1 = 1500$ m, and a downstream section, $D_2 = 0.5$ m and $L_2 = 1000$ m. The pipes are cast-iron and are connected end-to-end with a sudden reduction of area. Assume the water temperature at 10°C. Compute the discharge capacity.

Solution

The energy equation can be written as follows:

$$\frac{V_A^2}{2g} + \frac{P_A}{\gamma} + h_A = \frac{V_B^2}{2g} + \frac{P_B}{\gamma} + h_B + h_L$$

or eliminating zero and small terms

$$h_L = h_A - h_B = 10 \text{ m}$$

Since the discharge is yet unknown, the velocity in each pipe can only be assumed to be V_1 and V_2, respectively. The total energy equation, as above, will contain these two assumed quantities. It cannot be solved directly so an iteration procedure is used. For water temperature at 10°C, $\nu = 1.31 \cdot 10^{-6}$ m²/s. The corresponding Reynolds numbers may be expressed as

$$N_{R_1} = \frac{V_1 D_1}{\nu} = \frac{V_1(0.75)}{1.31 \cdot 10^{-6}} = 5.73 \cdot 10^5 \cdot V_1 \quad (1)$$

$$N_{R_2} = \frac{V_2 D_2}{\nu} = \frac{V_2(0.5)}{1.31 \cdot 10^{-6}} = 3.82 \cdot 10^5 \cdot V_2 \quad (2)$$

From the continuity condition, $A_1 V_1 = A_2 V_2$, we have

$$\frac{\pi}{4}(0.75)^2 \cdot V_1 = \frac{\pi}{4}(0.5)^2 \cdot V_2$$

$$V_2 = 2.25 V_1 \quad (3)$$

Substituting Equation (3) into Equation (2), we get

$$N_{R_2} = (3.82 \cdot 10^5) \cdot (2.25 V_1) = 8.60 \cdot 10^5 \cdot V_1$$

The energy equation may be written as

$$\begin{aligned} 10 &= \left[0.5 + f_1\left(\frac{1500}{0.75}\right)\right]\frac{V_1^2}{2g} + \left[0.24 + f_2\left(\frac{1000}{0.5}\right) + 1\right]\frac{V_2^2}{2g} \\ 10 &= (0.35 + 101.94\, f_1 + 516.05\, f_2)V_1^2 \end{aligned} \quad (4)$$

From Table 3.1, $e/D_1 = 0.00035$ and $e/D_2 = 0.00052$. As a first trial, we assume $f_1 = 0.015$ and $f_2 = 0.016$.

$$10 = [0.35 + 101.94(0.015) + 516.05(0.016)]V_1^2$$

$$V_1 = 0.99 \text{ m/sec}$$

$$N_{R_1} = 5.73 \cdot 10^5(0.99) = 5.67 \cdot 10^5$$

$$N_{R_2} = 8.60 \cdot 10^5(0.99) = 8.51 \cdot 10^5$$

From Figure 3.8, $f_1 = 0.0165$ and $f_2 = 0.0175$. These values do not agree with the values stated previously. For the second trial, assume that $f_1 = 0.0165$ and $f_2 = 0.0175$.

$$V_1 = 0.95 \text{ m/sec}; \qquad N_{R_1} = 5.44 \times 10^5; \qquad N_{R_2} = 8.17 \times 10^5$$

We obtain, from Figure 3.8, $f_1 = 0.0165$ and $f_2 = 0.0175$. These values are the same as the assumed values, suggesting that $V_1 = 0.95$ m/sec is the actual velocity in the upstream pipe. Therefore, the discharge is

$$Q = A_1 V_1 = \frac{\pi}{4}(0.75)^2(0.95) = 0.42 \text{ m}^3/\text{sec}$$

Example 4.3

A concrete pipeline is installed to deliver 6 m^3/sec of water between two reservoirs 17 km apart. If the elevation difference between the two reservoirs is 12 m, determine the diameter of the pipe.

Solution

As in the previous examples, the energy relationship between the two reservoirs is

$$\frac{V_A^2}{2g} + \frac{P_A}{\gamma} + h_A = \frac{V_B^2}{2g} + \frac{P_B}{\gamma} + h_B + h_L$$

Thus,

$$h_L = h_A - h_B = 12 \text{ m}$$

The mean velocity may be obtained by using the continuity condition, Equation (3.4):

$$V = \frac{Q}{A} = \frac{6}{\frac{\pi}{4}D^2} = \frac{7.64}{D^2}$$

and,

$$N_R = \frac{DV}{\nu} = \frac{D\left(\frac{7.64}{D^2}\right)}{1.31 \cdot 10^{-6}} = \frac{5.83 \cdot 10^6}{D}$$

Neglect the minor losses (for long pipes of $L/D >> 1000$, minor losses may be neglected). The energy loss contains only the friction loss term. Equation (3.16) gives

$$12 = f\left(\frac{L}{D}\right)\frac{V^2}{2g} = f\left(\frac{L}{D}\right)\frac{Q^2}{2gA^2} = f\left(\frac{17{,}000}{D}\right)\cdot\left(\frac{6^2}{2\cdot 9.81\cdot\left(\frac{\pi}{4}\right)^2\cdot D^4}\right)$$

Simplified,

$$0.000237 = \frac{f}{D^5} \qquad \text{(a)}$$

For concrete pipes, $e = 0.3$ mm to $e = 3.0$ mm (field construction), use $e =$ 0.36 mm, and assume that $D = 2.5$ m for the first trial.

$$\frac{e}{D} = \frac{0.36}{2500} = 0.00014$$

and $N_R = 2.33 \cdot 10^6$. From Figure 3.8 we obtain $f = 0.0135$. Substituting these values into Equation (a), we see that

$$D = \left(\frac{0.0135}{0.000237}\right)^{1/5} = 2.24$$

Hence a different pipe diameter must be used for the second iteration. Use $D = 2.24$ m. Hence, $e/D = 0.00016$ and $N_R = 2.6 \cdot 10^6$. From Figure 3.8, we obtain $f = 0.0136$. From Equation (a), we now have

$$D = \left(\frac{0.0136}{0.000237}\right)^{1/5} = 2.25 \text{ m}$$

The value on the right-hand side is considered close enough to that on the previous iteration and the pipe diameter of 2.25 m is selected.

PROBLEMS

4.1.1 Sketch the energy gradient line (EGL) and the hydraulic gradient line (HGL) for the compound pipe shown in Figure P4.1.1. Consider all the losses and changes of velocity and pressure heads.

4.1.2 A 40-cm straight pipe is used to carry water at 20°C from one reservoir to another reservoir that is 1.7 km away. There is an elevation difference of 9 m between the two reservoirs. Determine the discharge for the following pipes: (a) commercial steel, (b) cast iron, (c) aged cast iron, (d) smooth concrete, and (e) wood-stave pipe in best condition.

4.1.3 It is necessary to deliver 5 ℓ/min of a water-glycerol solution (sp.gr. = 1.1; $\nu = 1.03 \times 10^{-5}$ m^2/sec) under the pressure head of 50 mm Hg. A glass tube is used ($e = 0.003$ mm). Determine the tube diameter if the tube length is 2.5 m.

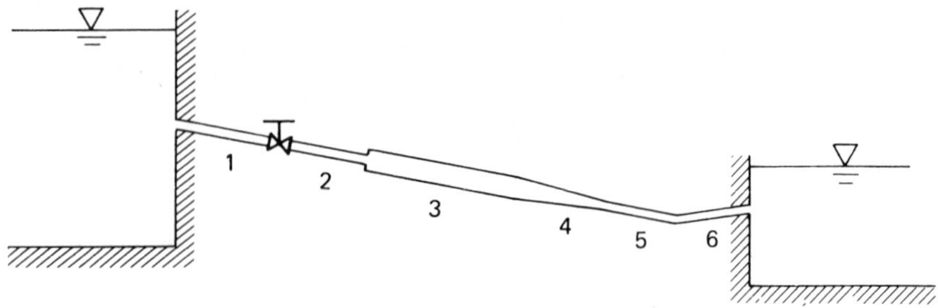

Figure P4.1.1

4.1.4 Water is taken from a head tank to a cooling pond by a 22-m-long, 25-cm-diameter steel pipe. If the elevation difference between the tank and the pond is 1.4 m, and if a globe valve is installed in the pipeline, determine the maximum discharge when the valve is fully opened. Assume square-end connections, and 10°C.

4.1.5 The elevation difference between two reservoirs, 30 km apart, is 100 m. Asphalted cast-iron pipes are used to transport water between the two reservoirs. (a) Determine the pipe diameter that will carry a discharge of 200 ℓ/sec. (b) Determine the elevation difference between the reservoirs if the flowrate increases to 250 ℓ/sec. Assume water temperature 10°C.

4.1.6 The 40-m-long, 4-in. commercial steel pipe connects reservoirs *A* and *B* as shown in Figure P4.1.6. The water temperature is 10°C. Determine the flowrate if reservoir *A* is subjected to atmospheric pressure and the globe valve is fully open.

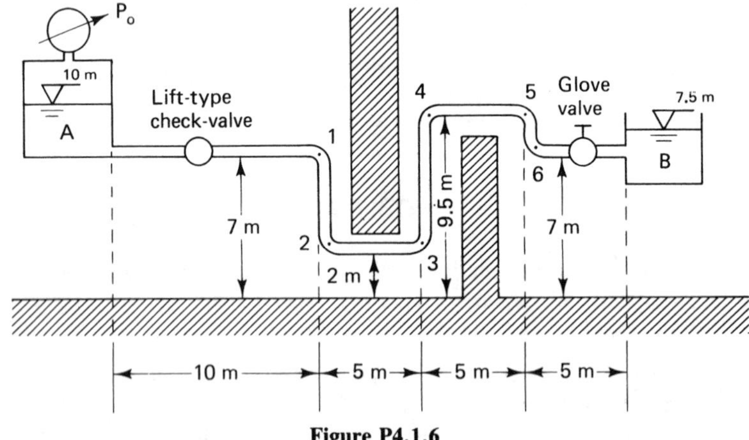

Figure P4.1.6

4.1.7 Determine the pressure at each point designated in Figure P4.1.6.

4.1.8 Determine the pressure at points 1 and 5 in Figure P4.1.6 if reservoir *A* is pressurized to 40,000 N/m^2 while reservoir *B* remains at atmospheric pressure.

4.1.9 Determine the minimum pressure P_0 that would keep the pressure head throughout the pipe positive in Figure P4.1.6.

4.1.10 Sketch the energy and hydraulic gradient lines for Problems 4.1.6 and 4.1.8.

4.1.11 Water flows through a 15-cm horizontal pipe at 20°C as shown in Figure P4.1.11. The pipe is cast iron, 200 m long, and discharges into the air. Determine the flowrate and sketch the EGL and the HGL.

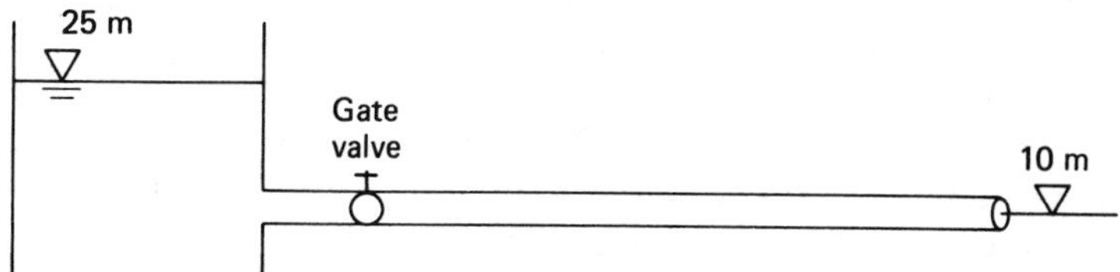

Figure P4.1.11

4.1.12 A 5-cm smooth nozzle is installed at the end of the pipe in Figure P4.1.11. (a) Compute the discharge and sketch the new EGL and HGL. (b) Comment on the change in discharge. Assume no energy loss at the nozzle.

4.2 PIPELINE WITH NEGATIVE PRESSURE OR PUMPS

Long pipelines laid to transport water from one reservoir to another over a large distance usually follow the natural contour of the land. Occasionally, a section of the pipeline may be raised to an elevation that is above the local hydraulic gradient line, as shown in Figure 4.2. As discussed in Chapter 3, the vertical distance measured between the energy gradient line and the hydraulic gradient line at any location along the pipeline is the velocity head, $V^2/2g$, at that location; and the vertical distance measured between the hydraulic gradient line and the pipeline is the local pressure head, P/γ. In the vicinity of the pipeline summit, S, the pressure head may take on negative values. This is because the total energy head

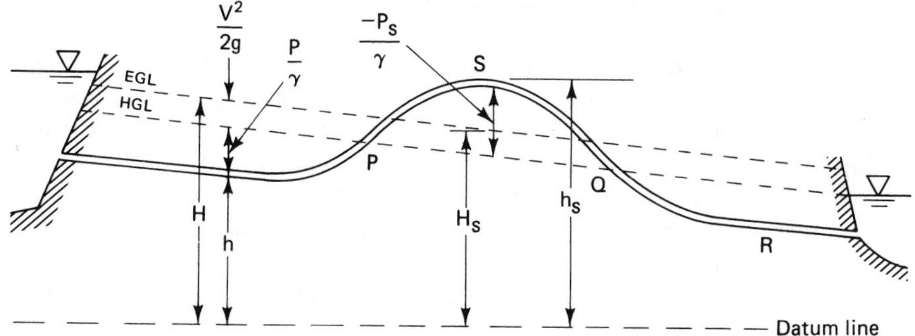

Figure 4.2 Elevated section in a pipeline.

$$H = \left(\frac{V^2}{2g} + h\right) + \frac{P}{\gamma}$$

must equal the vertical distance between the datum and the energy gradient line at any location in the pipeline. At the summit, for example, the elevation h_s is given and the velocity head $V^2/2g$ is also a fixed positive value. The sum $V^2/2g + h_s$ may become larger than H_s. Hence, the pressure head, P_s/γ, can only be negative. *Negative pressure* (in reference to atmospheric pressure as zero, $P_{atm} = 0$) *exists in the pipeline wherever the pipeline is raised above the hydraulic gradient line* (between P and Q). This negative pressure reaches a maximum value at the summit, $-(P_s/\gamma)$. Water flow from S to R must flow against the pressure gradient. In other words, it flows from a point of lower pressure toward a point of higher pressure. This is possible because water always flows from high-energy toward lower-energy locations, and in the closed conduit the elevation head converts to compensate the pressure head along the gradient line. For example, if a unit weight of water flowing from S to R experiences a pressure increase equal to 3 m of water column, the elevation of S must be at least 3 m higher than the elevation of R. The difference in elevation between S and R must equal exactly 3 m plus the loss of head between S and R. Or, more generally, the elevation difference between any two points 1 and 2 in a pipeline is

$$h_{12} = \left(\frac{P_2}{\gamma} - \frac{P_1}{\gamma}\right) + h_{f12} \tag{4.1}$$

However, it is important to maintain pressure at all points in a pipeline above the vapor pressure of water. As we discussed in Chapter 1, the vapor pressure of water is approximately equal to a negative water column of 10 m at 20°C. When the pressure in a pipe drops below this value, water will be vaporized locally to form vapor pockets that separate the water in the pipe. These vapor pockets collapse in regions of higher pressure downstream. The action of vapor collapse is most violent, causing vibrations and sound that can greatly damage the pipeline.

Theoretically, a pipeline may be designed to allow pressure to fall to the level of vapor pressure at certain sections in the pipeline. In practice, however, most of the time water contains dissolved gases that will come out from the water solution well before the vapor pressure point is reached. Such gases return into the solution very slowly. They usually move with the water in the form of large bubbles that reduce the pipe flow cross section and tend to disrupt the flow. For this reason, negative pressure shouldn't be allowed to exceed approximately two-thirds of barometric head in any section of the pipeline (ordinarily, 7 to 8 meters).

Example 4.4

A uniform steel pipeline 40 cm in diameter and 2000 m long carries water at 10°C between two reservoirs as shown in Figure 4.2. The two reservoirs have a water surface elevation difference of 30 m. At midlength the pipeline must be raised to carry the water over a small hill. Determine the maximum elevation to which the summit S may be raised above the lower reservoir water.

Solution

The total head loss between the two reservoirs is 30 m, which includes the entrance loss, h_e, the friction loss in the pipeline, h_f, and the discharge loss, h_d. We may write

$$h_A - h_B = 30 = \left(K_e + f\frac{L}{D} + K_d\right)\frac{V^2}{2g}$$

From Equations (3.32), (3.16), (3.35) assuming square entrance and that $f =$ 0.025, we have

$$30 = \left(0.5 + 0.025 \cdot \frac{2000}{0.4} + 1\right) \cdot \frac{V^2}{2(9.81)}$$

Hence, $V = 2.157$ m/sec. We must check the assumed friction factor f against this velocity. For commercial steel, $e/D = 0.00011$, and the Reynolds number is

$$N_R = \frac{VD}{\nu} = \frac{2.157 \cdot 0.4}{1.31 \cdot 10^{-6}} = 6.6 \cdot 10^5$$

The Moody chart (Figure 3.8) gives $f = 0.014$, which is significantly different from the assumed value. The energy relationship must be rewritten as follows:

$$30 = \left(0.5 + 0.014 \cdot \frac{2000}{0.4} + 1\right)\frac{V^2}{2(9.81)}$$

From which $V = 2.87$ m/sec. Here, $N_R = 8.76 \cdot 10^5$, and the Moody chart gives $f = 0.0138$, which is close enough to the second approximation, and the velocity of 2.87 m/sec is acceptable for the following computations.

The pressure is atmospheric at the downstream reservoir, and its water surface can be selected as the reference datum. The velocity head at the reservoir can be neglected. At 10°C the water vapor pressure head is −10.2 m.

The energy equation between the summit point S and the lower reservoir may be expressed as

$$\frac{V^2}{2g} - 10.2 + h_S = 0 + 0 + 0 + \left(f\frac{L}{D} + K_d\right)\frac{V^2}{2g}$$

The maximum elevation that the pipeline summit, S, may be raised, h_S, is therefore

$$h_s = (0 + 10.2) + \left(1 + f\frac{L}{D}\right)\frac{V^2}{2g} - \frac{V^2}{2g}$$

$$= 10.2 + \left[(0.014)\frac{1000}{0.4}\right]\frac{2.87^2}{2(9.81)} = 24.9 \text{ m}$$

Pumps may be needed in a pipeline to lift water from a lower elevation or simply to boost the rate of flow. Pump operation adds energy to water in the pipeline by boosting the pressure head. The details of pump design and selection will be discussed in Chapter 5, but the analysis of pressure head and energy head provided by a pump to a pipeline system are discussed here.

The computation for pump installation in a pipeline is usually carried out by separating the pipeline system into two sequential parts, the *suction side* and the *discharge side*.

Figure 4.3 shows a typical pump installation. The suction side of the system from the supply reservoir (1–1) to the inlet of the pump (2–2) is subjected to negative pressure, while the discharge side from the outlet of the pump (3–3) to the receiving reservoir (4–4) is subjected to positive pressure. A pump is a device that adds extra energy into the flow, mainly in the form of pressure head.

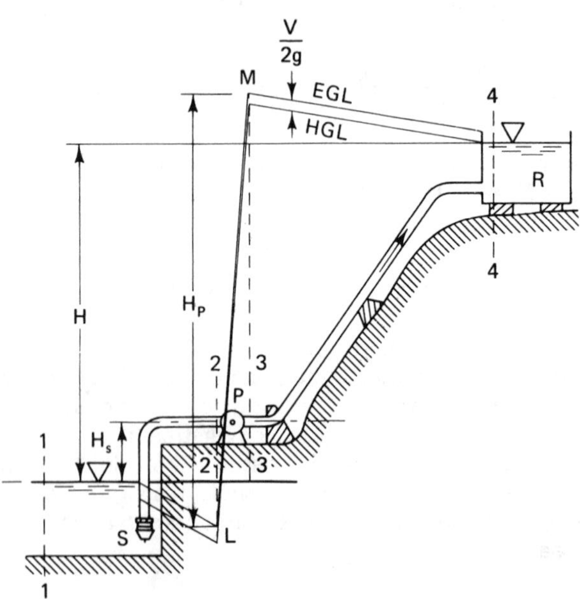

Figure 4.3 Energy gradient line and hydraulic gradient line of a pumping station.

The total energy head provided by the pump to the system, H_P, is represented by the vertical distance between the low point, L, and the high point, M, in the energy gradient line, at the inlet and outlet of the pump, respectively. The elevation of M represents the total energy at the outlet of the pump that delivers the water to the receiving reservoir, R. An energy equation can be written between the supply reservoir, S, and the receiving reservoir, R, as

$$H_S + H_P = H_R + h_L \tag{4.2}$$

where H_S and H_R are the total energy in unit weight of water (energy head) in the supply and receiving reservoirs, respectively. H_P is the energy head added by the pump, and h_L is the total energy loss in the system.

Example 4.5

An emergency pump is installed in the pipeline system of Example 4.4 at a distance of 500 m from reservoir A. The pump is used to boost the flow rate when needed. Determine the pressure head the pump must develop in order to double the flow rate.

Solution

The rate of flow in the pipeline without the pump, from Example 4.4 is

$$Q_0 = AV = \frac{\pi}{4}(0.4)^2(2.87) = 0.36 \text{ m}^3/\text{sec}$$

The pump operation will double the flow rate

$$Q = 2Q_0 = 2 \cdot 0.36 = 0.72 \text{ m}^3/\text{sec}$$

And the corresponding Reynolds number is

$$N_R = 2N_{R_0} = 1.75 \times 10^6$$

For which, the friction factor $f = 0.013$.
The energy loss between reservoir A and the pump is

$$h_{f_{AP}} = \left(0.5 + 0.013 \cdot \frac{500}{0.4}\right)\frac{V^2}{2g} \tag{1}$$

The head loss between the pump and reservoir B is

$$h_{f_{PB}} = \left(0.013 \cdot \frac{1500}{0.4} + 1\right)\frac{V^2}{2g} \tag{2}$$

The corresponding velocity is

$$V = \frac{Q}{A} = 5.73 \text{ m/sec}$$

Substituting this value into Equations (1) and (2), we have

$$h_{f_{AP}} = \left(0.5 + 0.013 \cdot \frac{500}{0.4}\right)\frac{5.73^2}{2(9.81)} = 28.03 \text{ m}$$

$$h_{f_{PB}} = \left(0.013 \cdot \frac{1500}{0.4} + 1\right)\frac{5.73^2}{2(9.81)} = 83.25 \text{ m}$$

The minimum pressure head the pump must provide to double the flow rate is

$$H_P = 28.03 + 83.25 - 30 = 81.28 \text{ m}$$

In addition to this value, a certain amount of pressure head must also be provided to compensate the energy loss that occurs at the pump when it is in operation.

PROBLEMS

4.2.1 A 10-cm tube is 12 m long and is used to siphon water from the reservoir and discharge it into the air, as shown in Figure P4.2.1. If the total head loss between the intake end of the tube and the summit, S, is 1.2 m and that between S and the discharge end is 2.0 m, determine the discharge of the tube and the pressure at S.

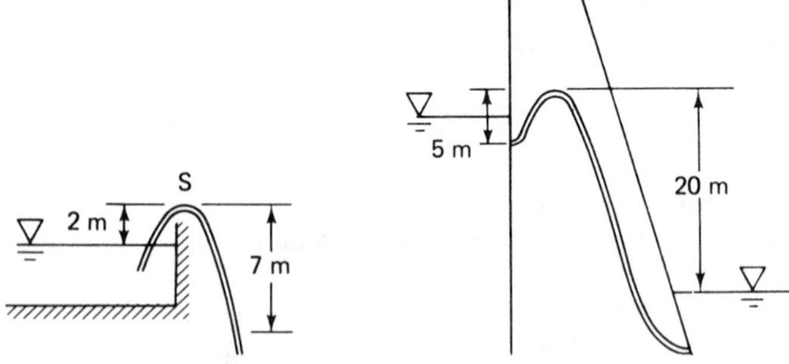

Figure P4.2.1 **Figure P4.2.2**

4.2.2 A siphon spillway with a square cross section of 1 m^2 is used to discharge water to a downstream reservoir 20 m below the crest of the spillway (see Figure P4.2.2). If the frictional head loss is equal to twice the velocity head and is evenly distributed throughout the spillway length, determine the range of water surface level within which the spillway may operate once it is primed. The crest of the spillway is located at one third the length from the entrance.

4.2.3 A pump draws water from reservoir *A* and lifts it to a higher reservoir *B*, as shown in Figure P4.2.3. The head loss from *A* to the pump is four times the velocity head in the 10-cm pipe, and the head loss from the pump to *B* is seven times the velocity head. Compute the pressure head the pump must deliver. The pressure head at the inlet of the pump is −6 m.

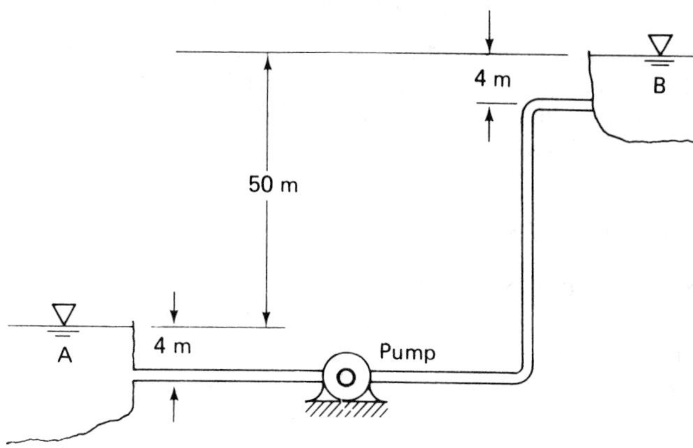

Figure P4.2.3

4.2.4 Construct the energy gradient line and the hydraulic gradient line for the pipeline system in Problem 4.2.3.

4.2.5 A pump installed at elevation 3.0 m delivers 220 ℓ/sec of water through a horizontal pipe system to a pressurized tank. The water surface elevation in the tank is 6.0 m. The pressure at the 30 cm-diameter suction side of the pump is $-15{,}000$ N/m^2 and at the 15 cm-diameter discharge side the pressure is 580,000 N/m^2. The cast-iron discharge pipe is 40 m long. Determine the pressure at the water surface in the tank. Construct a diagram to show the EGL and the HGL.

4.2.6 Water flows in a new 20-cm, 300-m-long galvanized iron pipe between reservoirs *A* and *B*, as shown in Figure P4.2.6. The pipe is elevated at *S*, which is 150 m downstream from reservoir *A*. The water surface in reservoir *B* is 25 m below the water surface in reservoir *A*. If the maximum allowable negative pressure head in the pipeline is −6 m, determine the distance Δs.

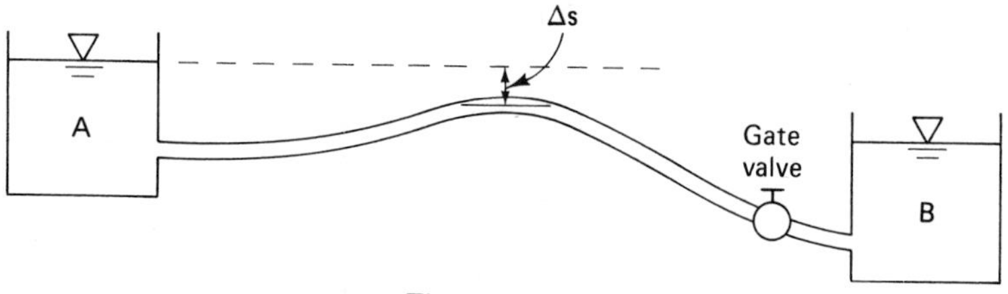

Figure P4.2.6

4.2.7 Determine the head loss through the gate valve in Figure P4.2.6 if Δs is 3 m.

4.2.8 In the system described in Problem 4.2.6 a booster pump is installed 100 m downstream from reservoir *A* and provides an energy head of 30 m. Determine the flowrate and the pressure head at the summit of the pipe if Δs is 1 m.

4.2.9 Determine the energy head that should be provided by a booster pump 100 m downstream from reservoir *A* if Δs is 3 m. (See Figure P4.2.6). Determine the flowrate and construct the EGL and the HGL.

4.2.10 A smooth nozzle is installed in a 30-cm pipe. Immediately upstream from the nozzle the water pressure is 78,000 N/m^2. Determine the minimum diameter of the nozzle that will keep the pressure head in the nozzle above -8 m when the flowrate is 240 ℓ/sec.

4.3 PIPE BRANCHINGS

Water is brought by pipes to a junction where more than two pipes meet. This system must simultaneously satisfy two basic conditions. First, the total amount of water brought by pipes to a junction must always be equal to that carried away from the junction by the other pipes. Second, all pipes that meet at the junction must share the same pressure at the junction.

The hydraulics of pipe branchings at a junction *J* can be best demonstrated by the classical *three-reservoirs problem,* in which three reservoirs of different elevation are connected to a common junction *J* as shown in Figure 4.4. Given the lengths, diameters, and materials of all pipes involved, and given the water elevation in each of the three reservoirs, the discharges to or from each reservoir, Q_1, Q_2, and Q_3, can be determined. If an open-ended vertical tube (*piezometer*) is installed at the junction, the water elevation in the tube will rise to the elevation *P*. The vertical distance between *P* and *J* is a direct reading of the pressure head at the junction. The difference in elevations between *P* and reservoir *A* represents the energy head loss by transporting water from *A* to *J*, as indicated by h_{f_1}. The elevation difference between *B* and *P*, h_{f_2}, represents the head loss by transporting water from *B* to *J*; h_{f_3} represents the head loss by transporting water from *J* to *C*.

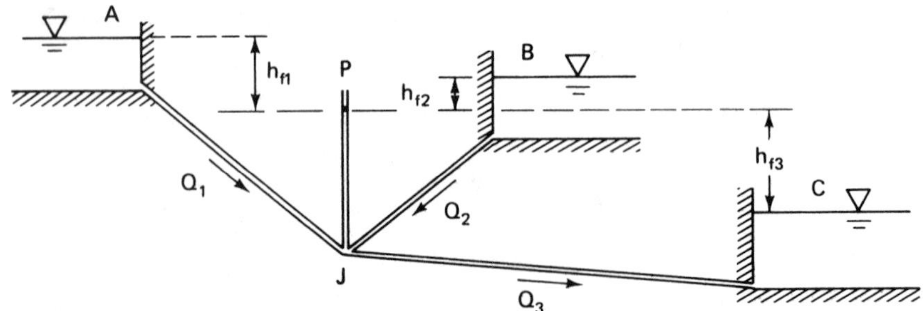

Figure 4.4 Branching pipes connecting three reservoirs.

Since the same quantity of water must be taken away from the junction as is brought to the junction at the same time, we may simply write

$$Q_3 = Q_1 + Q_2 \tag{4.3}$$

or,

$$\Sigma Q = 0$$

at the junction.

These types of problems are most conveniently solved by trial. Not knowing the discharge in each pipe, we may first assume a piezometric surface elevation, P, at the junction. This assumed elevation gives the head losses h_{f_1}, h_{f_2}, and h_{f_3} for each of the three pipes. From this set of head losses and the given pipe diameters, lengths, and materials, the trial computation gives a set of values for discharges Q_1, Q_2, and Q_3. If the assumed elevation P is correct, the computed Q's should satisfy the condition stated above, i.e.,

$$\Sigma Q = Q_1 + Q_2 - Q_3 = 0 \tag{4.4}$$

Otherwise, a new elevation P is assumed for the second trial. The computation of another set of Q's is performed until the above condition is satisfied. The correct values of discharge in each pipe are thus obtained.

Note that if the assumed water surface elevation in the piezometer tube is higher than the elevation in reservoir B, Q_1 should be equal to $Q_2 + Q_3$; if it is lower, $Q_1 + Q_2 = Q_3$. The error in the trial Q's indicates the direction in which the assumed piezometric elevation P should be set for the next trial.

It is helpful to plot the computed trial values of P against ΣQ. The resulting difference may be either plus or minus for each trial. However, with values obtained from three trials, a curve may be plotted as shown in Figure 4.5 (see page 108). The correct discharge is indicated by the intersection of the curve with the vertical axis. The computation procedure is demonstrated in the following example.

Example 4.6

In Figure 4.5 the three reservoirs *A, B,* and *C* are connected by pipes to a common joint *J*. Pipe *AJ* is 1000 m long and 30 cm in diameter; pipe *BJ* is 4000 m long and 50 cm in diameter; and pipe *CJ* is 2000 m long and 40 cm in diameter. The pipes are made of concrete for which $e = 0.6$ mm may be assumed. Determine the discharge in each pipe if the water temperature is 20°C. (Neglect minor losses.)

Solution

Let subscripts 1, 2, and 3 represent, respectively, pipes *AJ, BJ,* and *CJ*. As a first trial, we assume the pressure head at *J*, represented by the elevation *P*, at 110 m. Q_1, Q_2, and Q_3 may be calculated as follows.

For reservoir *A*:

$$h_1 = 120 - 110 = f_1\left(\frac{L_1}{D_1}\right)\frac{V_1}{2g} = f_1\left(\frac{1000}{0.3}\right)\frac{V_1^2}{2(9.81)}$$

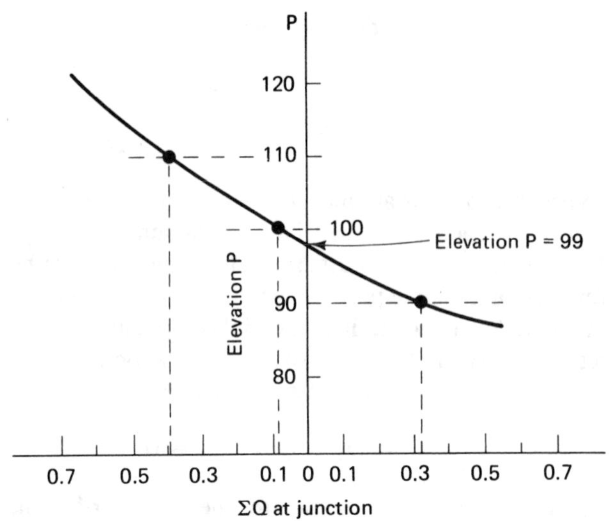

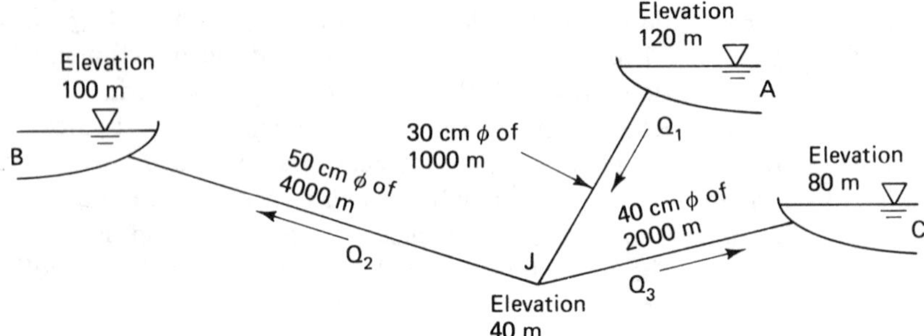

Figure 4.5 Trial for three-reservoir problem.

assuming fully developed turbulent flow,

$$\frac{e_1}{D_1} = \frac{0.6}{300} = 0.002 \therefore f_1 = 0.024$$

which yield $V_1 = 1.566$ m/sec. Hence, $N_R = V_1 D_1/\nu = 4.7 \times 10^5$, and $f_1 = 0.024$ is correct, therefore,

$$Q_1 = V_1 A_1 = 1.566 \frac{\pi}{4}(0.3)^2 = 0.111 \text{ m}^3/\text{sec}$$

Similarly, for reservoir *B*,

$$h_2 = 110 - 100 = f_2\left(\frac{V_2^2}{D_2}\right)\frac{2}{2g} = f_2\left(\frac{4000}{0.5}\right)\frac{V_2^2}{2(9.81)}$$

$$\frac{e_2}{D_2} = \frac{0.6}{500} = 0.0012 \qquad \therefore f_2 = 0.0205$$

which yields $V_2 = 1.094$ m/sec and $N_R = 5.5 \times 10^5$. From the Moody chart, a better estimation for f is obtained, $f = 0.021$. This yields $V_2 = 1.081$ m/sec and $N_R = 5.4 \times 10^5$. Since no further approximation is necessary, we may write

$$Q_2 = V_2A_2 = 1.081\frac{\pi}{4}(0.5)^2 = 0.212 \text{ m}^3/\text{sec}$$

For reservoir C,

$$h_3 = 110 - 80 = f_3\left(\frac{L_3}{D_3}\right)\frac{V_3^2}{2g} = f_3\left(\frac{2000}{0.4}\right)\frac{V_3^2}{2(9.81)}$$

$$\frac{e_3}{D_3} = \frac{0.6}{400} = 0.0015 \qquad \therefore f_3 = 0.022$$

which yields $V_3 = 2.313$ m/sec and $N_R = 9.3 \times 10^5$. No further approximations are needed for f_3, so

$$Q_3 = V_3A_3 = 2.313\frac{\pi}{4}(0.4)^2 = 0.291 \text{ m}^3/\text{sec}$$

Hence, the summation of flows into the joint J is

$$\Sigma Q = Q_1 - (Q_2 + Q_3) = -0.392 \text{ m}^3/\text{sec}$$

At the second trial we assume $P = 100$ m. Similar computations are repeated to obtain

$$Q_1 = 0.157 \text{ m}^3/\text{sec}$$
$$Q_2 = 0.0 \text{ m}^3/\text{sec}$$
$$Q_3 = 0.237 \text{ m}^3/\text{sec}$$

Hence,

$$\Sigma Q = (Q_1 + Q_2) - Q_3 = -0.080 \text{ m}^3/\text{sec}$$

The computations are repeated again for $P = 90$ m to obtain

$$Q_1 = 0.192 \text{ m}^3/\text{sec}$$
$$Q_2 = 0.300 \text{ m}^3/\text{sec}$$
$$Q_3 = 0.168 \text{ m}^3/\text{sec}$$

Hence,

$$\Sigma Q = (Q_1 + Q_2) - Q_3 = 0.324 \text{ m}^3/\text{sec}$$

With the values computed above, a small diagram may be constructed with P plotted against the corresponding values of ΣQ, as shown in Figure 4.5. The curve intersects the $\Sigma Q = 0$ line at $P = 99.0$ m, which is used to compute the final set of discharges. We obtain

$$Q_1 = 0.160 \text{ m}^3/\text{sec}$$

$$Q_2 = 0.067 \text{ m}^3/\text{sec}$$

$$Q_3 = 0.231 \text{ m}^3/\text{sec}$$

Hence, the condition $\Sigma Q = (Q_1 + Q_2) - Q_3 = 0$ is satisfied.

Example 4.7

A horizontal, galvanized iron pipe system consists of a 20-cm-diameter, 4-m-long main pipe between the two joints 1 and 2. A gate valve is installed at the downstream end immediately before 2. The branch pipe is 12 cm in diameter and 6.4 m long. It consists of two 90° elbows ($R/D = 2.0$) and a globe valve. The system carries a total discharge of 0.26 m³/sec water at 10°C. Determine the discharge in each of the pipes when the valves are both fully opened. (See Figure 4.6.)

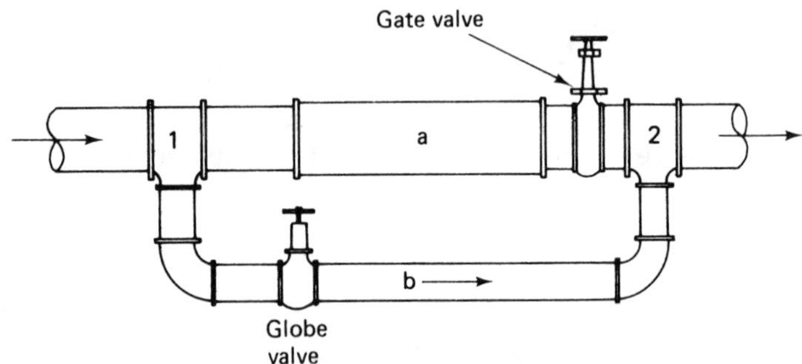

Figure 4.6

Solution

The cross-sectional area of pipes a and b are, respectively,

$$A_a = \pi\left(\frac{0.2}{2}\right)^2 = 0.0314 \text{ m}^2; \qquad A_b = \pi\left(\frac{0.12}{2}\right)^2 = 0.0113 \text{ m}^2$$

Continuity condition requires that

$$0.26 \text{ m}^3/\text{s} = A_a V_a + A_b V_b = 0.0314 V_a + 0.0113 V_b \tag{a}$$

where V_a and V_b are the velocities in pipes a and b, respectively. The head drop between 1 and 2 along the main pipe is h_a

$$h_a = f_a\left(\frac{L_a}{D_a}\right)\frac{V_a^2}{2g} + 0.15\frac{V_a^2}{2g}$$

The second term is from Table 3.5 for fully opened gate valves. The head drop between 1 and 2 along the branch pipe is h_b

$$h_b = f_b\left(\frac{L_b}{D_b}\right)\frac{V_b^2}{2g} + 2(0.19)\frac{V_b^2}{2g} + 10\frac{V_b^2}{2g}$$

where the second term is for the elbow losses and the third term is for fully opened globe valves from Table 3.5.

Since the pressure head losses through both pipes must be the same, $h_a = h_b$, we have,

$$\left[f_a\left(\frac{4}{0.2}\right) + 0.15\right]\frac{V_a^2}{2g} = \left[f_b\left(\frac{6.4}{0.12}\right) + 0.38 + 10\right]\frac{V_b^2}{2g}$$

or,

$$(20f_a + 0.15)V_a^2 = (53.33f_b + 10.38)V_b^2 \qquad \text{(b)}$$

Equations (a) and (b) are solved simultaneously for V_a and V_b. For galvanized iron, Table 3.1 gives

$$\left(\frac{e}{D}\right)_a = \frac{0.15}{200} = 0.00075$$

and

$$\left(\frac{e}{D}\right)_b = \frac{0.15}{120} = 0.00125$$

Assuming large Reynolds numbers, we may obtain from the Moody chart the following values:

$$f_a = 0.0185, \qquad \text{and} \qquad f_b = 0.021$$

as a first approximation.
Substituting the above values into Equation (b), we have

$$(20 \times 0.0185 + 0.15)V_a^2 = (53.33 \times 0.021 + 10.38)V_b^2$$

$$0.52V_a^2 = 11.50V_b^2$$

$$V_a = \sqrt{\frac{11.50}{0.52}}V_b = 4.70V_b$$

Substituting V_a into Equation (a) we find

$$0.26 = 0.0314 \cdot (4.70 V_b) + 0.0113 V_b = 0.159 V_b,$$

$$V_b = \frac{0.26}{0.159} = 1.636 \text{ m/sec}$$

Hence, $V_a = 4.70 V_b = 4.70 \cdot 1.636 = 7.818$ m/sec. Corresponding Reynolds numbers are calculated to check the above friction factors. For pipe a,

$$N_{R_a} = \frac{V_a D_a}{\nu} = \frac{7.818 \cdot 0.2}{1.31 \cdot 10^{-6}} = 1.19 \cdot 10^6$$

The Moody chart gives $f = 0.0185$, which is correct.
For pipe b,

$$N_{R_b} = \frac{V_b D_b}{\nu} = \frac{1.636 \cdot 0.12}{1.31 \cdot 10^{-6}} = 1.40 \cdot 10^5$$

The Moody chart gives $f_b = 0.0225 \neq 0.021$.
Equations (a) and (b) are solved again for the new value of f_b.

$$(20 \cdot 0.0185 + 0.15) V_a^2 = (53.33 \cdot 0.0225 + 10.38) V_b^2$$

$$V_a = 4.719 V_b$$

Substituting V_a into Equation (3.35),

$$0.26 = 0.0314(4.719 V_b) + 0.0113 V_b$$

$$V_b = 1.630 \text{ m/sec}, \quad \text{and} \quad V_a = 1.630 \cdot 4.719 = 7.693 \text{ m/sec.}$$

Therefore, the discharges are

$$Q_a = A_a V_a = 0.0314 \cdot 7.693 = 0.242 \text{ m}^3/\text{sec}$$

and

$$Q_b = A_b V_b = 0.0113 \cdot 1.630 = 0.018 \text{ m}^3/\text{sec}$$

Pipe branches connecting more than three reservoirs to a junction are not common in hydraulic engineering. However, problems of multiple (more than three) reservoirs can be solved by following the same principle (see Figure 4.7).

Assume that the water elevation in the piezometer reaches the level P at the junction J. The differences in water surface elevations between reservoirs A and B, A and C, and A and D are, respectively, H_1, H_2, and H_3. The head losses between reservoirs A, B, C, and D and the junction are, respectively, h_{f1}, h_{f2},

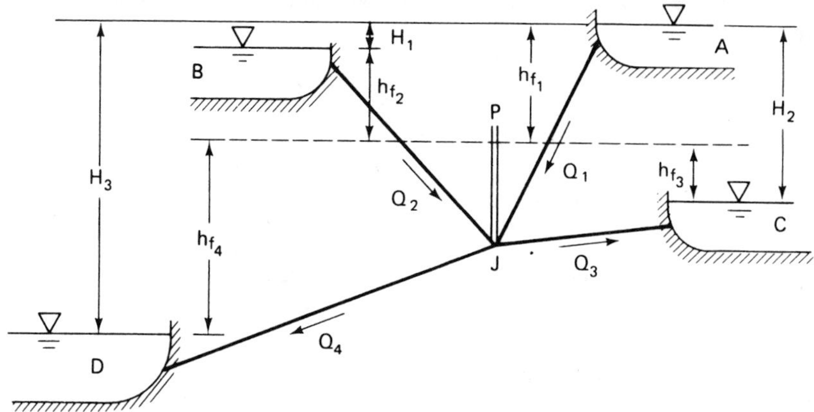

Figure 4.7 Multiple reservoirs connected at a junction.

h_{f3}, and h_{f4}, as shown in Figure 4.7. A set of four independent equations may be written in the following general form for the four reservoirs:

$$H_1 = h_{f_1} - h_{f_2} \tag{4.5}$$

$$H_2 = h_{f_1} + h_{f_3} \tag{4.6}$$

$$H_3 = h_{f_1} + h_{f_4} \tag{4.7}$$

In addition, Equation (4.4) requires that

$$\Sigma Q_j = 0 \tag{4.8}$$

For each of the pipe branches, the head loss may be expressed in the form of the Darcy-Weisbach equation,* Equation (3.16), as follows:

$$h_{f_1} = f_1\left(\frac{L_1}{D_1}\right)\frac{V_1^2}{2g} = f_1\left(\frac{L_1}{D_1}\right)\frac{Q_1^2}{2gA_1^2}$$

$$h_{f_2} = f_2\left(\frac{L_2}{D_2}\right)\frac{V_2^2}{2g} = f_2\left(\frac{L_2}{D_2}\right)\frac{Q_2^2}{2gA_2^2}$$

$$h_{f_3} = f_3\left(\frac{L_3}{D_3}\right)\frac{V_3^2}{2g} = f_3\left(\frac{L_3}{D_3}\right)\frac{Q_3^2}{2gA_3^2}$$

$$h_{f_4} = f_4\left(\frac{L_4}{D_4}\right)\frac{V_4^2}{2g} = f_4\left(\frac{L_4}{D_4}\right)\frac{Q_4^2}{2gA_4^2}$$

* Other empirical formulas, such as the Manning formula and the Hazen-Williams formula, may be applied using corresponding expressions.

Substituting the previous relationship into Equations (4.5), (4.6), and (4.7), we have

$$H_1 = \frac{1}{2g} \cdot \left(f_1 \frac{L_1}{D_1} \frac{Q_1^2}{A_1^2} - f_2 \frac{L_2}{D_2} \frac{Q_2^2}{A_2^2} \right) \tag{4.9}$$

$$H_2 = \frac{1}{2g} \cdot \left(f_1 \frac{L_1}{D_1} \frac{Q_1^2}{A_1^2} + f_3 \frac{L_3}{D_3} \frac{Q_3^2}{A_3^2} \right) \tag{4.10}$$

$$H_3 = \frac{1}{2g} \cdot \left(f_1 \frac{L_1}{D_1} \frac{Q_1^2}{A_1^2} + f_4 \frac{L_4}{D_4} \frac{Q_4^2}{A_4^2} \right) \tag{4.11}$$

and

$$\Sigma Q_j = 0 \tag{4.8}$$

Equations (4.9), (4.10), (4.11), and (4.8) can then be solved simultaneously for the four unknowns Q_1, Q_2, Q_3, and Q_4. These values are the discharge for each of the pipe branches shown. This procedure could be applied to any number of reservoirs connected to a common junction.

PROBLEMS

4.3.1 A 1.3-m concrete pipe 2000 m long carries water at 5 m³/sec from reservoir *A* to a junction and discharges the water into reservoir *B* and reservoir *C*. All connecting pipes are 1 m in diameter. The distances from the junction to reservoirs *B* and *C* are each 1500 m, and the elevation difference between *A* and *B* is 8 m. Determine the surface elevation at *C*.

4.3.2 Determine the flow into or out of each reservoir in Figure P4.3.2 if the connecting pipes all have a Hazen-Williams coefficient of 120.

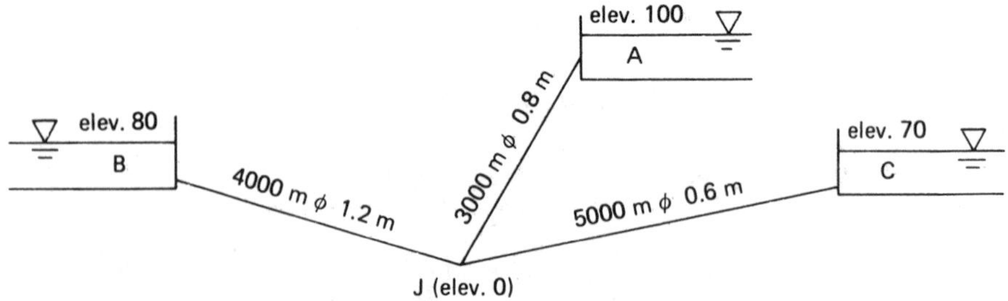

Figure P4.3.2

4.3.3 Determine the flow into or out of each reservoir in Figure P4.3.2 if the connecting pipes are made of the same material with e = 0.05 mm and water temperature at 20°C.

4.3.4 A cast-iron pipe 750 m long and 120 cm in diameter connects reservoir A to the junction J. The cast-iron pipe from reservoir B is 1800 m long and 100 cm in diameter, and the cast-iron pipe from reservoir C is 3400 m long and 80 cm in diameter. Determine the flow in or out of each reservoir if the water temperature is 10°C and the elevations are as sketched in Figure P4.3.4.

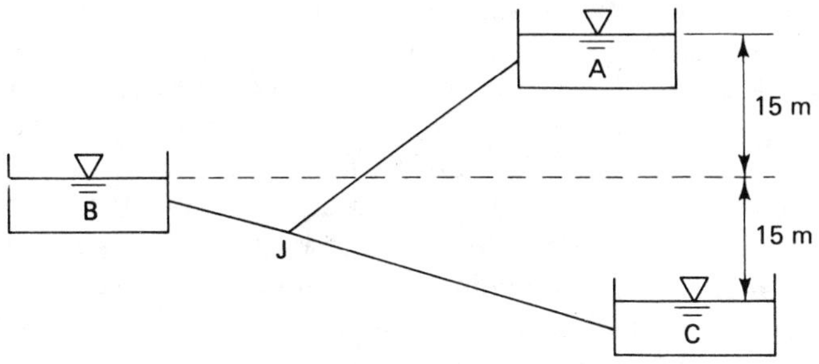

Figure P4.3.4

4.3.5 Solve Problem 4.3.4 using

(a) Hazen-Williams formula

(b) Manning's formula

4.3.6 Reservoir A discharges water through a 1000 m long pipe to a junction, J. From the junction, water flows into three reservoirs, B, C, and D. The discharge into reservoir B is 1.0 m^3/sec through a 500 m long pipe. Discharge into reservoir C, which is 25 m below reservoir A, is 0.6 m^3/sec through a 1200 m long pipe. Flow into reservoir D is 0.6 m^3/sec through a 900 m long pipe. All pipes are cast iron and the elevation difference between reservoirs A and B is 20 m, and between A and D, 30 m. Determine the four diameters that will minimize the total cost of the system if it can be assumed that the cost of each pipe is related to its length and diameter as $KLD^{1.2}$ where K is a constant multiplier.

4.4 PIPE NETWORKS

The water supply distribution system in a municipal district is usually constructed of a large number of pipes connected together to form loops and branches. Although the calculations of water flow in a network involve a large number of pipes and may become tedious, the solution to the problem is based on the same principles that govern the pipelines and pipe branchings previously discussed. In general, a series of simultaneous equations can be written for the network. These equations are written to satisfy the following conditions:

1. At any junction, $\Sigma Q = 0$ (junction equation).
2. Between any two junctions the total head loss is independent of the path taken (loop equation).

Depending on the number of unknowns, it is usually possible to set up a sufficient number of independent equations to solve the problem. A typical problem would, for example, be to determine the flow distribution in each pipe of a network when the total inflows (say, Q_1 and Q_2) and outflows (say, Q_3 and Q_4) are given. These equations may then be solved simultaneously.

For the simple network shown in Figure 4.8, a set of 12 independent equations (eight junction equations and four loop equations) is needed to solve for the flow distribution in the 12 pipes. As a general rule, a network with m loops and n junctions provides a total of $m + (n - 1)$ independent equations. For more complex networks, the number of equations increases proportionally. At a certain point it will be obvious that algebraic solution of the network equations becomes impractical. In such a case, the solutions may feasibly be obtained by means of programming the equations for a digital computer. A commonly used computer program for a pipe network is developed from the *Hardy-Cross method*, which is outlined below.

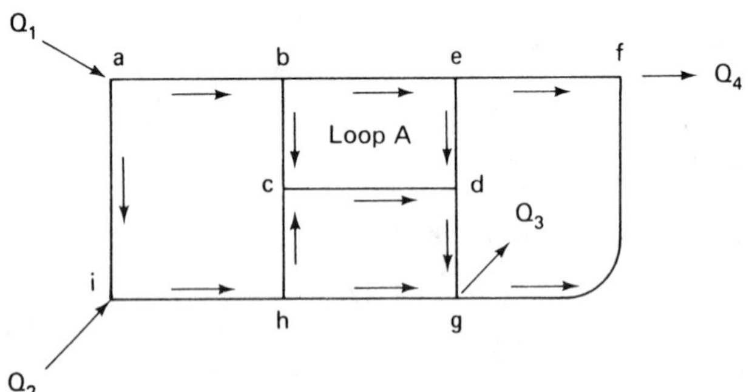

Figure 4.8 Schematics of a pipe network.

The Hardy-Cross method is a method of successive approximations based on the two conditions stated above for each junction and loop in the network system. In the elemental loop A shown in Figure 4.8, for example, the arrowheads indicate the estimated direction of flow. This loop must satisfy the conditions mentioned previously

1. At each junction b, c, d, or e the total inflow must equal the total ouflow.
2. The loss of head due to flow in the counterclockwise direction along pipes bc and cd must be equal to that of flow in the clockwise direction along pipes be and ed.

As a first approach, a distribution of flows in each pipe is estimated such that the total inflow must be equal to the total outflow at each junction throughout the network system. For a network with n junctions, $(n - 1)$ junction equations can be established to determine the flowrates in the system. The estimated flows, together with the given pipe diameters and lengths, friction coefficients, and other descriptions of the network, are the input to the computation program. The program computes the loss of head due to the estimated flowrates in all pipes in the network.

The chance for such an assumed flow distribution to satisfy the m loop equations is small. Usually, the estimated pipe flows must be corrected until head losses in the clockwise direction and in the counterclockwise direction are equal within each loop. The successive computational procedure basically uses the loop equations, one at a time, to correct the assumed pipeflows and thus equalize the head losses in the loop. Since the flow balance at each junction must be maintained, a given correction to the flow in a pipe, (for example, pipe *be*) in the clockwise direction, requires a corresponding correction of the same magnitude of flow in the clockwise direction in the other pipes (pipes *bc* and *ed*). The successive flow corrections to equalize the head loss are discussed as follows.

Knowing the diameter, length, and roughness of a pipe, we see that the head loss in the pipe is a function of the flow rate Q. Applying the Darcy-Weisbach formula [Equation (3.16)], we may write

$$h_f = f\left(\frac{L}{D}\right)\frac{V^2}{2g} = \left[f\left(\frac{L}{D}\right) \cdot \frac{1}{2gA^2}\right] \cdot Q^2 = KQ^2 \tag{4.12}$$

In any elementary loop, such as loop A, the total head loss in the clockwise direction (subscript c) is the sum of head losses in all pipes that carry flow in the clockwise direction around the loop.

$$\Sigma h_{fc} = \Sigma K_c Q_c^2 \tag{4.13}$$

Similarly, the loss of head in the counterclockwise direction (subscript cc) is

$$\Sigma h_{fcc} = \Sigma K_{cc} Q_{cc}^2 \tag{4.14}$$

With the assumed flow rates, Q's, it is not expected that these two values will be equal during the first trial, as mentioned previously. The difference,

$$\Sigma K_c Q_c^2 - \Sigma K_{cc} Q_{cc}^2$$

is the *closure error* of the first trial.

We need to determine a flow correction ΔQ which, when subtracted from Q_c and added to Q_{cc}, will equalize the two head losses. Thus, the correction ΔQ must satisfy the following equation:

$$\Sigma K_c (Q_c - \Delta Q)^2 = \Sigma K_{cc} (Q_{cc} + \Delta Q)^2$$

Expanding the terms in the parentheses on both sides of the equation, we have

$$\Sigma K_c(Q_c^2 - 2Q_c\Delta Q + \Delta Q^2) = \Sigma K_{cc}(Q_{cc}^2 + 2Q_{cc}\Delta Q + \Delta Q^2)$$

Assuming that the correction term is small compared to both Q_c and Q_{cc}, we may simplify the previous expression by dropping the last term on each side of the equation and write

$$\Sigma K_c(Q_c^2 - 2Q_c\Delta Q) = \Sigma K_{cc}(Q_{cc}^2 + 2Q_{cc}\Delta Q)$$

From this relationship we may solve for ΔQ

$$\Delta Q = \frac{\Sigma K_c Q_c^2 - \Sigma K_{cc} Q_{cc}^2}{2(\Sigma K_c Q_c + \Sigma K_{cc} Q_{cc})} \tag{4.15}$$

If we take Equation (4.12) and divide it by Q on both sides, we have

$$KQ = \frac{h_f}{Q} \tag{4.16}$$

Equations (4.13), (4.14), and (4.16) can be substituted into Equation (4.15) to obtain

$$\Delta Q = \frac{\Sigma h_{fc} - \Sigma h_{fcc}}{2\left(\Sigma \frac{h_{fc}}{Q_c} + \Sigma \frac{h_{fcc}}{Q_{cc}}\right)} \tag{4.17a)*$$

The second approximation involves this correction for a new estimated flow rate distribution. The computation results from the second approximation are expected to give a closer match of the two head losses along the *c* and *cc* directions in loop *A*. Note that pipes *bc*, *cd*, and *ed* in loop *A* are each common to two loops and, therefore, need to be subjected to double corrections from each loop. The successive computational procedure is repeated until each loop in the entire network is relaxed and the corrections become negligibly small.

The Hardy-Cross method can be conveniently handled by the digital computer. A typical program and an example problem of application are provided next.

* When the Hazen-Williams formula (Equation 3.27) is used, the equation should be,

$$\Delta Q = \frac{\Sigma h_{fc} - \Sigma h_{fcc}}{1.85\left(\Sigma \frac{h_{fc}}{Q_c} + \Sigma \frac{h_{fcc}}{Q_{cc}}\right)} \tag{4.17b}$$

```
C     THIS PROGRAM READS AN ARBITRARY PIPE NETWORK WITH ESTIMATED FLOWS.
C     THE HAZEN WILLIAMS FORMULA IS USED TO COMPUTE THE HEAD LOSSES FOR
C     EACH PIPE. EACH LOOP IS BALANCED BY THE HARDY CROSS METHOD.
C     PRESSURES AT EACH JUNCTION ARE CALCULATED BASED ON THE PRESSURE AT
C     A GIVEN JUNCTION AND THE ELEVATION OF EACH JUNCTION.
C
      DIMENSION LOOP(15,10),NR(15),P(100),Q(100),JI(100),JF(100),IR(100)
     1,FRIC(100),TE(100),HLOSS(100)
      REAL LGTH
      DO 10 I=1,100
   10 P(I)=999.
C
C     **READ NETWORK DATA. NUMBER OF REACHES AND JUNCTIONS,TOLERANCES ETC.
C
      READ(5,12)NRS,NODES,LOOPS,ITER,TOLQ,TOLH,PKNOW,NODE
   12 FORMAT(4I5,3F10.0,I5)
      WRITE(6,13)NRS,NODES,LOOPS,ITER,TOLQ,TOLH,NODE,PKNOW
   13 FORMAT(2X,'SOLUTION OF A PIPE NETWORK BY THE HARDY CROSS METHOD'//
     1/2X,'DATA INPUT'//5X,'NUMBER OF REACHES   =',I3/5X,'NUMBER OF NODES
     2      =',I3/5X,'NUMBER OF LOOPS      =',I3/5X,'ITERATION LIMIT      =',I
     33/5X,'DISCHARGE TOLERANCE=',F7.3/5X,'HEAD LOSS TOLERANCE=',F7.3/5X
     4,'PRESSURE AT JUNC.',I2,'=',F6.2/)
C
C     **READ LOOPS AND CORRESPONDENT REACHES
C
      DO 16 NL=1,LOOPS
      READ(5,14)JZ,(LOOP(NL,I),I=1,JZ)
   14 FORMAT(11I5)
      WRITE(6,15)JZ,NL,(LOOP(NL,I),I=1,JZ)
   15 FORMAT(5X,I2,' REACHES IN LOOP',I3,' ——',10I5)
   16 NR(NL)=JZ
```

Figure 4.10(a) A typical program for pipe network computations.

```
            P(NODE)=PKNOW
            DO 17 IP=1,NRS
     17     Q(IP)=0.
C
C           **READ THE DESCRIPTION OF THE REACHES,NUMBER,INITIAL AND FINAL JUNCTION . . .
C
            WRITE(6,18)
     18     FORMAT(/5X,'REACHES INFORMATION'/7X,'REACH     IJUNC     FJUNC     L(M)
          1          D(M)          Q(CMS)          C')
            DO 20 I=1,NRS
            READ(5,58)L,JI(L),JF(L),C,LGTH,D,Q(L)
     58     FORMAT(3I5,F5.0,3F10.0,I5)
            WRITE(6,19)L,JI(L),JF(L),LGTH,D,Q(L),C
     19     FORMAT(3X,3I8,F10.2,F9.2,F11.3,F8.0)
            IR(I)=L
     20     FRIC(L)=10.63*LGTH/((C**1.85)*(D**4.87))
C
C           **BEGIN BALANCING PROCEDURE
C
            DO 32 IT=1,ITER
            LOK=0
            DO 31 K=1,LOOPS
            SUMH=0.
            SUMZ=0.
            IWX=NR(K)
C
C           **CALCULATE THE CORRECTION FACTOR FOR EACH LOOP
C
            DO 25 IRL=1,IWX
            S1=1
            S2=1
```

Figure 4.10(a) Cont.

```
      L=LOOP(K,IRL)
      IF(L)21,25,22
   21 S1=-1
      L=-L
   22 FLOWR=Q(L)
      IF(FLOWR)23,24,24
   23 S2=-1
      FLOWR=-FLOWR
   24 H=S1*S2*FRIC(L)*FLOWR**1.85
      Z=1.85*FRIC(L)*FLOWR**0.85
      SUMH=SUMH+H
      SUMZ=SUMZ+Z
   25 CONTINUE
      FCORR=SUMH/SUMZ
C
C     **CORRECT DISCHARGES FOR THIS LOOP
C
      DO 28 IRL=1,IWX
      L=LOOP(K,IRL)
      IF(L)26,28,27
   26 L=-L
      Q(L)=Q(L)+FCORR
      GO TO 28
   27 Q(L)=Q(L)-FCORR
   28 CONTINUE
C
C     **TEST CORRECTION FACTORS AGAINST TOLERANCES
C
      TEST=ABS(FCORR)-TOLQ
      IF(TEST)30,30,29
   29 TEST=ABS(SUMH)-TOLH
```

Figure 4.10(a) Cont.

```
      IF(TEST)30,30,31
   30 LOK=LOK+1
   31 CONTINUE
      IF(LOOPS .LE. LOK)GO TO 33
   32 CONTINUE
C
C     **READ THE TOPOGRAPHY AND CALCULATE THE PRESSURES
   33 WRITE(6,34)
   34 FORMAT(/5X,'JUNCTIONS INFORMATION'/7X,'JUNC. ELEVATION')
      DO 36 I=1,NODES
      READ(5,35)N,TE(N)
   35 FORMAT(I5,F10.0)
   36 WRITE(6,37)N,TE(N)
   37 FORMAT(7X,I3,4X,F8.4)
      DO 38 I=1,NRS
      L=IR(I)
      S=1
      IF(Q(L) .LT. 0.0)S=-1
   38 HLOSS(L)=S*FRIC(L)*(ABS(Q(L)))**1.85
   39 IFLAG=0
      DO 50 K=1,NRS
      I=IR(K)
      JUPS=JI(I)
      JDOWN=JF(I)
      IF(P(JUPS)-999.)40,42,42
   40 IF(P(JDOWN)-999.)50,41,41
   41 P(JDOWN)=P(JUPS)-HLOSS(I)+TE(JUPS)-TE(JDOWN)
      GO TO 50
   42 IF(P(JDOWN)-999.)44,43,43
   43 IFLAG=1
      GO TO 50
```

Figure 4.10(a) Cont.

```
44   P(JUPS)=P(JDOWN)+HLOSS(I)-TE(JUPS)+TE(JDOWN)
50   CONTINUE
     IF(IFLAG .EQ. 1)GO TO 39
     WRITE(6,51)IT
51   FORMAT(//5X,'FLOWS AFTER',I3,' ITERATIONS ARE'/7X,'REACH     Q(CMS)
    1.        HL(M)')
     DO 52 I=1,NRS
52   WRITE(6,53)I,Q(I),HLOSS(I)
53   FORMAT(8X,13,4X,F6.3,4X,F7.3)
     WRITE(6,54)
54   FORMAT(/5X,'PRESSURE AT EACH JUNCTION IN COLUMN OF WATER'/8X,'JUNC
    1.        P(M)')
     DO 55 I=1,NODES
55   WRITE(6,56)I,P(I)
56   FORMAT(8X,I3,3X,F7.3)
99   STOP
     END
```

Figure 4.10(a) Cont.

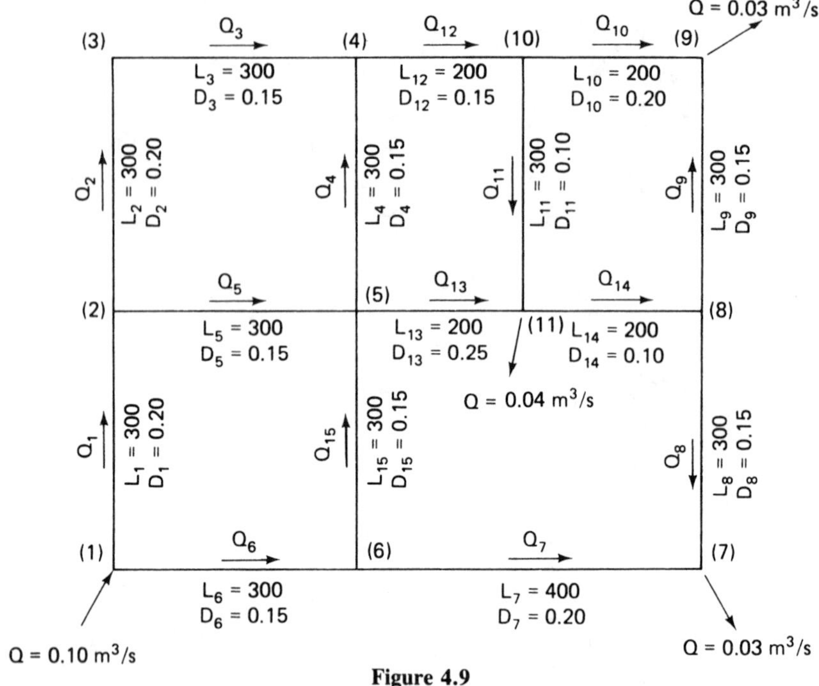

Figure 4.9

Example 4.8

Determine the discharge in each pipe in the pipe network shown in Figure 4.9. If the pressure head measured at junction 1 is 100 m, determine the pressure head at each of the junctions in the network. All pipes are new, cast iron and in good condition.

Solution

The Hardy-Cross method is used to solve this problem by using the computer program.

The computer program, as shown in Figure 4.10(a), recognizes positive (+) and negative (−) instead of clockwise (*c*) and counterclockwise (*cc*), respectively. The Hazen-Williams formula [Equation (3.27)] is used to compute the head loss in each pipe based on the assumed flow rates. The Hardy-Cross method is used to balance the flowrate-pressure relationship in each loop. The program is provided for general application of any numbers and combinations of loops and junctions in a network of given pipe material (Hazen-Williams coefficients).

For the first approximation, a set of discharges is arbitrarily assumed to satisfy the continuity condition at each joint, i.e., $\Sigma Q = 0$.

$Q_1 = 0.05$ $Q_2 = 0.03$ $Q_3 = 0.03$ $Q_4 = 0.00$

$Q_5 = 0.02$ $Q_6 = 0.05$ $Q_7 = 0.02$ $Q_8 = 0.01$

$Q_9 = 0.01 \quad Q_{10} = 0.02 \quad Q_{11} = 0.01 \quad Q_{12} = 0.03$

$Q_{13} = 0.05 \quad Q_{14} = 0.02 \quad Q_{15} = 0.03$

The program results show that five consecutive iterations were made in the computation. The results are shown in Figure 4.10(b).

```
DATA INPUT
  NUMBER OF REACHES    = 15
  NUMBER OF NODES      = 11
  NUMBER OF LOOPS      =  5
  ITERATION LIMIT      = 50
  DISCHARGE TOLERANCE=  0.001
  HEAD LOSS TOLERANCE=  0.005
  PRESSURE AT JUNC. 1  =100.00
   4 REACHES IN LOOP  1 ——     1    5  -15   -6
   4 REACHES IN LOOP  2 ——     2    3   -4   -5
   4 REACHES IN LOOP  3 ——     4   12   11  -13
   4 REACHES IN LOOP  4 ——   -11   10   -9  -14
   5 REACHES IN LOOP  5 ——    15   13   14    8   -7
  REACHES INFORMATION
   REACH  IJUNC  FJUNC    L(M)       D(M)    Q(CMS)      C
     1      1      2     300.00      0.20    0.050     130.
     2      2      3     300.00      0.20    0.030     130.
     3      3      4     300.00      0.15    0.030     130.
     4      5      4     300.00      0.15    0.        130.
     5      2      5     300.00      0.15    0.020     130.
     6      1      6     300.00      0.15    0.050     130.
     7      6      7     400.00      0.20    0.020     130.
     8      8      7     300.00      0.15    0.010     130.
     9      8      9     300.00      0.15    0.010     130.
    10     10      9     200.00      0.20    0.020     130.
    11     10     11     300.00      0.10    0.010     130.
    12      4     10     200.00      0.15    0.030     130.
    13      5     11     200.00      0.25    0.050     130.
    14     11      8     200.00      0.10    0.020     130.
    15      6      5     300.00      0.15    0.030     130.
  JUNCTIONS INFORMATION
   JUNC.  ELEVATION
     1      0.
     2      0.
     3      0.
     4      0.
     5      0.
     6      0.
     7      0.
     8      0.
     9      0.
    10      0.
    11      0.
```

Figure 4.10(b) Solution of a pipe network by the Hardy-Cross method.

```
FLOWS AFTER  5 ITERATIONS ARE
  REACH  Q(CMS)
      1      0.059
      2      0.027
      3      0.027
      4     -0.007
      5      0.031
      6      0.041
      7      0.032
      8     -0.002
      9      0.006
     10      0.024
     11     -0.003
     12      0.021
     13      0.047
     14      0.004
     15      0.009
PRESSURE AT EACH JUNCTION IN COLUMN OF WATER
   JUNC.  P(M)
      1   100.000
      2    94.770
      3    93.506
      4    88.378
      5    88.007
      6    88.902
      7    86.632
      8    86.592
      9    86.250
     10    86.902
     11    87.549
```

Figure 4.10(b) Cont.

PROBLEMS

4.4.1 The total discharge from *A* to *B* in Figure P4.4.1 is 12 ℓ/sec. Determine the head loss between *A* and *B* and the flow rate in each pipe if cast-iron pipes are used at 10°C.

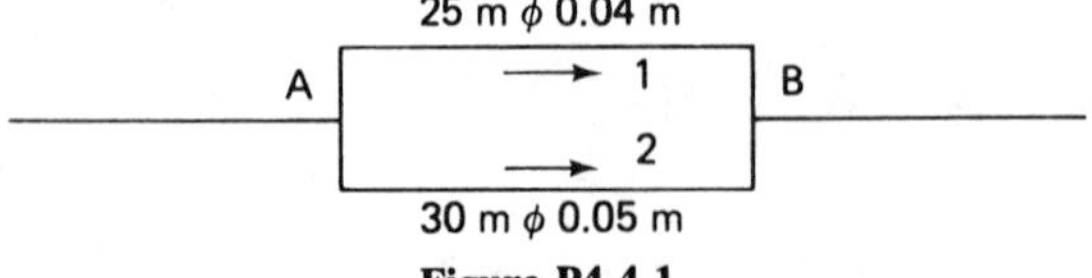

Figure P4.4.1

4.4.2 The system shown in Figure P4.4.2 carries a total discharge of 1.3 m^3/sec. All pipes are commercial steel. Determine the head loss between *A* and *D* at 10°C.

4.4.3 Parallel cast-iron pipes 1, 2, and 3 in Figure P4.4.3 carry a total discharge of 0.8 m^3/sec. Determine the flow rate in each pipe. Neglect minor losses.

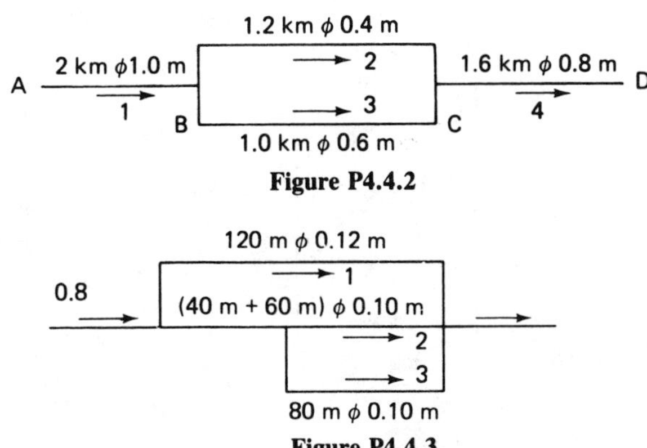

Figure P4.4.2

Figure P4.4.3

4.4.4 Determine the flow in each cast-iron pipe in the network shown in Figure P4.4.4 at 10°C.

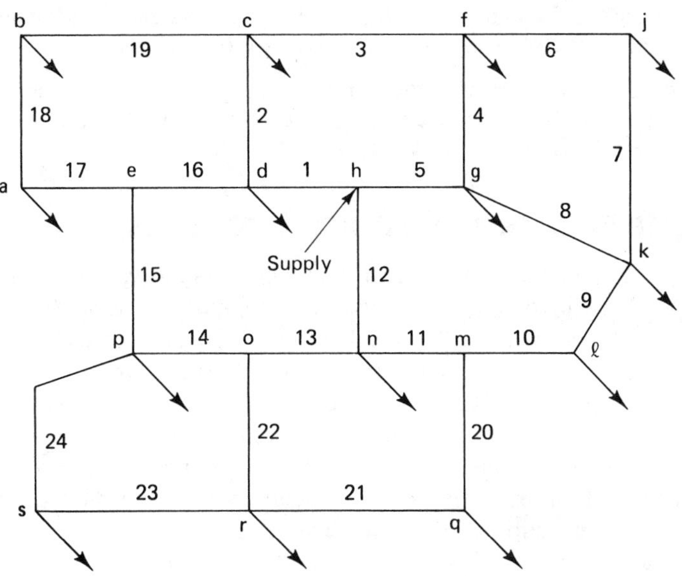

Q = 100 ℓ/sec at a, b, c, d, f, g, j, k, n, r, and s

Q_ℓ = 80 ℓ/sec; $Q_p = Q_q$ = 130 ℓ/sec

L = 400 m for 2, 4, 6, 12, 15, 18, 20, and 22

L = 550 m for 3, 7, 8, 19, 21, 23, and 24

L = 275 m for 1, 5, 9, 10, 11, 13, 14, 16, and 17

D = 60 cm for 1, 5, and 12

D = 40 cm for 2, 4, 8, 11, 13, and 16

D = 30 cm for 14, 15, 17, and 22

D = 20 cm for 3, 6, 7, 9, 10, 18, 19, 20, 21, 23, and 24

Figure P4.4.6

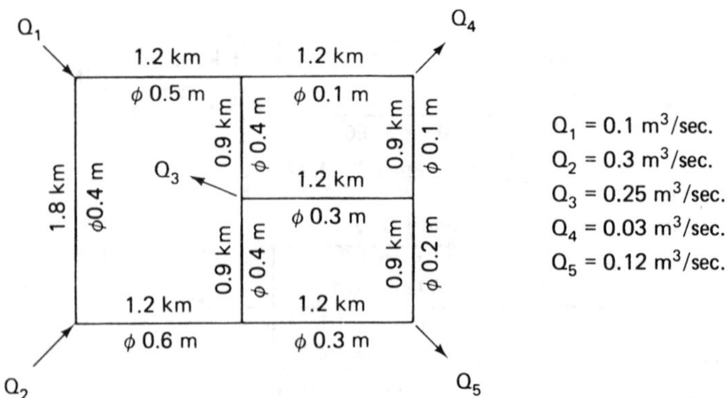

Figure P4.4.4

4.4.5 In Problem 4.4.4. use the Hazen-Williams formula instead of the Darcy-Weisbach equation to solve for the flows.

4.4.6 Determine the flow through each pipe in the network shown in Figure P4.4.6 assuming Hazen-Williams coefficient $C_{HW} = 120$ for all pipes.

4.4.7 For the network described in Problem 4.4.6. the pressure head measured at node *h* is 40 m. Determine diameters for all pipes in the network if the pressure head at all nodes must be kept above 20 m.

4.5 WATER HAMMER PHENOMENON IN PIPELINES

A sudden change of flow rate in a large pipeline (due to valve closure, pump turnoff, etc.) may involve a great mass of water moving inside the pipe. The force resulting from changing the speed of the water mass may cause a pressure rise in the pipe with a magnitude several times greater than the normal static pressure in the pipe. This phenomenon is commonly known as the *water hammer phenomenon*. The excessive pressure may fracture the pipe walls or cause other damage to the pipeline system. The possible occurrence of such pressure, its magnitude, and the propagation of the pressure wave form must be carefully investigated in connection with the pipeline design.

The sudden change of pressure due to a valve closure may be viewed as the result of the force developed in the pipe necessary to stop the flowing water column. The column has a total mass M and is changing its velocity at the rate of dV/dt. According to Newton's second law of motion,

$$F = m\frac{dV}{dt} \tag{4.18}$$

If the velocity of the entire water column could be reduced to zero instantly, Equation 4.18 would become

$$F = \frac{m(V_0 - 0)}{0} = \frac{mV_0}{0} = \infty$$

The resulting force (hence, pressure) would be infinite. Fortunately, such an instantaneous change is almost impossible because a mechanical valve requires a certain amount of time to complete a closure operation. In addition, neither the pipe walls nor the water column involved are perfectly rigid under large pressure. The elasticities of both the pipe walls and the water column play very important roles in the water hammer phenomenon.

Consider a pipe length L with inside diameter D, wall thickness t, and the modulus of elasticity E_p. Immediately following the valve closure, the water in close proximity to the valve is brought to rest. The sudden change of velocity in the water mass causes a local pressure increase. As a result of this pressure increase, the water column in this section is somewhat compressed, and the pipe walls expand slightly due to the corresponding increase of stress in the walls. Both of these phenomena help provide a little extra volume, allowing water to enter the section continuously until it comes to a complete stop.

The next section immediately upstream is involved in the same procedure an instant later. In this manner, a wave of increased pressure propagates up the pipe toward the reservoir, as shown in Figure 4.11(b). When this pressure wave reaches the upstream reservoir, the entire pipe is expanded and the water column within is compressed by the increased pressure. At this very instant the entire water column within the pipe comes to a complete halt.

Obviously, this transient state cannot be maintained because the pressure in the pipe is much higher than the pressure in the open reservoir. The halted water in the pipe begins to flow back into the reservoir as soon as the pressure wave reaches the reservoir. This process starts at the reservoir end of the pipe and a decreased pressure wave travels downstream toward the valve, as shown in Figure 4.11(d). During this period, the water behind the wave front moves in the upstream direction as the pipe continuously contracts and the column decompresses. The time required for the pressure wave to return to the valve is $2L/C$, where C is the speed of the wave travel along the pipe. It is also known as *celerity*.

The speed of pressure wave travel in a pipe depends on the modulus of elasticity of water E_b, and the modulus of elasticity of the pipe wall material E_p. The relationship may be expressed as

$$C = \sqrt{\frac{E_c}{\rho}} \tag{4.19}$$

where E_c is the composite modulus of elasticity of the water pipe system and ρ is the density of water. E_c is composed of the elasticity of the pipe walls and the elasticity of the fluid within. It may be calculated by the following relationship:

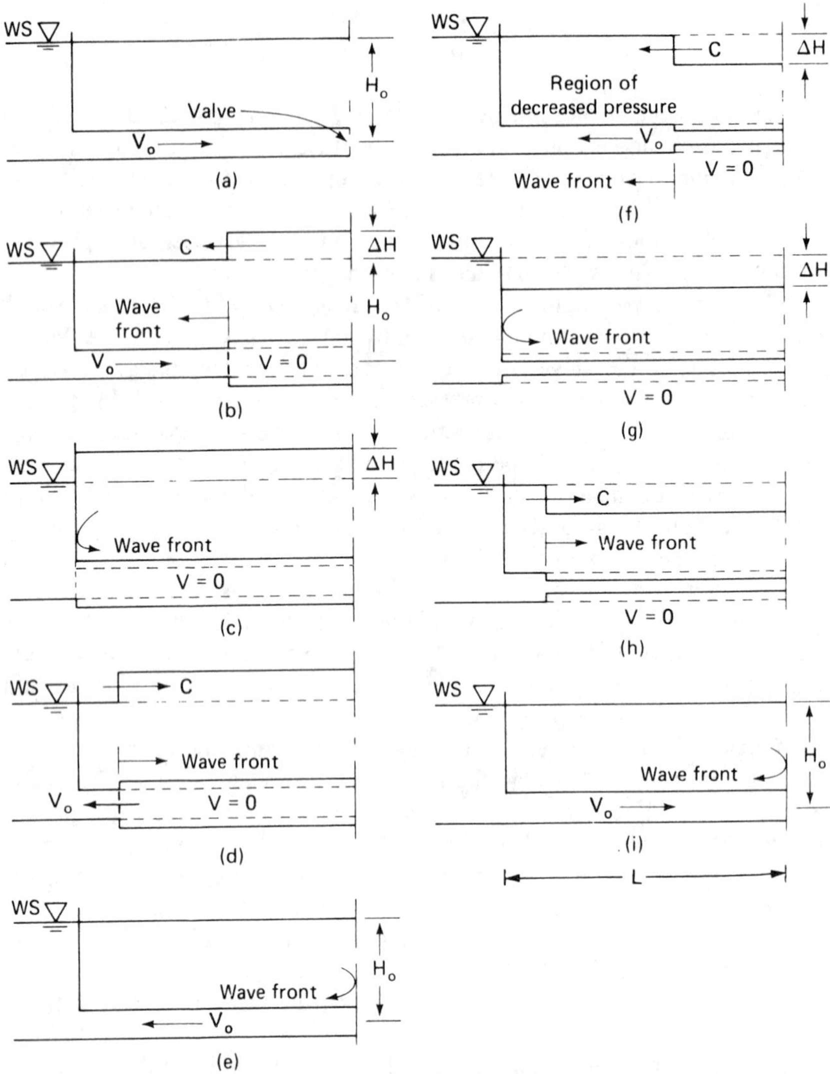

Figure 4.11 Propagation of water hammer pressure waves (friction in pipe neglected). (a) Steady-state condition previous to valve movement. (b) Transient conditions at $t < L/C$. (c) Transient conditions at $t = L/C$. (d) Transient conditions at $L/C < t < 2L/C$. (e) Transient conditions at $t = 2L/C$. (f) Transient conditions at $2L/C < t < 3L/C$. (g) Transient conditions at $t = 3L/C$. (h) Transient conditions at $4L/C > t > 3L/C$. (i) Transient conditions at $t = 4L/C$. Note: After $t = 4L/C$ the cycle repeats and continues indefinitely if the friction in the pipe is zero. The symbol ↻ or ↺ is used to denote the direction of reflection of the wave front.

$$\frac{1}{E_c} = \frac{1}{E_b} + \frac{Dk}{E_p th} \tag{4.20a}$$

The modulus of elasticity of water, E_b, and the density of water are given in Chapter 1. The modulus of elasticity of various common pipe materials is listed in Table 4.1, k is a constant depending on the method of pipeline anchoring, and th is the thickness of the pipe walls.

TABLE 4.1 Modulus of Elasticity, E_p, of Common Pipe Materials

Pipe Material	E_p *(dyn/cm²)*	E_p *(psi)*
Aluminum	$7 \cdot 10^{11}$	$10 \cdot 10^6$
Brass, Bronze	$9 \cdot 10^{11}$	$13 \cdot 10^6$
Cast-iron, gray	$11 \cdot 10^{11}$	$16 \cdot 10^6$
Cast-iron, malleable	$16 \cdot 10^{11}$	$23 \cdot 10^6$
Concrete, reinforced	$16 \cdot 10^{11}$	$25 \cdot 10^6$
Glass	$7 \cdot 10^{11}$	$10 \cdot 10^6$
Lead	$31 \cdot 10^8$	$4.5 \cdot 10^4$
Lucite	$28 \cdot 10^8$	$4 \cdot 10^4$
Copper	$97 \cdot 10^{10}$	$14 \cdot 10^6$
Rubber, vulcanized	$14 \cdot 10^{10}$	$2 \cdot 10^6$
Steel	$19 \cdot 10^{11}$	$28 \cdot 10^6$

$k = \left(\frac{5}{4} - \epsilon\right)$, for pipes free to move longitudinally;

$k = (1 - \epsilon^2)$, for pipes anchored at both ends against longitudinal movement;

$k = (1 - 0.5\epsilon)$, for pipes with expansion joints,

where ϵ is the Poisson's ratio of the pipe wall material. It may take on the value $\epsilon = 0.25$ for common pipe materials.

If the longitudinal stress in a pipe can be neglected, $k = 1.0$, and Equation (4.20a) can be simplified

$$\frac{1}{E_c} = \frac{1}{E_b} + \frac{D}{E_p th} \tag{4.20b}$$

Figure 4.11(e) shows that by the time the decreased pressure wave arrives at the valve, the entire column of water within the pipe is in motion in the upstream direction. This motion cannot pass the already closed valve and must be stopped when the wave arrives at the valve. The inertia of this moving water mass causes the pressure at the valve to drop below the normal static pressure. A third oscillation period begins as the wave of negative pressure propagates up the pipe toward the reservoir, as shown in Figure 4.11(f). At the instant the negative pressure reaches the reservoir the water column within the pipe again comes to a complete standstill and the pressure in the pipe is less than the static pressure. Since the pressure at the entrance to the pipe is less than that in the reservoir, water again flows into the pipe, starting a fourth period of oscillation.

The fourth period is marked by a wave of normal static pressure moving downstream toward the valve, as in Figure 4.11(h). The water mass behind the wave front also moves in the downstream direction. This fourth-period wave arrives at the valve at time $4L/C$, with the entire pipe returning to normal static pressure and water moving in the downstream direction. For an instant the conditions throughout the pipe are somewhat similar to that at the time of the valve closure (the beginning of the first-period wave), except that the water velocity in the pipe is less than what is was at the time of valve closure. This is a result of energy losses to heat due to friction and the viscoelastic behavior of the pipe walls and the water column.

Another cycle begins instantly. The four sequential waves travel up and down the pipe in exactly the same manner as the first cycle described above, except that the corresponding pressure waves are much smaller in magnitude. The pressure wave oscillation continues with each set of waves successively diminishing, until finally the waves die out.

As mentioned previously, the closure of a valve usually requires a certain period of time t to complete. If t is less than $2L/C$ or if the valve closure is completed before the first pressure wave could return to the valve, the resulting rise in pressure should be the same as if the valve had an instantaneous closure. However, if t is greater than $2L/C$, then the first pressure wave returns to the valve before the valve is completely closed. The returned negative pressure wave compensates the pressure rise resulting from the final closure of the valve.

The maximum pressure created by the water hammer phenomenon may be calculated. The formulas for water hammer pressure are derived as follows.

Consider a pipe with a rapidly closing valve ($t \leqq 2L/C$); the extra volume of water ΔVol that enters the pipe during the first period ($t = L/C$) [Figure 4.11(c)] is

$$\Delta\text{Vol} = V_0 A\left(\frac{L}{C}\right) \tag{4.21}$$

where V_0 is the initial velocity of water flowing in the pipe and A is the pipe cross-sectional area. The increased pressure ΔP is related to this extra volume by

$$\Delta P = E_c \cdot \frac{\Delta\text{Vol}}{\text{Vol}} = \frac{E_c \cdot \Delta\text{Vol}}{AL} \tag{4.22}$$

where Vol is the original volume of the water column in the pipe and E_c is the combined modulus of elasticity, as defined by Equation (4.20a). Substituting Equation (4.21) into Equation (4.22), we may write

$$\Delta P = \frac{E_c}{AL} \cdot \left[V_0 \cdot A \cdot \left(\frac{L}{C} \right) \right] = \frac{E_c V_0}{C} \tag{4.23}$$

As the pressure wave propagates upstream along the pipe at speed C, the water behind the wave front is immediately brought to a stop from the initial velocity of V_0. The total mass of water involved in this sudden change of speed from V_0 to zero in time Δt is $m = \rho \cdot AC \cdot \Delta t$. Applying Newton's second law to this mass, we have

$$\Delta P \cdot A = m \frac{\Delta V}{\Delta t} = \rho \cdot A \cdot C \cdot \Delta t \frac{(V_0 - 0)}{\Delta t} = \rho A C V_0$$

or

$$C = \frac{\Delta P}{\rho V_0}$$

and Equation 4.19 results. Substituting the above value of C into Equation (4.23), we have

$$\Delta P = E_c V_0 \frac{\rho V_0}{\Delta P}$$

$$(\Delta P)^2 = \rho E_c V_0^2$$

or

$$\Delta H = \frac{\Delta P}{\rho g} = \frac{V_0}{g} \sqrt{\frac{E_c}{\rho}} = \frac{V_0}{g} C \tag{4.24}$$

which is the pressure head caused by the water hammer. The formula is applicable to rapid valve movement ($t \leqq 2L/C$).

For closure time $t > 2L/C$, ΔP will not develop fully because the reflected negative wave arriving at the valve will compensate for the pressure rise. For such slow valve closures, the maximum water hammer pressure may be calculated by the Allievi formula.*

*L. Allievi, *The Theory of Water Hammer* (translated by E. E. Halmos), *Trans. ASME*, 1929. The maximum water hammer pressure calculated by the Allievi formula is

$$\Delta P = P_0 \left(\frac{N}{2} + \sqrt{\frac{N^2}{4} + N} \right) \tag{4.25}$$

where P_0 is the static-state pressure in the pipe, and

$$N = \left(\frac{\rho L V_0}{P_0 t} \right)^2 \tag{4.26}$$

The total pressure experienced by the pipe is

$$P = \Delta P + P_0$$

Applying the water hammer formula to pipe flow problems, we shall first determine the energy gradient line and the hydraulic gradient line for the pipe system under steady flow conditions, as shown in Figure 4.12.

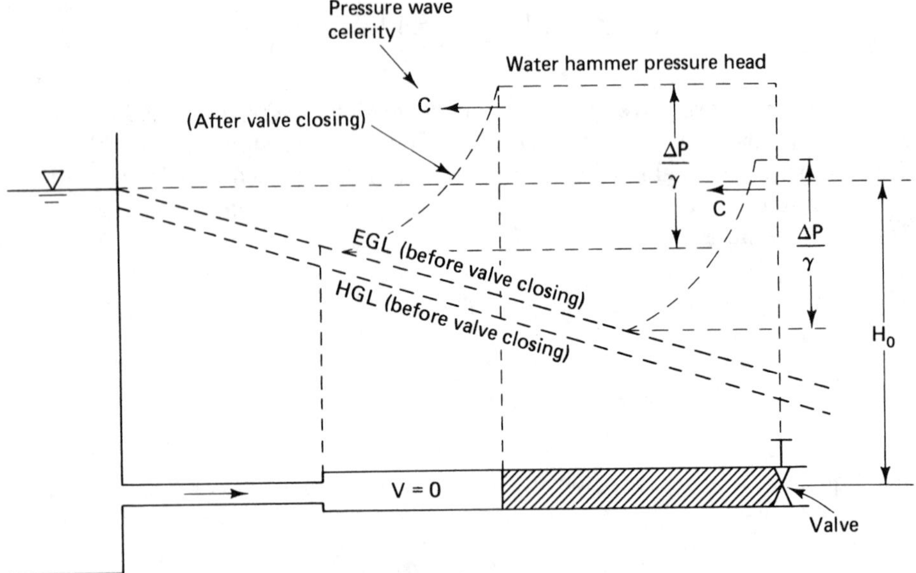

Figure 4.12 Water hammer pressure in a pipeline.

As the pressure wave travels up the pipeline, energy is being stored in the form of pressure energy in the pipe behind the wave front. Maximum pressure is reached when the wave front arrives at the reservoir,

$$P_{max} = \gamma H_0 + \Delta P$$

where H_0 is the total energy head prior to the valve closure, as indicated by the water elevation in the reservoir in Figure 4.12.

Example 4.9

A steel pipe 1500 m long laid on a uniform slope has a 0.5-m diameter and a 5-cm wall thickness. The pipe carries water from a reservoir and discharges it into the air at an elevation 50 m below the reservoir free surface. A valve installed at the downstream end of the pipe allows a flow rate of 0.8 m^3/sec. If the valve is completely closed in 1.4 sec, calculate the maximum water hammer pressure at the valve. Neglect longitudinal stresses.

Solution

From Equation (4.20b),

$$\frac{1}{E_c} = \frac{1}{E_b} + \frac{D}{E_p th}$$

where $E_b = 2.2 \cdot 10^9$ N/m^2, and $E_p = 1.9 \cdot 10^{11}$ N/m^2, as shown in Table 4.1. The above equation may thus be written as

$$\frac{1}{E_c} = \frac{1}{2.2 \cdot 10^9} + \frac{0.5}{(1.9 \cdot 10^{11}) \cdot 0.05}$$

Hence,

$$E_c = 1.97 \cdot 10^9 \text{ N/m}^2$$

From Equation (4.19) we have the speed of wave propagation along the pipe

$$C = \sqrt{\frac{E_c}{\rho}} = \sqrt{\frac{1.97 \cdot 10^9}{998}} = 1404 \text{ m/sec}$$

The time required for the wave to return to the valve is t

$$t = \frac{2L}{C} = \frac{2 \cdot 1500}{1404} = 2.14 \text{ sec}$$

Since the water velocity in the pipe before valve closure is

$$V_0 = \frac{0.80}{\frac{\pi}{4} \cdot (0.5)^2} = 4.07 \text{ m/sec}$$

the maximum water hammer pressure at the valve can be calculated.

$$\Delta P = \rho V_0 C = 998 \cdot 4.07 \cdot 1404 = 5.7 \cdot 10^6 (\text{N/m}^2)$$

Example 4.10

A cast-iron pipe 20 cm in diameter and with 15-mm thick walls is carrying water when the outlet is suddenly closed. If the design discharge is 40 ℓ/sec, calculate the water hammer pressure rise if

(a) the pipe wall is rigid;
(b) the longitudinal stress is neglected;
(c) the pipeline has expansion joints throughout its length.

Solution

$$A = \frac{\pi}{4} \cdot 0.2^2 = 0.0314$$

hence,

$$V = \frac{Q}{A} = \frac{0.04}{0.0314} = 1.274 \text{ m/sec}$$

(a) For rigid pipe wall, $Dk/E_p th = 0$, Equation (4.20a) gives the following relation

$$\frac{1}{E_c} = \frac{1}{E_b}, \quad \text{or} \quad E_c = E_b = 2.2 \cdot 10^9 \text{ N/m}^2$$

$$\rho = 998 \text{ kg/m}^3$$

From Equation (4.19), we can calculate the speed of pressure wave,

$$C = \sqrt{\frac{E_c}{\rho}} = \sqrt{\frac{2.2 \cdot 10^9}{998}} = 1485 \text{ m/sec}$$

From Equation (4.24), we can calculate the rise of water hammer pressure

$$H = \frac{VC}{g} = \frac{1.274 \cdot 1485}{9.81} = 193 \text{ m } (H_2O)$$

$$\therefore P = \gamma H = 1.89 \cdot 10^6 \text{ N/m}^2$$

(b) For pipes with no longitudinal stress, $k = 1$, we may use Equation (4.20b),

$$E_c = \frac{1}{\left(\dfrac{1}{E_b} + \dfrac{D}{E_p th}\right)} = \frac{1}{\left(\dfrac{1}{2.2 \cdot 10^9} + \dfrac{0.2}{(1.6 \cdot 10^{11})(0.015)}\right)} = 1.86 \cdot 10^9$$

and

$$C = \sqrt{\frac{E_c}{\rho}} = 1365 \text{ m/sec}$$

Hence, the rise of water hammer pressure can be calculated

$$H = \frac{VC}{g} = \frac{1.274 \cdot 1365}{9.81} = 177 \text{ m } (H_2O)$$

and

$$P = \gamma H = 1.74 \cdot 10^6 \text{ N/m}^2$$

(c) For pipes with expansion joints, $k = (1 - 0.5 \cdot 0.25) = 0.875$. From Equation (4.20a)

$$E_c = \frac{1}{\dfrac{1}{E_b} + \dfrac{0.875D}{E_p th}} = \frac{1}{\left(\dfrac{1}{2.2 \cdot 10^9} + \dfrac{(0.875)(0.2)}{(1.6 \cdot 10^{11})(0.015)}\right)} = 1.90 \cdot 10^9$$

and

$$C = \sqrt{\frac{E_c}{\rho}} = 1378 \text{ m/sec}$$

Equation (3.57) is used to calculate the rise of water hammer pressure

$$H = \frac{VC}{g} = \frac{1.274 \cdot 1378}{9.81} = 179 \text{ m } (H_2O)$$

Hence,

$$P = \gamma H = 1.76 \cdot 10^6 \text{ N/m}^2$$

In water hammer analysis the time history of pressure oscillation in the pipeline is determined. Because of the friction between the oscillatory water mass and the pipe wall, the pressure–time pattern is modified, and the oscillation gradually dies out. A typical pressure oscillation is shown in Figure 4.13.

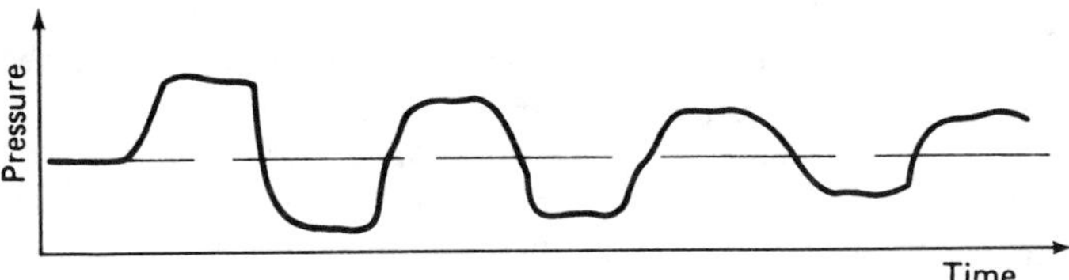

Figure 4.13 Friction effect on pressure-time pattern.

In reality, a valve cannot be closed completely in zero time. The time required for closure of a valve is a certain period, t_c. The water hammer pressure increases gradually with the rate of the closure of the valve. The typical valve closure curve is shown in Figure 4.14.

If t_c is smaller than the time required for the wave front to make a round trip along the pipeline and return to the valve site ($t_c < 2L/C$), the operation is defined as *rapid closure*. The shock pressure will reach its maximum value. The computation of a quick closure operation is the same as that of an instantaneous closure. In order to keep the water hammer pressure within manageable limits, valves are commonly designed with closure times considerably greater than $2L/C$. For slow closure operation, however, $t_c > 2L/C$, the pressure wave

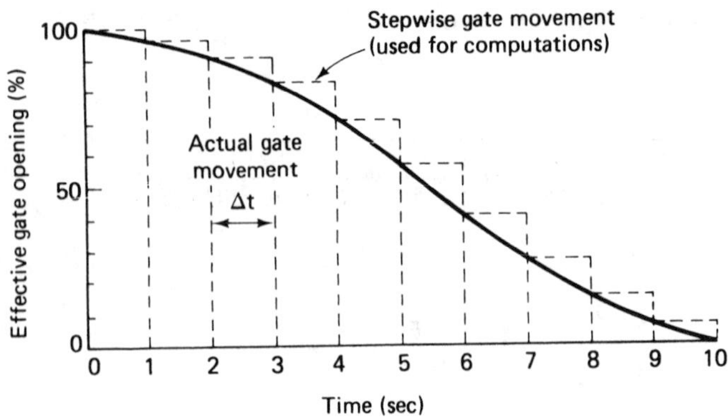

Figure 4.14 Typical valve closure curve.

returns to the valve site before the closure is completed. A certain amount of water continuously passes through the valve when the pressure wave returns. As a result, the pressure wave pattern will be altered. A complete treatment of the water hammer phenomenon, with consideration of friction and slow valve closure operation, may be found in Wylie, 1978* and Pickford 1969.†

PROBLEMS

4.5.1 A 500-m-long steel pipe carries water from a reservoir that has a water elevation at 55 m above the exit of the pipe. The pipe has a diameter of 40 cm and a wall thickness of 2 cm. A gate valve at the downstream end is slowly closed during a 30-sec period. Determine the maximum water hammer pressure.

4.5.2 A horizontal pipe 30 cm in diameter and 420 m long has a wall thickness of 1 cm. The pipe is steel and carries water from a reservoir to a level 100 m below to discharge into open air. A rotary valve is installed at the downstream end. Determine the discharge when the valve is fully opened. Calculate the maximum pressure on the valve if it closes in a 0.5-sec period (neglect longitudinal stresses).

4.5.3 Calculate the water hammer pressure and the speed of the pressure wave in Problem 4.5.2. Assume that

(a) the pipe is free to move longitudinally;
(b) the pipe is anchored against longitudinal motion;
(c) the pipe has expansion joints.

4.5.4 Determine the wall thickness of a steel pipe 2 m in diameter and 700 m long conveying water from a reservoir to a generator. The total head is 150 m. A valve

* E. B. Wylie and V. L. Streeter, *Fluid Transients* (New York: McGraw-Hill, 1978).
† John Pickford, *Analysis of Water Surge* (New York: Gordon and Beach Scientific Publisher, 1969).

is installed at the downstream end of the pipe. The design requires that we consider that the valve can suddenly close. Use Laplace equation ($P \cdot D = 2\tau \cdot$ thickness), with $\tau = 1.1 \times 10^8$ N/m^2 (neglect longitudinal stresses).

4.5.5 Determine the wall thickness in Problem 4.5.4, if (a) expansion joints were used, or (b) the valve closes in 60 sec.

4.5.6 For the pipe described in Problem 4.5.4, compute the maximum pressure if the wall thickness is 10 cm.

4.6 SURGE TANKS

In the case of valve closure, the water hammer pressure may be alleviated by the provision of relief valves or diverters. Although this may be the most simple solution to the problem, it results in a waste of water.

The inclusion of a surge tank near the control station (Figure 4.15) in the pipeline system may give rise to the forces required to retard the mass of water in the pipeline.

A surge tank is defined as a stand pipe or storage reservoir placed at the downstream end of a long pipeline to prevent sudden pressure increases in case of a sudden rejection (valve closing) or demand (valve opening) of the water load. As the valve is being closed, the large mass of water in the long pipeline retards slowly. The difference between flow in the pipeline and that allowed to pass the closing valve causes a rise of water level in the surge tank. As the water rises above the level of the reservoir, a counter pressure is produced so that the water in the pipeline flows back towards the reservoir and the water level in the surge tank drops. The cycle is repeated with *mass oscillation* of water in the pipeline and the surge tank until it is gradually damped out by friction.

Newton's second law can be applied to analyze the effect of the surge tank on the water column *AB*, between the two ends of the pipeline. At any time during the closure or opening of the valve, the acceleration of the water mass is always equal to the forces acting on it. That is

$$\begin{aligned}\rho LA \frac{dV}{dt} = {} & \text{(the pressure force on the column at } A\text{)} \\ & + \text{(the weight component of the column} \\ & \quad \text{in the direction of the pipeline)} \\ & - \text{(the force acting on the column at } B\text{)} \\ & + \text{(friction losses)}\end{aligned}$$

The pressure force at *A* is due to the elevation difference between the water surface in the reservoir and the pipeline inlet, modified by the entrance loss. The force acting on the column at *B* depends on the elevation of the water surface in the surge tank, also modified by the losses occurring at the entry (may be a restrictive throttle) to the tank. Hence,

$$\rho LA \frac{dV}{dt} = \rho g A[(H_A \pm \text{entrance loss}) + (H_B - H_A) - (H_B + y \pm \text{throttle loss}) \pm (\text{pipeline losses})] \tag{4.27}$$

The sign of the pipeline losses depends on the direction of the flow. The losses always occur in the direction of the flow.

If we introduce the modulus form, $h_L = K_f V \cdot |V|$ and $H_T = K_T U \cdot |U|$, where

$$U = \frac{dy}{dt} \tag{4.28}$$

which is the upward velocity of the water surface in the tank, the sign of the losses would always be correct. Here K_f is the pipeline friction factor $K_f = fL/2gD$ and h_L is the total head loss in the pipeline between A and B. H_T is the throttle loss.

Substituting these values into Equation (4.27) and simplifying, we have the dynamic equation for the surge tank

$$\frac{L}{g}\frac{dV}{dt} + y + K_f V \cdot |V| + K_T U \cdot |U| = 0 \tag{4.29}$$

In addition, the continuity condition at B must be satisfied

$$VA = UA_s + Q \tag{4.30}$$

where Q is the discharge allowed to pass the closing valve at any given time t.

The combination of Equations (4.28), (4.29), and (4.30) gives second order differential equations that can only be solved for special cases. A special solution may be obtained by what is frequently called the *logarithmic method*.* The method provides simple theoretical analysis of surge heights that are close to those observed in practice, if the cross-sectional area A_s remains constant.

The solution for a simple (unrestricted) constant-area surge tank (Figure 4.15) may be expressed as

$$\frac{y_{max} + h_L}{\beta} = \ln\left(\frac{\beta}{\beta - y_{max}}\right) \tag{4.31}$$

where β is the damping factor, defined as

$$\beta = \frac{LA}{2gK_f A_s} \tag{4.32}$$

* John Pickford, *Analysis of Water Surge* (New York: Gordon and Beach, 1969), pp. 111–124.

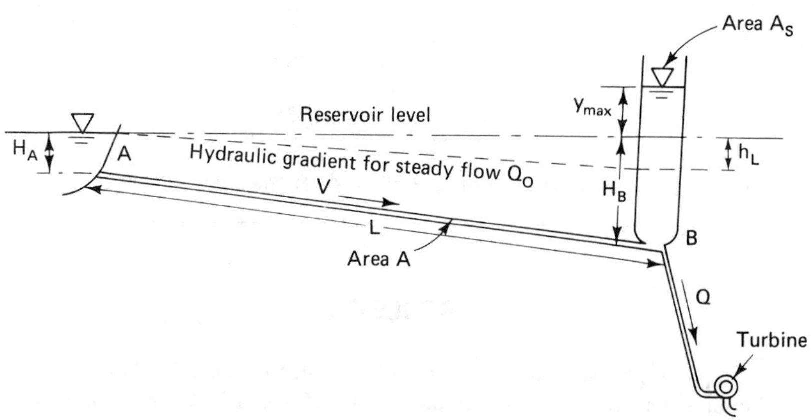

Figure 4.15 Surge tank.

Equation (4.31) may be solved by trial and error for the surge height, y, as demonstrated in the following example.

Example 4.11

A simple surge tank 8 m in diameter is located at the downstream end of a 1500 m long pipe, 2.2 m in diameter. The head loss between the upstream reservoir and the surge tank is 15.13 m when the flowrate is 20 m³/sec. Determine the maximum elevation of the water in the surge tank if a valve downstream suddenly closes.

Solution

For a smooth entrance where the head loss may be neglected, we may write

$$h_L \cong h_f = K_f V^2$$

or

$$K_f = \frac{h_L}{V^2} = \frac{15.13}{(5.26)^2} = 0.5466$$

and the damping factor, from Equation (4.32), is

$$\beta = \frac{LA}{2gK_f A_s} = \frac{(1500)(3.80)}{2(9.81)(0.5466)(50.27)} = 10.58$$

Applying Equation (4.31)

$$\frac{y_{max} + 15.13}{10.58} = \ln\left(\frac{10.58}{10.58 - y_{max}}\right)$$

the solution is obtained by trial and error.

y_{max}	*LHS*	*RHS*
9.5	2.33	2.28
9.6	2.34	2.38
9.55	2.33	2.33

The maximum elevation for water is 9.55 m over the reservoir level.

PROBLEMS

4.6.1 A simple surge tank is installed to retard the water mass in a 1050 m long pipeline, 2 m in diameter. The design discharge is 13 m^3/sec at 10°C and the pipe material is smooth concrete. Determine the diameter of the surge tank if the water in the tank is allowed to rise to an elevation 5 m above the feeding reservoir after the flow is suddenly stopped.

4.6.2 If the surge tank in Problem 4.6.1 was 6 m in diameter, determine the maximum water elevation if the pipeline is discharging 10 m^3/sec at 10°C when the flow is interrupted.

4.6.3 A simple surge tank is installed in a pipeline to protect an electric generator. The circular concrete tunnel between the reservoir and the surge tank is 1600 m long and 1.5 m in diameter. The maximum flow is 6 m^3/sec at 10°C. Compute the maximum water rise if the surge tank is 6 m in diameter.

4.6.4 Determine the minimum diameter of the surge tank in Problem 4.6.3 if the allowable water surface rise is 3 m over the feeding reservoir water level.

PROJECT PROBLEM

4.6.5 From a reservoir 1200 m away a 15-cm-diameter main pipe supplies water to six multistorey buildings in an industrial park. The reservoir is 80 m above the datum elevation. The positions of the buildings are shown in Figure P4.6.5. The height and the water demand of each building are as follows:

Buildings	*A*	*B*	*C*	*D*	*E*	*F*
Height (m)	9.4	8.1	3.2	6.0	9.6	4.5
Water Demand (ℓ/sec)	5.0	6.0	3.5	8.8	8.0	10.0

If commercial steel pipes are used for the network (downstream from the junction J), determine the size of each pipeline. A gate valve ($K = 0.15$ when fully opened) is installed in the main pipe immediately upstream from the junction J. Determine the material to be used for the main pipe. Determine the water hammer pressure if the valve is suddenly closed. What must be the minimum wall thickness of the pipe in order to withstand the pressure?

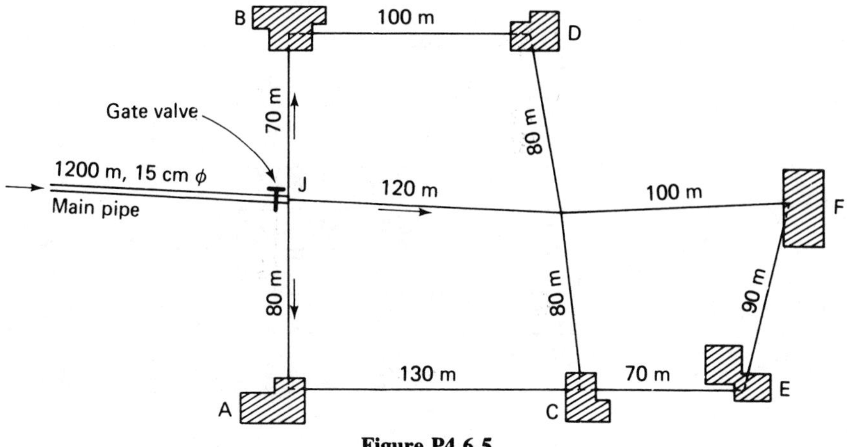

Figure P4.6.5

WATER PUMPS

Water pumps are devices designed to convert mechanical energy to hydraulic energy. All forms of water pumps may be classified into two basic categories

1. turbo-hydraulic pumps,
2. positive-displacement pumps.

Turbo-hydraulic pumps include mainly the centrifugal pumps, propeller pumps, and jet pumps. Analysis of the turbo-hydraulic machines is a problem involving fundamental principles of hydraulics. Positive-displacement pumps move fluid strictly by precise machine displacements such as a gear system rotating within a closed housing (screw pumps) or a piston moving in a sealed cylinder (reciprocal pumps). Analysis of the positive-displacement pumps involves purely mechanical concepts and does not require the detailed knowledge of hydraulics. This chapter will only treat the first category, which constitutes most of the water pumps used in modern hydraulic engineering systems.

5.1 CENTRIFUGAL PUMPS

The fundamental principle of the centrifugal pump was first demonstrated by Demour in 1730. The simple pump, as shown in Figure 5.1 consisted of two straight pipes in the form of a tee. If the tee is primed with a liquid and the lower end of the tee is submerged, the rotation of the horizontal arms develops a centrifugal force which, when it overcomes the gravitational force on the liquid, moves the liquid upward through the vertical shaft and discharges it through the two ends of the horizontal arms.

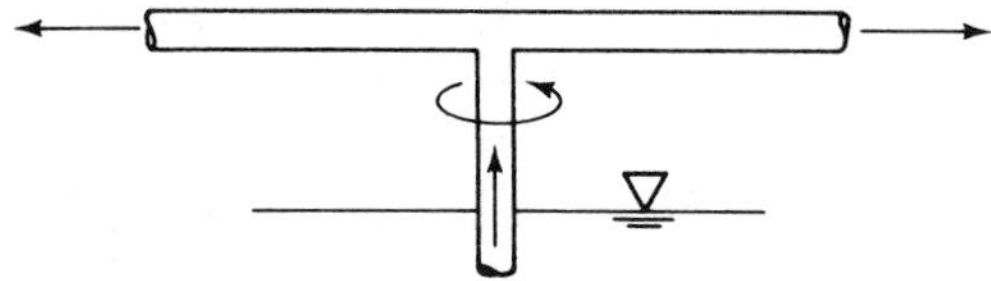

Figure 5.1 Demour's centrifugal pump.

Modern centrifugal pumps are constructed with this same hydraulic principle, but with new configurations designed to improve the efficiency. Modern centrifugal pumps basically consist of two parts

1. The rotating element that is commonly called the *impeller*.
2. The *housing* that encloses the rotating element and seals the pressurized liquid inside.

The power of a pump is supplied by a motor to the shaft of the impeller. The rotary motion of the impeller creates a centrifugal force that enables the liquid to enter the pump at the low-pressure region near the center (*eye*) of the impeller and to move along the direction of the impeller vanes toward the higher-pressure region near the outside of the housing surrounding the impeller, as shown in Figure 5.2. The housing is designed with a gradually expanding spiral shape so that the entering liquid is led toward the discharge pipe with minimum loss while the kinetic energy in the liquid is converted into pressure energy.

The theory of centrifugal pumps is based on the *principle of angular momentum conservation*. Physically, the term momentum, which usually refers to the linear momentum, is defined as the product of a mass and its velocity, or

$$\text{momentum} = (\text{mass})(\text{velocity})$$

The angular momentum (or moment of momentum) with respect to a fixed axis of rotation may thus be defined as the moment of the linear momentum with respect to the axis

$$\begin{aligned}\text{angular momentum} &= (\text{radius}) \cdot (\text{momentum}) \\ &= (\text{radius}) \cdot (\text{mass})(\text{velocity})\end{aligned}$$

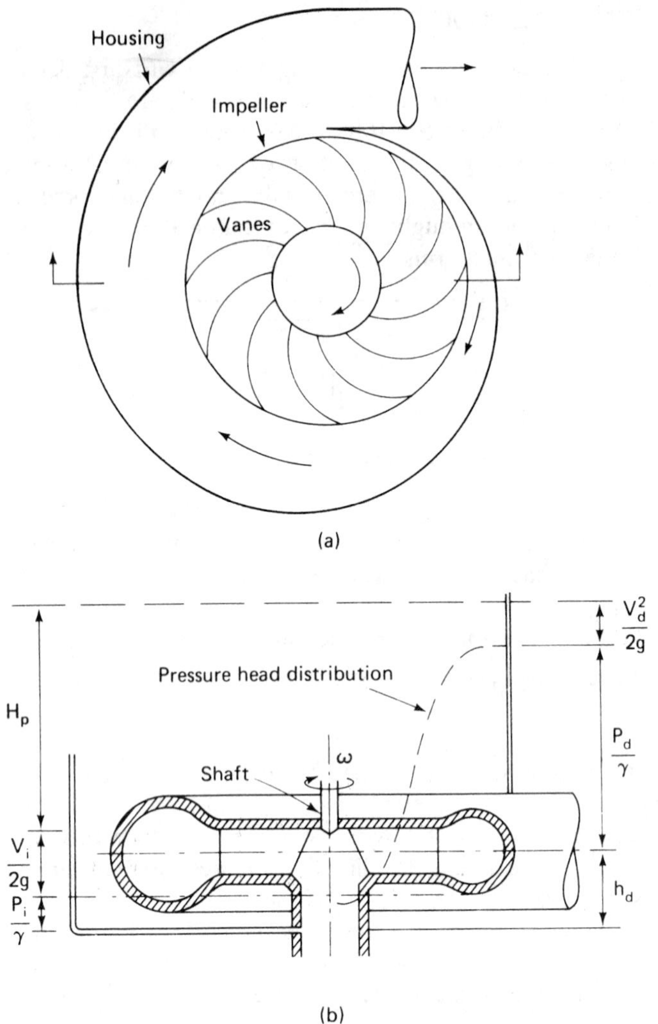

Figure 5.2 Cross sections of a centrifugal pump.

The principle of conservation of angular momentum requires that *the time rate of change of angular momentum in a body of fluid be equal to the torque resulting from the external force acting on the body*. This relationship may be expressed as

$$\text{torque} = \frac{(\text{radius})\cdot(\text{mass})(\text{velocity})}{\text{time}} = (\text{radius})\cdot\rho\left(\frac{\text{volume}}{\text{time}}\right)\cdot(\text{velocity})$$

The diagram in Figure 5.3 can be used to analyze this relationship.

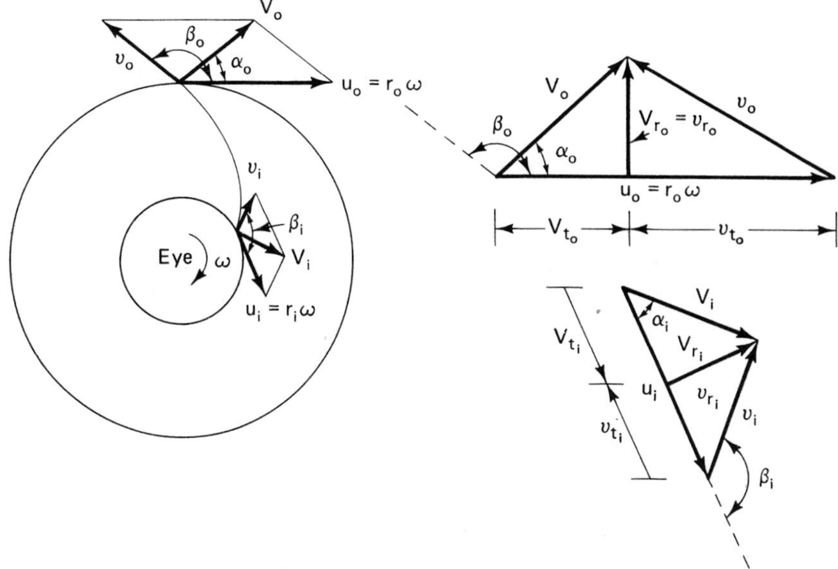

Figure 5.3 Velocity diagram of a pump impeller. u is the speed of the impeller vane ($u = r\omega$); v is the relative velocity of the liquid with respect to the vane; V is the absolute velocity of the liquid, a vector sum of u and v; β_0 is the vane angle at the exit; β_i is the vane angle at the entrance; $r = r_i$ is the radius of the impeller eye at the entrance; and $r = r_0$ is the radius of the impeller at the exit.

The angular momentum (or moment of momentum) for a small fluid mass per unit time ($\rho\, dQ$) is

$$(\rho\, dQ)(V \cos \alpha)(r)$$

where $V \cos \alpha$ is the tangential component of the absolute velocity. For the total fluid mass that enters the pump in unit time, the angular momentum can be evaluated by the following integral:

$$\rho \int_Q rV \cos \alpha\, dQ$$

The torque applied to a pump impeller must equal the difference of angular momentum at the inlet to and outlet from the impeller. It may be expressed as follows:

$$T = \rho \int_Q r_o V_o \cos \alpha_o\, dQ - \rho \int_Q r_i V_i \cos \alpha_i\, dQ \tag{5.1}$$

For steady flow and uniform conditions around the impeller, $r_o V_o \cos \alpha_o\, dQ$ and $r_i V_i \cos \alpha_i\, dQ$ have constant values. Equation (5.1) may be simplified to

$$T = \rho Q (r_o V_o \cos \alpha_o - r_i V_i \cos \alpha_i) \quad (5.2)$$

Let ω be the angular velocity of the impeller. The power input to the pump can be computed as

$$\mathbf{P}_i = \omega T = \rho Q \omega (r_o V_o \cos \alpha_o - r_i V_i \cos \alpha_i) \quad (5.3)$$

The output power of a pump is usually expressed in terms of the pump discharge and the total energy head that the pump imparts to the liquid, H_P. As previously discussed, the energy head in fluid can usually be expressed as the sum of the three forms of hydraulic energy:

1. kinetic energy, $V^2/2g$,
2. pressure energy, P/γ,
3. elevation energy, h.

$$H = \frac{V^2}{2g} + \frac{P}{\gamma} + h$$

The *pump output power* may be expressed as

$$\mathbf{P}_o = \gamma Q H_P = \gamma Q \left(\frac{V^2}{2g} + \frac{P}{\gamma} + h \right) \quad (5.4)$$

where V, P, and h are, respectively, the velocity, pressure, and elevation added to the liquid by the pump.

The *polar vector diagram* (Figure 5.3) is generally used in analyzing the vane geometry and its relationship to the flow. As designated previously, the subscripts i and o are used, respectively, for the inlet and exit flow conditions. u represents the peripheral velocity of the impeller, v represents the water velocity relative to the vane blade (hence, in the direction of the blade), and V is the absolute water velocity. V_t is the tangential component of the absolute velocity, and V_r is the radial component. Theoretically, the energy loss at the inlet reaches its minimum value when water enters the impeller without whirl. This is achieved when the impeller is operated at such a speed that the absolute water velocity at the inlet is in the radial direction.

The efficiency of a centrifugal pump depends largely on the particular design of the vane blades and the pump housing. It also depends on the conditions under which the pump operates.

The *efficiency* of a pump is defined by the ratio of the output power to the input power of the pump.

$$e_P = \frac{\mathbf{P}_o}{\mathbf{P}_i} = \frac{\gamma Q H_P}{\omega T} \quad (5.5)$$

A hydraulic pump is usually driven by a motor. The efficiency of the motor is defined by the ratio of the power applied to the pump by the motor, P_i, to the power input to the motor, P_m.

$$e_m = \frac{P_i}{P_m} \tag{5.6}$$

The *overall efficiency* of the pump system is thus

$$e = e_P e_m = \left(\frac{P_o}{P_i}\right) \cdot \left(\frac{P_i}{P_m}\right) = \frac{P_o}{P_m} \tag{5.7}$$

or

$$P_o = e \cdot P_m \tag{5.8}$$

The values of the e's are always less than unity due to friction and other energy losses that occur in the system.

In Figure 5.2 the total energy head at the entrance to the pump is represented by

$$H_i = \frac{P_i}{\gamma} + \frac{V_i^2}{2g}$$

and the total energy head at the discharge is

$$H_d = h_d + \frac{P_d}{\gamma} + \frac{V_d^2}{2g}$$

The difference between the two is the amount of energy that the pump imparted to the liquid.

$$H_P = H_d - H_i = \left(h_d + \frac{P_d}{\gamma} + \frac{V_d^2}{2g}\right) - \left(\frac{P_i}{\gamma} + \frac{V_i^2}{2g}\right) \tag{5.9}$$

Example 5.1

A centrifugal pump has the following characteristics: $r_i = 12$ cm, $r_o = 40$ cm, $\beta_i = 118°$, $\beta_o = 140°$. The width of the impeller vanes is 10 cm and is uniform throughout. At the angular speed of 550 rpm the pump delivers 0.98m^3/sec of water between two reservoirs with a 25-m elevation difference. If a 500-kw motor is used to drive the centrifugal pump, determine the efficiency of the pump and the overall efficiency of the system at this stage of operation.

Solution

The peripheral speeds of the vanes at the entrance and at the exit of the impeller are, respectively,

$$u_i = \omega r_i = 2\pi \cdot \frac{550}{60} \cdot 0.12 \text{ m} = 6.91 \text{ m/sec}$$

$$u_o = \omega r_o = 2\pi \cdot \frac{550}{60} \cdot 0.40 \text{ m} = 23.04 \text{ m/sec}$$

and the radial velocity of the water may be obtained by applying the continuity equation; $Q = A_i V_{r_i} = A_o V_{r_o}$, where $A_i = 2\pi r_i B$ and $A_o = 2\pi r_o B$. It can be easily shown that $V_{r_i} = v_{r_i}$ and $V_{r_o} = v_{ro}$.

$$v_{r_i} = \frac{Q}{A_i} = \frac{Q}{2\pi r_i B} = \frac{0.98}{2\pi \cdot 0.12 \cdot 0.1} = 13.00 \text{ m/sec}$$

$$v_{r_o} = \frac{Q}{A} = \frac{Q}{2\pi r_o B} = \frac{0.98}{2\pi \cdot 0.4 \cdot 0.1} = 3.90 \text{ m/sec}$$

From the velocity vector diagram (see Figure 5.3)

$$v_{t_i} = \frac{v_{r_i}}{\tan \beta_i} = \frac{13.00}{\tan 118^\circ} = -6.91 \text{ m/sec}$$

$$v_{t_o} = \frac{v_{r_o}}{\tan \beta_o} = \frac{3.90}{\tan 140^\circ} = -4.65 \text{ m/sec}$$

and

$$V_i = \sqrt{v_{r_i}^2 + (u_i + v_{t_i})^2} = \sqrt{(13.00)^2 + (0.00)^2} = 13.00 \text{ m/sec}$$

$$\alpha_i = \tan^{-1} \frac{v_{r_i}}{(u_i + v_{t_i})} = \tan^{-1}\left(\frac{13.00}{0.00}\right) = 90^\circ$$

$$\cos \alpha_i = 0$$

$$V_o = \sqrt{v_{r_o}^2 + (u_o + v_{t_o})^2} = \sqrt{(3.90)^2 + (18.39)^2} = 18.80 \text{ m/sec}$$

$$\alpha_o = \tan^{-1} \frac{v_{r_o}}{u_o + v_{t_o}} = \tan^{-1}\left(\frac{3.90}{18.93}\right) = 11.97^\circ$$

$$\cos \alpha_o = 0.978$$

Applying Equation (5.3), we get

$$P_i = \rho Q \omega (r_o V_o \cos \alpha_o - r_i V_i \cos \alpha_i)$$

$$P_i = 1000 \cdot 0.98 \cdot 2\pi \cdot \frac{550}{60} \cdot (0.40 \cdot 18.8 \cdot 0.978 - 0) = 415{,}120 \text{ w}$$

$$= 415.12 \text{ kw}$$

Applying Equation (5.4) and assuming the only considerable head is the elevation, we get $H_P = h$

$$P_o = \gamma Q H_P = 9.81(\text{kN/m}^3)\cdot 0.98(\text{m}^3/\text{sec})\cdot 25(\text{m}) = 240.35 \text{ kw}$$

From Equation (5.5), the efficiency of the pump is

$$e_P = \frac{P_o}{P_i} = \frac{240.35}{415.12} = 0.579 \cong 58\%$$

From Equation (5.6), the overall efficiency of the system is

$$e = e_P e_m = \left(\frac{P_o}{P_i}\right)\left(\frac{P_i}{P_m}\right) = (0.58)\cdot\left(\frac{415.12}{500}\right)$$
$$= 0.48 = 48\%$$

PROBLEMS

5.1.1 A centrifugal pump delivers 2.5 m^3/sec against an energy head of 20 m. If the pump operates at 85% efficiency, determine the input power at the pump shaft.

5.1.2 The pump in Problem 5.1.1 is driven by a 750-kw electric motor. Determine the efficiency of the motor and the efficiency of the pump–motor system.

5.1.3 A 25,000-w pump–motor system delivers 75 ℓ/sec against a head of 22 m. Determine the system's overall efficiency.

5.1.4 A 24-kw motor with 82% efficiency drives a centrifugal pump delivering 72 ℓ/sec against a head of 21 m. Compare the efficiency of this system to the one described in Problem 5.1.3

5.1.5 A centrifugal pump with uniform impeller thickness 0.1 m, inlet radius r_i = 30 cm, outlet radius r_o = 75 cm, $\beta_i = 120°$, $\beta_o = 135°$ delivers a discharge of 2 m^3/sec against a head of 10 m. If the pump rotates at such a speed that no tangential velocity component exists at the inlet, what is the rotational speed? Calculate the efficiency of the pump.

5.1.6 At the outlet of a centrifugal pump impeller the radius is 60 cm and the width is 15 cm. If the velocity of water at the outlet (the absolute velocity) is measured to be 40 m/sec at an angle of 60° measured with respect to the radial line, calculate the torque exerted on the impeller by the outlet flow.

5.1.7 A centrifugal pump impeller has an inlet diameter of 50 cm and outlet diameter of 150 cm. With $\beta_i = 135°$ and $\beta_o = 150°$, the pump is rotating at an angular velocity of 100 rad/sec. The impeller has a uniform thickness of 30 cm. If the radial velocity component v_{r_i} is the same magnitude as the tangential velocity component V_{t_i}, calculate the discharge of the pump and the power input to the shaft of the pump.

5.1.8 If the efficiency of the pump in Problem 5.1.7 is at 85%, calculate the head against which the discharge is delivered.

5.2 PROPELLER (AXIAL-FLOW) PUMPS

A rigorous mathematical analysis for designing propellers based strictly on the energy-momentum relationship is not available. However, the application of the basic *principle of impulse-momentum* provides a simple means of describing their operations.

Linear impulse is defined as the integral of the product of the force and the time, dt, from t' to t'' during which the force acts on the body

$$I = \int_{t'}^{t''} F \, dt$$

If a constant force is involved during the time period, T, the impulse may be simplified to

$$(\text{impulse}) = (\text{force})(\text{time})$$

The principle of impulse-momentum requires that *the linear impulse of a force (or force system) acting on a body during a time interval be equal to the change in linear momentum in the body during that time.*

$$(\text{force})(\text{time}) = (\text{mass})(\text{velocity change})$$

or

$$(\text{force}) = \frac{(\text{mass})(\text{velocity change})}{(\text{time})} \tag{5.10}$$

The relationship may be applied to a body of fluid in steady motion by taking a volume between any two sections as shown in Figure 5.4. The factor, (mass)/(time), may be expressed as the mass involved per unit time, or,

$$\frac{(\text{mass})}{(\text{time})} = \frac{(\text{density})(\text{volume})}{(\text{time})} = (\text{density})(\text{discharge}) = \rho Q$$

and the change in velocity is thus the change of the fluid velocity between the two ends of the control volume.

$$(\text{velocity change}) = V_i - V_f$$

Substituting the above relationships into Equation (5.10), we have

$$\Sigma F = (\text{force}) = \rho Q (V_i - V_f) \tag{5.11}$$

Figure 5.4 schematically shows a propeller pump installed in a horizontal position. Four sections are selected along the pump tube to demonstrate the energy relationship. The system can be readily analyzed by the principle of impulse-momentum.

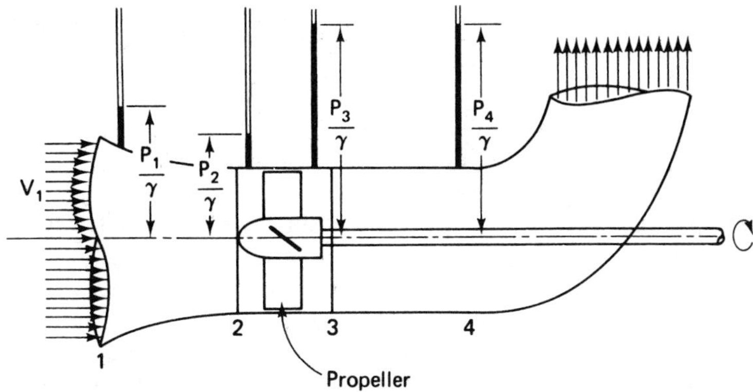

Figure 5.4 Propeller pump.

While the fluid moves from section 1 to section 2, the velocity increases and the pressure drops according to the Bernoulli principle of energy balance

$$\frac{P_1}{\gamma} + \frac{V_1^2}{2g} = \frac{P_2}{\gamma} + \frac{V_2^2}{2g}$$

Between the two adjacent sections 2 and 3, energy is added to the fluid by the propeller. The energy is added to the fluid in the form of pressure head, which results in a higher pressure immediately downstream from the propeller. Further downstream at the exit end of the pump (section 4), the flow condition is more stable, and a slight drop in pressure head may result from both the head loss between sections 3 and 4 and a slight increase in mean stream velocity.

Applying the impulse-momentum relationship, Equation (5.11), between sections 1 and 4, we may write the following equation:

$$P_1A_1 + F - P_4A_4 = \rho Q(V_4 - V_1) \tag{5.12}$$

where F is the force exerted to the fluid by the propeller. The right-hand side of Equation (5.12) drops out when the pump is installed in a conduit of uniform diameter, for which $V_1 = V_4$ and

$$F = (P_1 - P_4)A$$

In this case, the force imparted by the pump is totally used to generate pressure.

$$\frac{P_1}{\gamma} + \frac{V_1^2}{2g} = \frac{P_2}{\gamma} + \frac{V_2^2}{2g} \tag{5.13}$$

and between sections 3 and 4, we may write

$$\frac{P_3}{\gamma} + \frac{V_3^2}{2g} = \frac{P_4}{\gamma} + \frac{V_4^2}{2g} \tag{5.14}$$

Subtracting Equation (5.13) from Equation (5.14), and noting that $V_2 = V_3$ for the same cross-sectional area, we have

$$\frac{P_3 - P_2}{\gamma} = \left(\frac{P_4}{\gamma} + \frac{V_4^2}{2g}\right) - \left(\frac{P_1}{\gamma} + \frac{V_1^2}{2g}\right) = H_P \tag{5.15}$$

where H_P is the total energy imparted to the fluid by the pump. The total power output from the pump is, therefore,

$$P_o = \gamma Q H_P = Q(P_3 - P_2) \tag{5.16}$$

The efficiency of the pump may be computed by the ratio of the output power of the pump to the input power of the motor.

Propeller pumps are usually used for low head (under 12 m), high-capacity (above 20 ℓ/sec) applications. Figure 5.8 on page 157 shows the relative regions of application for different types of pumps.

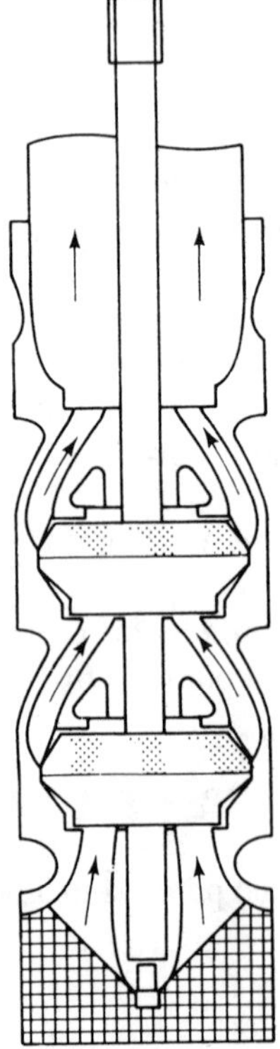

Figure 5.5 Multistage propeller pump.

More than one set of propeller blades may be mounted on the same axis of rotation in a common housing to form a *multistage propeller pump*, as shown in Figure 5.5. These pumps are capable of delivering a large quantity of water over a great elevation difference. The pumps are usually designed for self-priming operations and are used most frequently for pumping deep-water wells.

Example 5.2

A 3-m-diameter propeller pump is installed to deliver a large quantity of water over a 2.6-m elevation head. If the shaft power supplied to the pump is 1500 kw and if the pump operates at 80% efficiency, determine the discharge rate.

Solution

The energy imparted to the flow by the pump is given by Equation 5.5

$$P_o = e_P P_i = 0.8 \cdot 1500 \text{ kw} = 1200 \text{ kw} = 1{,}200{,}000 \text{ w}$$

also,

$$P_o = \gamma Q H_P = \gamma Q\left(h + \Sigma k \frac{V^2}{2g}\right)$$

For $k_{in} = 0.5$ and $k_{out} = 1.0$, we have

$$P_o = \gamma Q\left(h + 1.5\frac{Q^2}{2gA^2}\right); \qquad \gamma = 9810 \text{ N/m}^3$$

and

$$1{,}200{,}000 = 9810\ Q\left(2.6 + 1.5\frac{Q^2}{(1.5\pi)^2 \cdot 2 \cdot 9.81}\right)$$

By trial, the above equation may be solved to obtain

$$Q = 30.45 \text{ m}^3/\text{sec}$$

5.3 JET (MIXED-FLOW) PUMPS

Jet pumps utilize a high-pressure stream. The pressurized fluid ejects from a nozzle at high speed into the surrounding fluid, transferring its energy to the surrounding fluid and bringing it to delivery. Jet pumps are usually used in combination with a centrifugal pump, which supplies the high-pressure stream, and can be used to lift liquid in deep wells. They are usually compact in size and light in weight. They are sometimes used in construction work for dewatering the field. Since the energy loss during the mixing procedure is heavy, the efficiency of the jet pump is normally very low (rarely more than 25%). A schematic diagram of a jet pump is shown in Figure 5.6.

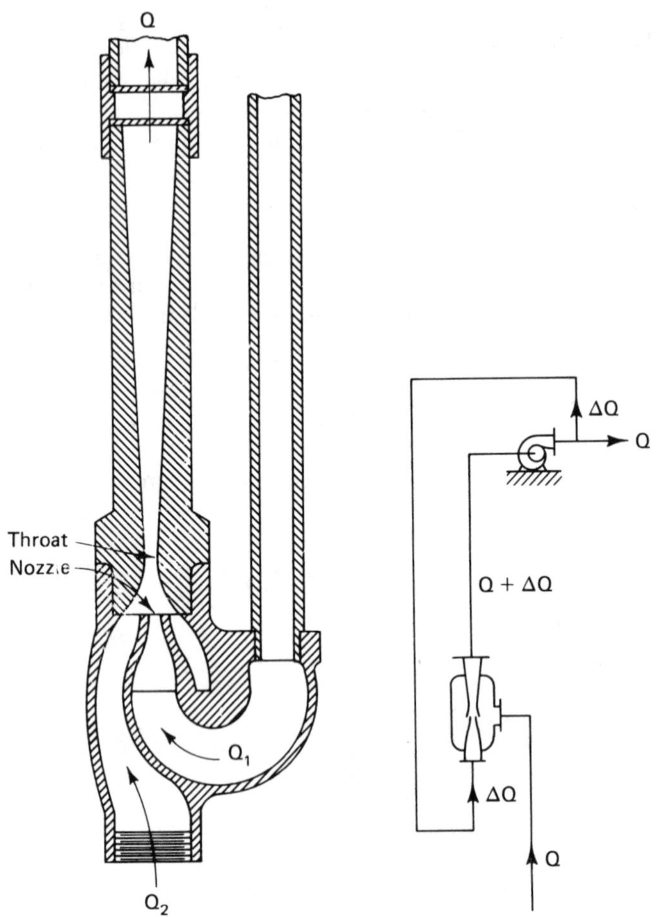

Figure 5.6 Jet pump.

Figure 5.7 Jet pump as a booster.

A jet pump can also be installed in series with a centrifugal pump system as a booster pump. The jet pump may be built into the casing of the centrifugal pump suction line to boost the water surface elevation at the inlet of the centrifugal pump as shown schematically in Figure 5.7. This arrangement avoids any unnecessary installation of moving parts in the well casing, which is usually buried deep below the ground surface.

5.4 SELECTION OF A PUMP

The efficiency of a pump depends on the discharge, head, and power requirement of the pump. The approximate ranges of application of each type of pump are indicated in Figure 5.8.

The total head that the pump delivers its discharge against includes the elevation head and the head losses incurred in the system. The friction loss and other minor losses in the pipeline depend on the velocity of water in the pipe

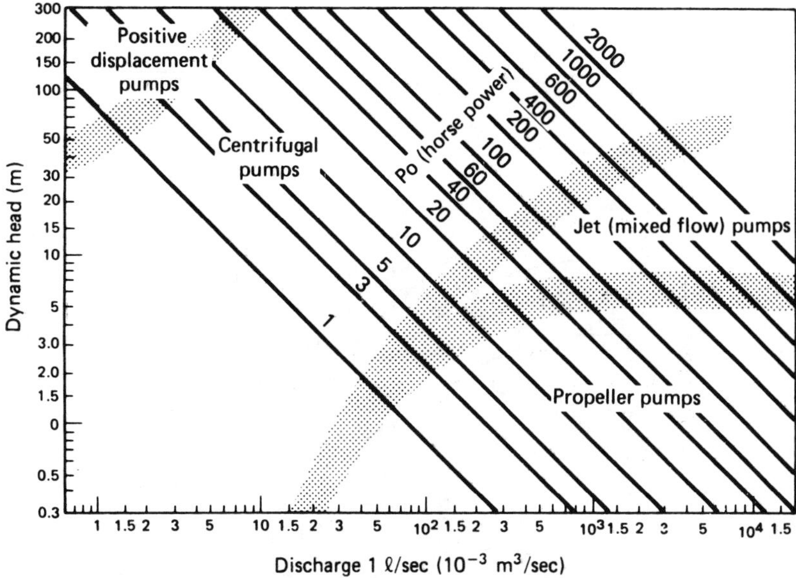

Figure 5.8 Head, discharge, and power requirement of different types of pumps.

(Chapter 3), and the total head loss can be related to the discharge rate. For a given pipeline system (including the pump), a unique H–Q curve can be plotted, as shown in Example 5.3, by computing the head losses for several discharges.

In selecting a particular pump for a given system, the design conditions are specified and a pump is selected for the range of applications. The H–Q curve is then matched to the pump performance chart (for example, Figures 5.9 and 5.10) provided by the manufacturer. The matching point, M, indicates the actual

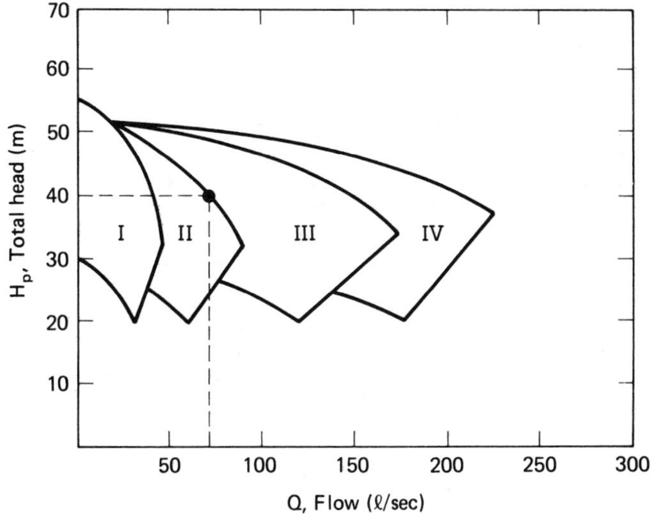

Figure 5.9 Pump selection chart.

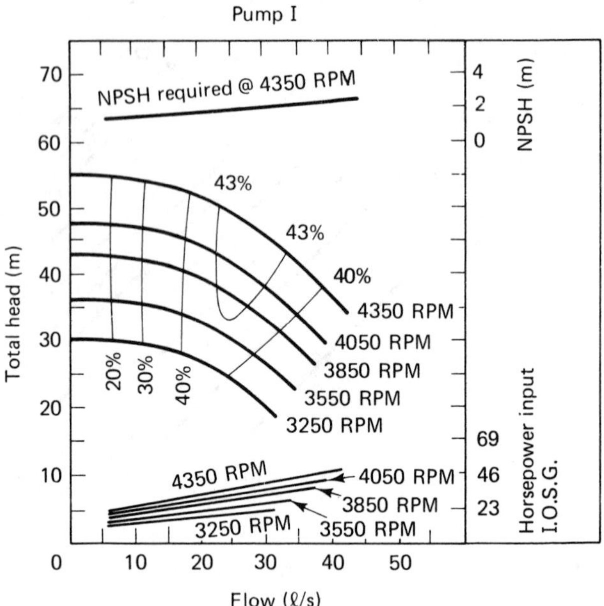

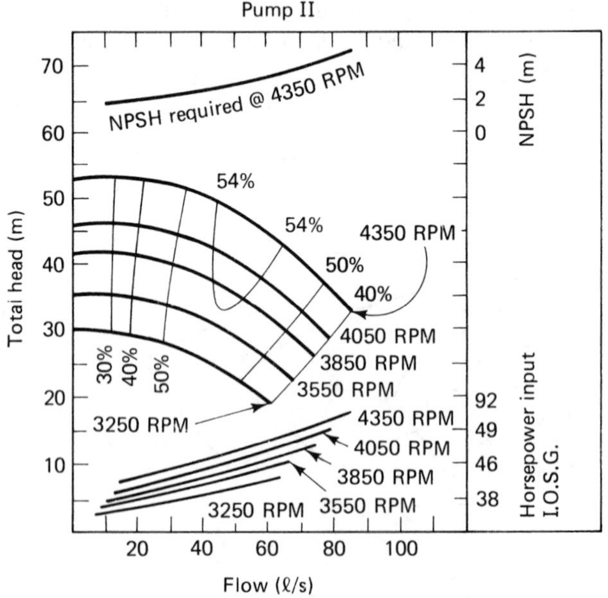

Figure 5.10 Characteristic curves for several pumps.

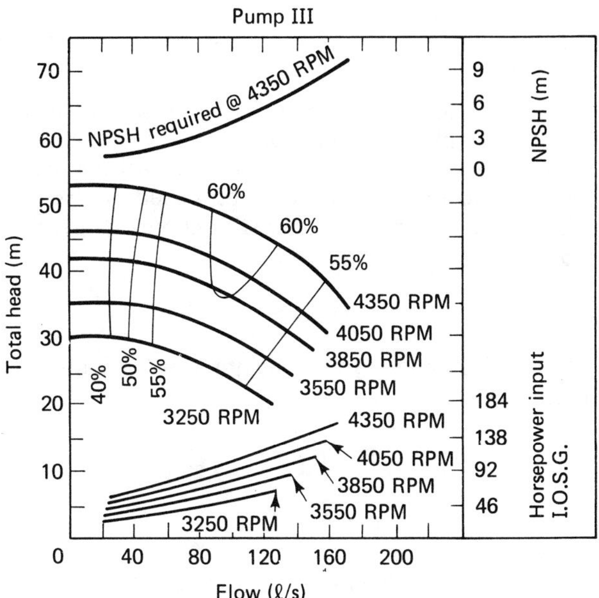

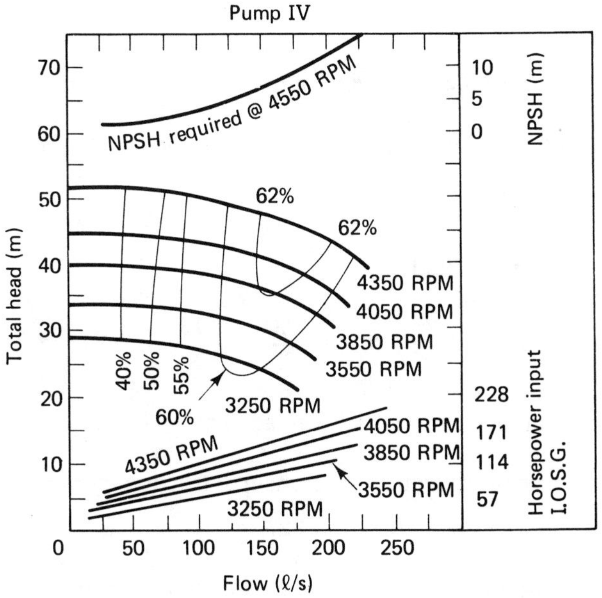

Figure 5.10 Cont.

working conditions. The selection process is demonstrated in the following example.

Example 5.3

A pump will be used to deliver the discharge of 70 ℓ/sec of water between two reservoirs 1000 m apart with an elevation difference of 20 m. If commercial steel pipes 20-cm in diameter are used for the project, select the proper pump and determine the working condition for the pump based on the pump selection chart (Figure 5.9) and the pump characteristics curves (H vs. Q, Figure 5.10), both provided by the manufacturer.

Solution

For commercial steel pipe, the roughness height, $e = 0.045$ mm (Table 3.1). The flow velocity in the pipe is

$$V = \frac{Q}{A} = \frac{0.070 \text{ m}^3/\text{sec}}{\left(\frac{\pi}{4} \cdot 0.2^2\right) \text{m}^2} = 2.23 \text{ m/sec}$$

and the corresponding Reynolds number at 20°C is

$$N_R = \frac{VD}{\nu} = \frac{2.23 \text{ m/sec} \cdot 0.2 \text{ m}}{1 \cdot 10^{-6} \text{ m}^2/\text{sec}} = 4.5 \cdot 10^5$$

and

$$e/D = 0.045 \text{ mm}/200 \text{ mm} = 2.3 \cdot 10^{-4} = 0.00023$$

The friction coefficient can be obtained from the Moody diagram (Figure 3.8).

$$f = 0.016$$

The pipe friction losses are, then,

$$h_f = f\left(\frac{L}{D}\right)\frac{V^2}{2g} = 0.016 \cdot \frac{1000}{0.2} \cdot \frac{2.23^2}{2 \cdot 9.81} = 20.27 \text{ m}$$

The total head that the pump must work against is

$$H_P = (\text{elevation difference}) + (\text{friction loss})$$
$$= 20 + 20.27 = 40.27 \text{ m}$$

From the pump selection chart provided by the manufacturer (see, for example, Figure 5.9), pumps II and III may be used for the project. The system curve (H–Q) should be determined before making a selection.

$Q(\ell/s)$	$V(m/s)$	N_R	f	H_f	H_P
50	1.59	$3.2 \cdot 10^5$	0.0165	10.65	30.65
60	1.91	$3.8 \cdot 10^5$	0.016	14.87	34.87
80	2.55	$5.1 \cdot 10^5$	0.0155	25.61	45.61

These values are plotted, with H_P vs. Q, as a curve. Matching the curve with the characteristic curves of pumps II and III as provided by the manufacturer (Figure 5.10), we find that the possibilities are as follows:

First trial. Pump II at 4350 rpm:

$$Q = 70\ \ell/\text{sec}, \qquad H_P = 40.3\ \text{m};$$

hence,

$$P_i = 71\ \text{hp and efficiency} = 52\%$$

Second trial. Pump III at 3850 rpm:

$$Q = 68\ \ell/\text{sec}, \qquad H_P = 39\ \text{m};$$

hence,

$$P_i = 61\ \text{hp and efficiency} = 57\%$$

Third trial. Pump III at 4050 rpm:

$$Q = 73\ \ell/\text{sec}, \qquad H_P = 42\ \text{m};$$

hence,

$$P_i = 70\ \text{hp and efficiency} = 58\%$$

Hydraulically, the selection should be pump II at 4350 rpm since it best fits the given conditions. However, one may notice that the second (pump III at 3850 rpm) and the third (pump III at 4050 rpm) trials also fit the conditions rather closely. In this case, the selection would best be made based on the considerations of the cost of the pump versus the cost of electricity.

PROBLEMS

5.4.1 Select a pump to be used to supply water to a reservoir at a flow rate of 30 ℓ/sec. The elevation head is 20 m and the distance between the supply and delivery is 100 m. A ball check valve is used with 15-cm diameter galvanized iron pipeline. Determine the working conditions.

5.4.2 Select a pump and determine its working conditions for a job that demands a minimum discharge of 80 ℓ/sec against an elevation head of 40 m. The distance between the supply and delivery is 150 m. A ball check valve will be used in the system that consists of commercial steel pipes. Also select the most economical diameter for the pipeline if the total cost can be expressed as

$$C = d^{1.75} + 0.75\boldsymbol{P} + 18$$

where d is the pipe diameter in cm and $\boldsymbol{P}$ is the power in horsepower.

5.4.3 The pumps in Figure 5.10 are used to pump water from a reservoir to an elevated water tank at elevation 25 m above the reservoir. If a 300-m-long, 25-cm-diameter asphalted cast-iron pipe is used to transport water at 10°C, determine the discharge, head, and efficiency of the appropriate pumps when operated at 3850 rpm.

5.4.4 A 70-kw motor is available to drive one of the pumps shown in Figure 5.10. The system is designed to deliver a minimum discharge of 80 ℓ/sec, over an elevation difference of 20 m. The system uses a wrought-iron pipe, 150 m long and 15 cm in diameter, to transport water at 10°C. Select the pump based on the consideration of lowest energy consumption.

5.5 PUMPS IN PARALLEL OR IN SERIES

As discussed in Section 5.1, the efficiency of a pump varies with the discharge rate of the pump and the height over which the delivery is made. The optimum efficiency of a pump can be obtained only over a limited range of operations (see Figure 5.10). To install a pumping station that can be effectively operated over a large range of fluctuations in both discharge and pressure, it may be advantageous to install several identical pumps at the station.

When several pumps are connected in parallel in a pipeline, the discharge is increased but the pressure head remains the same as with a single pump. It should be noted that two identical pumps operating in parallel may not double the discharge in a pipeline since the total head loss in a pipeline is proportional to the second power of discharge ($H_P \propto Q^2$). The additional resistance in the pipeline will cause a reduction in the total discharge. Figure 5.11 schematically shows the operation of two identical pumps in parallel. The joint discharge of the two pumps is always less than twice the discharge of a single pump.

Pumps connected in series in a pipeline will increase the total output pressure, but the discharge will remain approximately the same as that of a single pump. A typical performance curve for two pumps connected in series is shown by curve C in Figure 5.11.

The efficiency of two (or more) pumps operating in parallel or in series is almost the same as that of the single pump. The installation can be arranged with one separate motor for each pump or with one motor to operate two (or more) pumps. Multipump installations could be designed to perform either in-series or in-parallel operations with the same set of pumps. Figure 5.12 is a typical

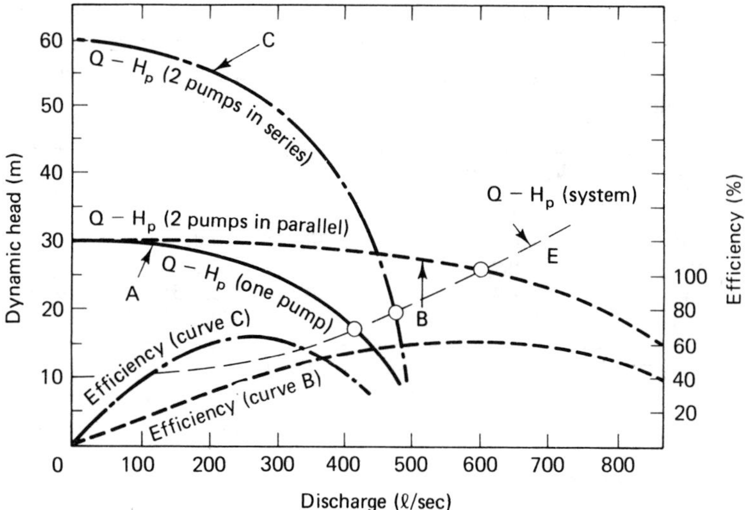

Figure 5.11 Typical performance curves of two pumps connected in parallel *B* and in series *C*.

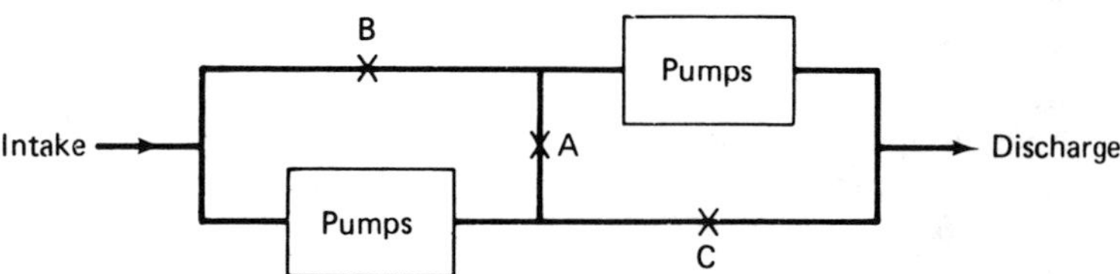

Figure 5.12 Schematic installation for operation either in series or in parallel.

schematic of such an installation. For series operations, valve *A* is opened and valves *B* and *C* are closed; for parallel operations, valve *A* is closed and valves *B* and *C* are open.

Example 5.4

Two reservoirs are connected by a 300-m-long asphalted cast-iron pipeline, 40 cm in diameter. The minor losses include the entrance, the exit, and a gate valve. The elevation difference between the reservoirs is 10 m and the water temperature is 10°C. Determine the discharge, head, and efficiency using (a) one pump, (b) two pumps in series, and (c) two pumps in parallel. Use the pump described in Figure 5.11.

Solution

To deliver the water, the pump system must provide a total energy head H_p

$$H_p = H + \left(f\frac{L}{D} + \Sigma K\right)\frac{V^2}{2g} = 10 + (750f + 1.65)\frac{V^2}{2g}$$

Using $\nu = 1.31 \times 10^{-6}$ m²/sec, and $e/D = 0.0003$, the following values are computed for a range of discharges within which each pump system may expect to operate. A Q vs. H_p curve is constructed based on the values computed in the following table (curve E, Figure 5.11).

$Q(\ell/sec)$	$V(m/sec)$	N_R	f	$H_p(m)$
0	0	—	—	10.0
100	0.796	2.4×10^5	0.0175	10.48
300	2.387	7.3×10^5	0.016	13.96
500	3.979	1.2×10^6	0.0155	20.71
700	5.570	1.7×10^6	0.0155	30.99

From Curve E in Figure 5.12, we obtain

(1) One pump
$Q = 420\ \ell/\text{sec}$
$H_p = 18$ m
$e_p = 40\%$

(2) Two pumps in series
$Q = 470\ \ell/\text{sec}$
$H_p = 20$ m
$e_p = 15\%$

(3) Two pumps in parallel
$Q = 590\ \ell/\text{sec}$
$H_p = 26$ m
$e_p = 62\%$

PROBLEMS

5.5.1 Two identical pumps have the characteristic curves shown in Figure 5.11. The pumps are connected in series and deliver water through a horizontal, 15-cm-diameter, 1000-m-long steel pipe into a reservoir in which the water level is 25 m above the pump. Neglect the minor losses in the system and determine the discharge.

5.5.2 Determine the discharge when the two pumps in Problem 5.5.1 are connected in parallel.

5.5.3 A pumping station is installed to deliver 10°C water from a reservoir to an elevated storage tank at a minimum required discharge of 300 ℓ/sec. The difference in elevations is 15 m, and a 1500-m-long, wrought-iron pipe that is 40 cm in diameter is used. Select the pumps from the set given in Figure 5.10. Determine the discharge, total head, and efficiency at which the pumps operate.

5.5.4 For the system in Problem 5.5.3 use three pumps (pump III in Figure 5.10), and determine the total input power to the pumps.

5.5.5 A set of two pumps is available to deliver water at 10°C from a reservoir to an elevated tank. The elevation difference is 25 m, and a 150-m-long, cast-iron pipe, 35 cm in diameter, is to be used. Determine the discharge and the power input to the pumps when the pumps are connected (a) in series, and (b) in parallel. Use pump III in Figure 5.10, at 3850 rpm.

5.5.6 If the system in Problem 5.5.5 must deliver 150 ℓ/sec at 10°C, determine which pump or set of pumps better satisfies the demand. Assume that only pumps I and II from Figure 5.10 are available.

5.6 CAVITATION IN WATER PUMPS

One of the important considerations in pump installation design is the relative elevation between the pump and the water surface in the supply reservoir. Water enters into the suction line through a *strainer* that is designed to keep out the trash and thus causes additional energy loss at the entrance.

The water in the suction line in a pumping installation is usually under pressure lower than atmospheric. The phenomenon of cavitation becomes a potential danger whenever the water pressure at any location in the pumping system drops substantially below atmospheric pressure. A common site of cavitation is near the tips of the impeller vanes where the velocity is very high. In regions of high velocities much of the pressure energy is converted to kinetic energy. This is added to the elevation difference between the pump and the supply reservoir, h_P, and to the inevitable energy loss in the pipeline between the reservoir and the pump, h_L. Those three items all contribute to the *total suction head,* H_s, in a pumping installation as shown schematically in Figure 5.13.

The value of H_s must be kept within a limit so that the pressure at any location in the pump is always above the vapor pressure of water; otherwise, the water will be vaporized and cavitation will occur. The vaporized water forms small vapor bubbles in the flow. These bubbles collapse when they reach the region of higher pressure in the pump. Violent vibrations may result from the collapse of vapor bubbles in water. Successive bubble breakup with considerable impact force may cause high local stresses on the metal surface of the vane blades and the housing. These stresses cause surface pitting and will rapidly damage the pump.

To prevent cavitation, the pump should be installed at an elevation so that the total suction head is less than the difference between the atmospheric head and the water vapor pressure head, or

$$\left(\frac{p_{\text{atm}}}{\gamma} - \frac{p_{\text{vapor}}}{\gamma}\right) > H_s$$

The maximum velocity near the tip of the impeller vanes is not assessable by the users. Pump manufacturers usually provide a value known commercially as the *net positive suction head* (NPSH), or H_s' as shown in Figure 5.10. NPSH

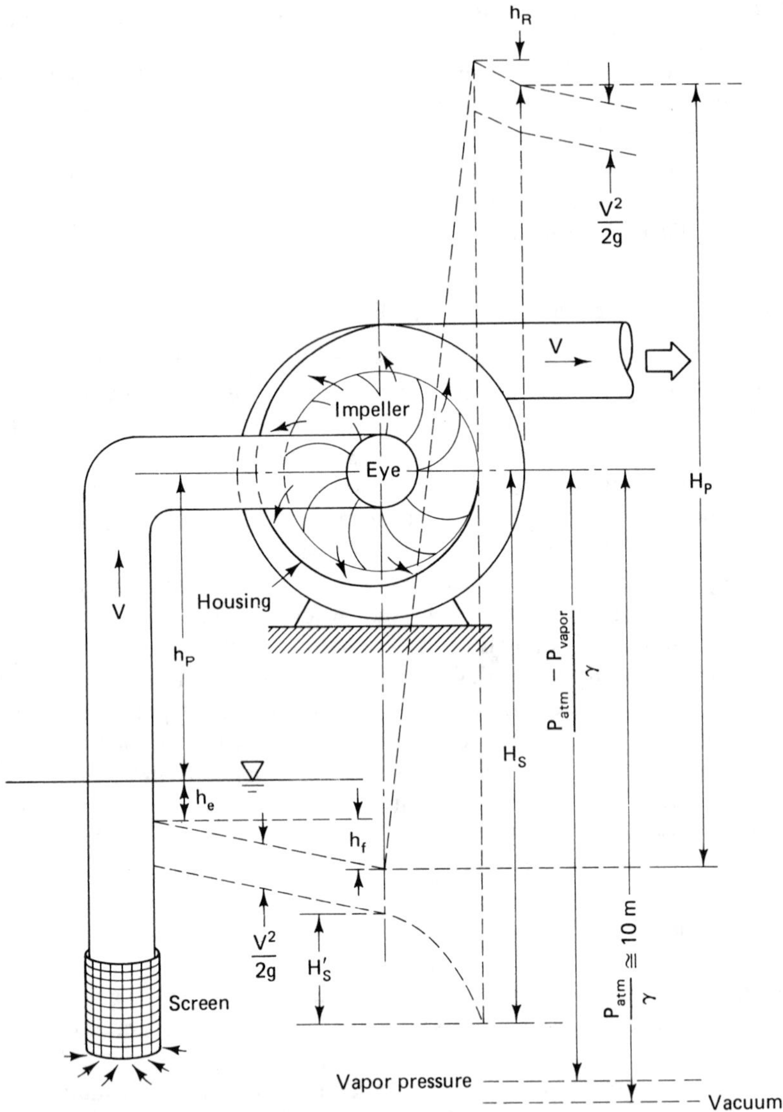

Figure 5.13 Energy and pressure relationship in a centrifugal pump.

represents the pressure drop between the eye of the pump and the tip of the impeller vanes. With the value of NPSH given, the maximum pump elevation above the supply reservoir can be easily determined

$$h_P \leq \frac{p_{\text{atm}}}{\gamma} - \left(\frac{p_{\text{vapor}}}{\gamma} + H_s' + \frac{V^2}{2g} + h_L\right) \tag{5.17}$$

where h_L is the total loss of energy in the suction side of the pump. It usually includes the entrance loss at the strainer, the friction loss in the pipe, and other minor losses.

The other parameter commonly used for preventing cavitation in a pump is described by the *cavitation parameter,* σ, which is defined as

$$\sigma = \frac{H_s'}{H_P} \tag{5.18}$$

where H_P is the total dynamic head developed by the pump and the numerator is the NPSH. The increase in velocity about the impeller vanes is already accounted for in the parameter σ. The value of σ for each type of pump is usually furnished by the manufacturer and is based on pump test data.

Applying Equation (5.18) and the relationships expressed in Figure 5.13, we may write

$$H_s' = \sigma H_P = \frac{p_{\text{atm}}}{\gamma} - \frac{p_{\text{vapor}}}{\gamma} - \left(\frac{V_i^2}{2g} + h_P + h_L\right) \tag{5.19}$$

where V_i is the speed of water at the entrance to the impeller. Rearranging Equation (5.19), we get

$$h_P = \frac{p_{\text{atm}}}{\gamma} - \frac{p_{\text{vapor}}}{\gamma} - \frac{V_i^2}{2g} - h_L - \sigma H_P \tag{5.20}$$

that will give the maximum allowable elevation of the intake (entrance to the impeller) above the surface of the supply reservoir. If the value determined by Equation (5.20) is negative, then the pump must be placed at an elevation below the water surface elevation in the supply reservoir.

Example 5.5

A pump is installed at sea level in a 15-cm, 300-m-long pipeline to pump 0.060 m^3/sec water at 20°C. The height is 25 m. If the pump has a 18-cm impeller intake diameter, a cavitation parameter of $\sigma = 0.12$, and experiences a total head loss of 1.3 m at the suction side, determine the relative elevation between the pump intake and the water surface elevation in the supply tank. Assume the pipeline has $C_{HW} = 120$.

Solution

The head loss in the pipeline can be determined by the chart provided in Figure 3.9 (Hazen-Williams formula). For $Q = 0.06$ m^3/sec, and $C_{HW} = 120$; $D = 0.15$ m and $L = 300$ m, we found the slope of the energy gradient line, $S = 0.086$ and

$h_f = L \cdot S = 25.8$. The only local loss is at the exit, for which we may assume $k_e = 1.0$, as discussed in Chapter 3.

The velocities in the intake pipe and in the conduction pipe line are, respectively,

$$V_i = \frac{Q}{A_i} = \frac{0.06}{\pi(0.09)^2} = 2.36 \text{ m/sec}$$

$$V_d = \frac{Q}{A_d} = \frac{0.06}{\pi(0.075)^2} = 3.40 \text{ m/sec}$$

The total dynamic head developed by the pump can be determined by applying Bernoulli's equation to the reservoirs at both ends of the system

$$\frac{V_1^2}{2g} + \frac{P_1}{\gamma} + h_1 + H_p = \frac{V_2^2}{2g} + \frac{P_2}{\gamma} + h_2 + h_L$$

where subscripts 1 and 2 refer to the reservoirs at the supply end and delivery end, respectively. At the surface level of the reservoirs, we may write, $V_1 \simeq V_2 \simeq 0$ and $P_1 = P_2 = P_{atm}$; so

$$H_p = (h_2 - h_1) + h_L = 25 + \left(1.3 + 25.8 + 1 \cdot \frac{3.40^2}{2g}\right) \simeq 52.7 \text{ m}$$

The vapor pressure is found in Table 1.1.

$$P_{vapor} = 0.023042 \text{ bar} = 2335 \text{ N/m}^2,$$

while

$$P_{atm} = 1 \text{ bar} = 101{,}357 \text{ N/m}^2$$

Finally, applying Equation (5.20), we obtain the maximum relative elevation of the pump

$$h_P = \frac{P_{atm}}{\gamma} - \frac{P_{vapor}}{\gamma} - \frac{V_i^2}{2g} - \Sigma H_{L_s} - \sigma H_P$$

$$= \frac{101{,}357}{9810} - \frac{2335}{9810} - \frac{(2.36)^2}{2 \cdot 9.81} - 1.3 - 0.12 \cdot 52.7 = 2.19 \text{ m}$$

PROBLEMS

5.6.1 Factory tests indicated that a cavitation parameter for a particular pump is $\sigma = 0.075$. The pump is installed to pump 60°C water at sea level. If the total head loss between the inlet and the suction side of the pump is 0.5 m, determine the allowable level of the pump intake relative to the supply reservoir water surface elevation. The discharge is 0.04 m^3/sec and the pipe system is the one used in Example 5.5.

5.6.2 The efficiency of a pump will drop suddenly if cavitation takes place in the pump. If this phenomenon is observed in a particular pump (with $\sigma = 0.08$) operating at sea level and the pump delivers 0.42 m^3/sec of water at 40°C, determine the sum of the gage pressure and the velocity head at the inlet. The dynamic head is 85 m and the suction pipe diameter is 30 cm.

5.6.3 In Problem 5.5.5, the suction pipe is 10 m long and the strainer and minor losses coefficients add up to 3.65. Determine the allowable elevation difference between the pump and the reservoir for (a) pumps in series, and (b) pumps in parallel.

5.6.4 Pump IV (in Figure 5.10) is operating at 4350 rpm to deliver water at 10°C between a reservoir and a water tank 20 m higher. The suction side consists of a strainer ($K_S = 2.5$), a 90° bend, and 10 m of cast-iron pipe, 25 cm in diameter. The discharge side includes a 160-m-long, cast-iron pipe, 20 cm in diameter, and a gate valve. Determine the allowable elevation difference between the pump and the reservoir water surface.

5.6.5 Solve Problem 5.6.4 for water temperature (a) 20°C, and (b) 50°C.

5.7 SPECIFIC SPEED AND PUMP SIMILARITY

The selection of a pump for a particular service is based on the required discharge rate and the head against which the discharge is delivered. To lift a large quantity of water over a relatively small elevation (for example, removing water from an irrigation canal onto a farm field) requires a high-capacity, low-stage pump. To pump a relatively small quantity of water against great heights (such as supplying water to a high-rise building) requires a low-capacity, high-stage pump. The designs of these two pumps are very different.

Generally speaking, impellers of relatively large radius and narrow flow passages transfer more kinetic energy head to the pressure head than impellers of smaller radius of large flow passages. Pumps designed with geometry that allows water to exit the impeller in a radial direction impart more centrifugal acceleration to the flow than those that allow water to exit axially or at an angle. Thus, the relative geometry of the impeller and the pump housing determine the performance and the field application of a specific centrifugal pump.

Dynamic analysis (Chapter 10) shows that centrifugal pumps built with identical proportions but different sizes have similar dynamic performance characteristics that are consolidated into one number called a *shape number*. The shape number of a particular pump design is a dimensionless number defined as

$$S = \frac{\omega\sqrt{Q}}{(gH_p)^{3/4}} \tag{5.21}$$

TABLE 5.1 Conversion of Specific Speed

Units	Unit Discharge	Unit Head	Unit Pump Speed	Formula	Symbol	Conversion	
United States	U.S. gal/min	ft	rev/min	(5.22)	N_{s_1}	$N_{s_1} = 45.6\ S$	$N_{s_1} = 51.6\ N_{s_3}$
English	Imp. gal/min	ft	rev/min	(5.22)	N_{s_2}	$N_{s_2} = 37.9\ S$	$N_{s_2} = 43.0\ N_{s_3}$
Metric	m^3/sec	m	rev/min	(5.22)	N_{s_3}	$N_{s_3} = 0.882\ S$	$N_{s_3} = 0.019\ N_{s_1}$
S.I.	m^3/sec	m	rad/sec	(5.21)	S	$S = 0.022\ N_{s_1}$	$S = 1.134\ N_{s_3}$

Note: $g = 9.81$ m/sec^2, 1 rad/sec = 9.52 rev/min

where ω is the angular velocity of the impeller in radians per second, Q is the discharge of the pump in cubic meters per second, g is the gravitational acceleration in meters per second squared, and H_P is the total dynamic head in meters that the pump develops.

In engineering practice, however, the dimensionless shape number is not commonly used. Instead, most of the commercial pumps are specified by the term *specific speed.* The specific speed of a specific design of centrifugal pump can be defined in two different ways. Some manufacturers define the specific speed of a specific pump design as the speed of a unit of the design series of such a magnitude that it delivers unit discharge at unit head. This way, the specific speed may be expressed as

$$N_s = \frac{\omega\sqrt{Q}}{H_p^{\,3/4}} \tag{5.22}$$

Other manufacturers define the specific speed of a specific design as the speed of a unit of the series of such a magnitude that it produces unit power with unit head. This way, the specific speed is expressed as

$$N_s = \frac{\omega\sqrt{P_i}}{H_p^{\,5/4}} \tag{5.23}$$

Most of the commercial pumps manufactured in the United States are presently specified with the United States conventional units [gallons per minute (gpm), brake horsepower (bhp), feet (ft), and revolutions per minute (rpm)]. In S.I. units, cubic meters per second, kilowatts, meters, and radians per second are usually used in the computations. The conversions of specific speed among the U.S. units, the English units, the metric units, and S.I. units are provided in Table 5.1.

Normally, the specific speed is defined, one way or the other, at the optimum point of operation efficiency. In practice, pumps with high specific speeds are generally used for large discharges at low-pressure heads, while pumps with low specific speeds are used to deliver small discharge at high-pressure heads. Centrifugal pumps with identical geometric proportions but different sizes have the same specific speed. Specific speed varies with impeller type. Its relationship to discharge and pump efficiency is shown in Figure 5.14.

Example 5.6

A centrifugal water pump operates at its optimum efficiency and delivers 2.5 m^3/sec over a height of 20 m. The pump has a 36-cm diameter impeller and rotates at 3000 rpm. Compute the specific speed of the pump (a) in terms of discharge and (b) in terms of power if the maximum efficiency of the pump is 80%.

Solution

The given conditions are Q = 2.5 m^3/sec, H_P = 20, and ω = 3000 rpm.
(a) Applying Equation (5.22), we get

$$N_s = \frac{3000\sqrt{2.5}}{(20)^{3/4}} = 500 \text{ rpm}$$

(b) At 80% efficiency the shaft power is

$$P_i = \frac{\gamma Q H_p}{0.80} = \frac{9810 \cdot 2.5 \cdot 20}{0.80} = 613{,}125 \text{ w} = 613 \text{ kw}$$

Applying Equation (5.23), we get

$$N_s = \frac{3000\sqrt{613}}{(20)^{5/4}} \simeq 1760 \text{ rpm}$$

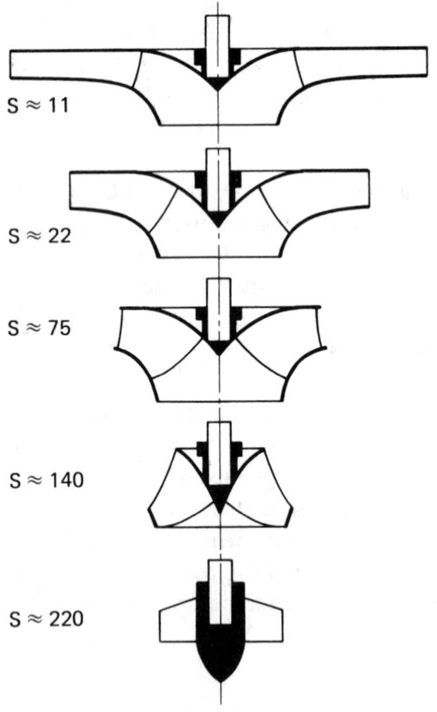

Figure 5.14 Relative impeller shapes and the approximate values of shape numbers, S, as defined in Table 5.1.

Example 5.7

The impeller of the pump in Example 5.6 has a diameter of 0.36 meters. What diameter should the impeller of a geometrically similar pump be for it to deliver one half of the water discharge at the same head? What is the speed of the pump?

Solution

Applying Equation 5.22, and from Example 5.6, we have

$$N_s = \frac{\omega\sqrt{\frac{1}{2} \times 2.5}}{(20)^{3/4}} = 500$$

The pump speed $\omega = \dfrac{500(20)^{3/4}}{(1.25)^{1/2}} = 4230$ rpm.

By definition, two pumps are geometrically similar if their specific speed, N_s, at optimal efficiency are identical. The following relationship will be satisfied.

$$\frac{Q_1}{\omega_1 D_1^3} = \frac{Q_2}{\omega_2 D_2^3} \tag{5.24}$$

Hence,

$$\frac{2.5}{3000 \cdot (0.36)^3} = \frac{1.25}{4230 \cdot (D_2)^3}$$

The diameter, $D_2 = 0.255$ m $= 25.5$ cm.

PROBLEMS

5.7.1 A homologous pump of the same design as in Example 5.6 has an impeller diameter of 72 cm and has the same efficiency when operated at 1800 rpm. Determine the discharge and shaft power required to operate the pump against a head of 30 m.

5.7.2 Two centrifugal pumps of the same design are operated at the same efficiency. Pump *A* is one fourth the size of pump *B* in the corresponding dimensions. When operating at 450 rpm, pump *B* delivers 2.4 m^3/sec against a 22-m head. With the rotational speed of 1800 rpm what will be the discharge of pump *A* when it is delivering water against the same head?

5.7.3 The design of a centrifugal water pump is studied by a 1/10 scale model in a hydraulic laboratory. At the optimum efficiency of 89% the model delivers

75.3 ℓ/sec of water against a 10-m head at 4500 rpm. If the prototype pump has a 72% efficiency at the rotational speed of 2250 rpm, determine the discharge and shaft power required to operate the pump under this condition.

PROJECT PROBLEM

5.7.4 A pumping system is designed to pump water from a 6-m-deep supply reservoir to a water tower 40 m above the ground. The system consists of a pump (or a combination of pumps), a 20-m long pipeline with one elbow on the suction side of the pump, a 60-m long pipeline, a gate valve, a check valve, and two elbows on the delivery side of the pump. The system is designed to pump 420 ℓ/sec of water, operating 350 days a year. Select a pump (or pumps) based on the characteristics provided in Figures 5.9 and 5.10, and a pipe size for optimum economy. Hazen-Williams coefficient 100 can be assumed for all pipe sizes listed below, and all elbows are 90° ($R/D = 2.0$). Power cost is \$0.04/kw-hr. Motor efficiency is 85% for all sizes listed.

Pump	*Cost* (\$)	*Motor* (*hp*)	*Cost* (\$)
I	700	60	200
II	800	95	250
III	900	180	300
IV	1020	250	340

	20 cm (\$)	*25 cm* (\$)	*30 cm* (\$)
Pipe (10 m)	120	150	180
Elbow	15	25	35
Gate valve	60	90	120
Check valve	80	105	130

WATER FLOW IN OPEN CHANNELS

Open channel flow differs from pipe flow in one important aspect: Pipe flow usually fills the entire conduit and open channel flow must have a free surface. Normally, a free water surface is subject to the atmospheric pressure, which approximately remains a constant value throughout the entire length of the channel. Pipe flow, while confined in a closed conduit, experiences only the hydraulic pressure, which varies from one section to another along the pipeline.

In Figure 6.1 an open channel flow is schematically compared to a pipe flow. Figure 6.1(a) shows a pipe flow with two open-ended vertical tubes (piezometric tubes) installed through the pipe wall at an upstream section, 1, and a downstream section, 2. The water level in each tube represents the pressure head, P/γ, in the pipe at the section. The difference in water surface elevations in the two tubes indicates the *hydraulic gradient line* over these sections. The velocity head at each section is represented in the familiar form, $V^2/2g$, where V is the mean velocity, $V = Q/A$, at the section. The total energy head at any section is equal to the sum of the elevation (potential) head, h, the pressure head,

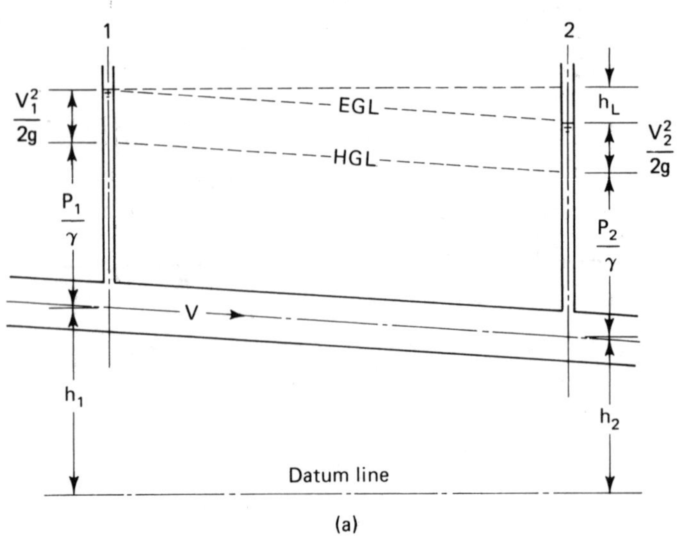

(a)

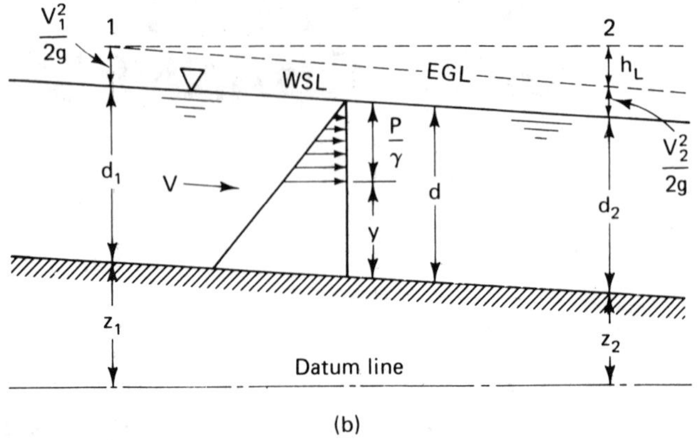

(b)

Figure 6.1 Comparison of: (a) a pipe flow, and (b) an open channel flow.

P/γ, and the velocity head, $V^2/2g$. The difference in total energy head between the two sections is indicated as the *energy gradient line*. The amount of energy loss when water flows from section 1 to section 2 is indicated by h_L.

Figure 6.1(b) shows an open channel flow. The free water surface is subjected to only atmospheric pressure, which is commonly referred to as the *zero pressure reference* in hydraulic engineering practice. The pressure distribution at any section is directly proportional to the depth measured from the free water surface. In this case, the water surface line corresponds to the hydraulic gradient line in pipe flow.

To solve problems of open channel flow, we must seek the interdependent relationships between the slope of the channel bottom, the discharge, the water depth, and other given channel conditions.

6.1 CLASSIFICATION OF OPEN CHANNEL FLOWS

Open channel flows may be classified by either the time criterion or the space criterion.

Based strictly on the time criterion, open channel flows may be classified into two categories

1. *steady flows,*
2. *unsteady flows.*

In a steady flow the discharge and water depth at any section in the reach do not change with time during the period of interest. In unsteady flow the discharge and the water depth at any section in the reach change with time.

Based on the space criterion, an open channel flow is said to be *uniform flow* if the disharge and the water depth remain the same in every section in the channel reach. Uniform flows in open channel are mostly steady (Figure 6.2a); unsteady uniform flows (Figure 6.2b) are very rare in nature. A *varied flow* in open channel is one in which the water depth and/or the discharge change along the length of the channel. Varied open channel flow may be either steady or unsteady. Steady flow over a spillway crest (Figure 6.2c) is a *varied steady flow*; a flood wave (Figure 6.2d) and a tidal surge (Figure 6.2e) are examples of *varied unsteady flows*.

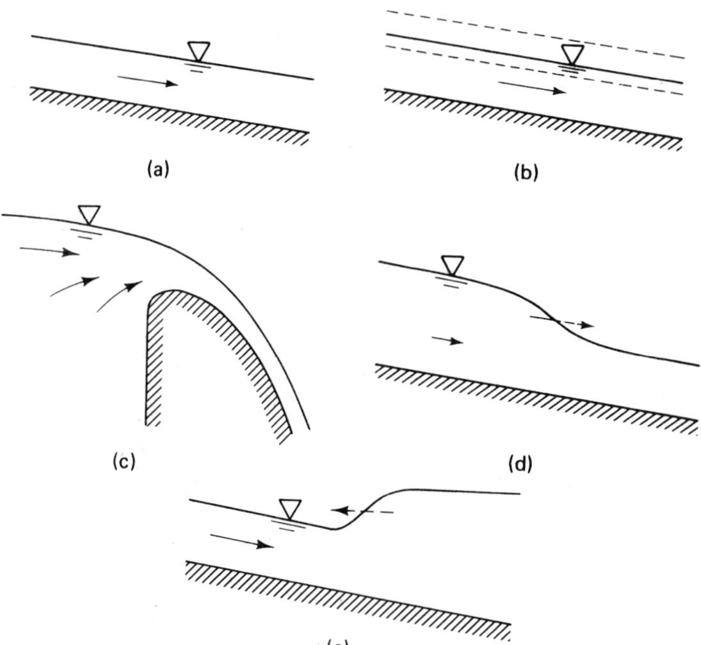

Figure 6.2 Classification of open channel flows: (a) uniform flow, (b) unsteady, uniform flow, (c) steady, varied flow, (d) unsteady, varied flow, (e) unsteady, varied flow.

PROBLEM

6.1.1 Using the time and space criterion, classify the following open channel flows during:

(a) The opening of a gate to let water into an otherwise uniform flow channel,
(b) Uniform rainfall over a sloped, long roof top,
(c) Flow in the gutter resulting from (b),
(d) Flow in the above gutter when the rainfall intensity increases with time.

6.2 UNIFORM FLOW IN OPEN CHANNELS

Uniform flow in an open channel must satisfy the following main features

1. The water depth, water area, discharge, and the velocity distribution at all sections throughout the entire channel reach must remain unchanged.
2. The energy gradient line, the water surface line, and the channel bottom line must be parallel to each other.

The slopes of these parallel lines are the same

$$S_e = S_{W.S.} = S_0$$

as shown in Figure 6.3.

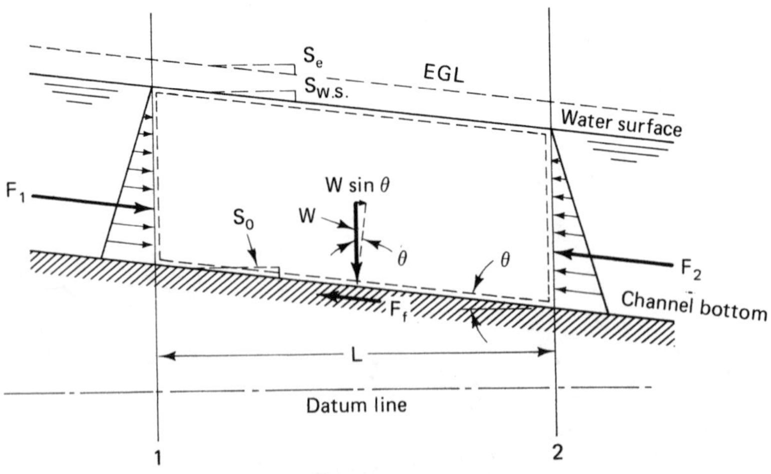

Figure 6.3 Force components in uniform open channel flow.

Water in an open channel can reach the state of uniform flow only if no acceleration (or deceleration) takes place between sections. This is possible when the gravity force component and the resistance to the flow are equal and opposite in direction in the reach.

In two arbitrary sections, 1 and 2, in a uniform open channel, a *free-body diagram* can be taken between the two sections to show the balance of the force components of gravity and resistance (Figure 6.3).

The forces acting upon the free body in the direction of the flow include

1. the hydrostatic pressure forces, F_1 and F_2, acting at each end of the free body;
2. the weight of the water body in the reach, W, has a component, $W \sin \theta$, in the direction of the flow;
3. the resistance force, F_f, exerted by the channel bed to the flow.

A summation of all these force components gives the following equation:

$$F_1 + W \sin \theta - F_2 - F_f = 0 \tag{6.1a}$$

This equation can be futher simplified, since there is no change in water depth in uniform flow between any two sections, 1 and 2. The hydrostatic forces at the two ends must be equal, $F_1 = F_2$. The total weight of the water body is

$$W = \gamma AL$$

where γ is the unit weight of water, A is the water cross-sectional area normal to the flow, and L is the length of the reach. In most open channels the channel slopes are small and the approximation, $\sin \theta = S_0$, is made. The slope S_0 is equal to the energy gradient line slope S_e. The gravity force component may thus be expressed as

$$W \sin \theta = \gamma ALS_0 \tag{6.1b}$$

The resistance force exerted by the channel boundaries may be expressed in terms of resisting force per unit area of the channel bed times the total channel bed area wetted by the flowing water. The wetted bed area is the product of the wetted perimeter, P, and the length of the channel reach, L.

In 1769 a French engineer, Antoine Chezy, assumed that the resisting force per unit area of the channel bed is proportional to the square of the mean velocity, KV^2, where K is a constant of proportionality. The total resistance force may thus be written as

$$F_f = \tau_0 PL = KV^2 PL \tag{6.1c}$$

where τ_0 is the resisting force per unit area of the channel bed, also known as the *wall shear stress*.

Substituting Equation (6.1b) and Equation (6.1c) into Equation (6.1a), we have

$$\gamma ALS_0 = KV^2PL$$

or

$$V = \sqrt{\left(\frac{\gamma}{K}\right)\left(\frac{A}{P}\right)S_0}$$

In this equation, $A/P = R_h$, and $\sqrt{(\gamma/K)}$ may be represented by a constant, C. For uniform flow, $S_0 = S_e$, the above equation may thus be simplified to

$$V = C\sqrt{R_h S_e} \tag{6.2}$$

in which R_h is the *hydraulic radius* of the channel cross section. The hydraulic radius is defined as the water area divided by the wetted perimeter for all shapes of open channel cross sections.

Equation (6.2) is the well-known *Chezy's formula* for open channel flow. Chezy's formula is probably the first formula derived for uniform flow. The constant C is commonly known as Chezy's resistance factor; it was found to vary in relation to both the conditions of the channel and the flow.

Over the past two centuries many attempts have been made to determine the value of Chezy's C. The simplest relationship and the one most widely applied in the United States is derived from the work of an Irish engineer, Robert Manning (1891–1895).* Using the analysis performed on his own experimental data and on those of others, Manning derived the following empirical relation:

$$C = \frac{1}{n}R_h^{1/6} \tag{6.3}$$

in which n is known as the Manning's coefficient of the channel roughness. Some typical values of Manning's coefficient are given in Table 6.1.

TABLE 6.1 Typical Values of Manning's *n*

Channel Surface	*n*
Smooth steel surface	0.012
Corrugated metal	0.024
Smooth concrete	0.011
Concrete culvert (with connection)	0.013
Glazed brick	0.013
Earth excavation, clean	0.022
Natural stream bed, clean, straight	0.030
Smooth rock cuts	0.035
Channels not maintained	0.050–0.1

*Robert Manning, "On the Flow of Water in Open Channels and Pipes," *Transactions, Institution of Civil Engineering of Ireland*, 10 (1891), 161–207; 24 (1895), 179–207.

Substituting Equation (6.3) into Equation (6.2), we have the *Manning's formula for uniform flow*

$$V = \frac{1}{n} R_h^{2/3} S_e^{1/2} \tag{6.4}$$

The discharge in a uniform flow channel may be determined by

$$Q = AV = \frac{1}{n} A R_h^{2/3} S_e^{1/2} \tag{6.5}$$

On the right-hand side of this equation, the water area, A, and the hydraulic radius R_h, are both functions of water depth, d, which is known as the *uniform depth* or *normal depth* when the flow is uniform.

The computation of uniform flow may be performed by the use of either Equation (6.5) or Equation (6.4) and basically involves six variables: the roughness coefficient, n; the channel slope, S; the normal discharge, Q; the channel geometry that includes the water area, A, and the hydraulic radius, R_h; the mean velocity, V; and the normal depth, d_n.

In practical applications, any four of the six variables are given, and the remaining two unknowns can be determined by using the equations.

Example 6.1

A 3-m-wide rectangular irrigation channel carries a discharge of 25.3 m³/sec at a uniform depth of 1.2 m. Determine the slope of the channel with Manning's coefficient, $n = 0.022$.

Solution

For a rectangular channel the wetted perimeter and the hydraulic radius are

$$A = bd = 3 \cdot 1.2 = 3.6 \text{ m}^2$$

$$P = b + 2d = 5.4 \text{ m}$$

$$R_h = \frac{A}{P} = \frac{3.6}{5.4} = \frac{2}{3} = 0.667 \text{ m}$$

Equation 6.5 can be rewritten as

$$S_0 = S_e = \left(\frac{Qn}{AR_h^{2/3}}\right)^2 = 0.041$$

Example 6.2

A trapezoidal open channel (see Figure 6.4) has a bottom width of 10 m and side slopes with inclination 1 : 2. The channel is paved with smooth cement surface.

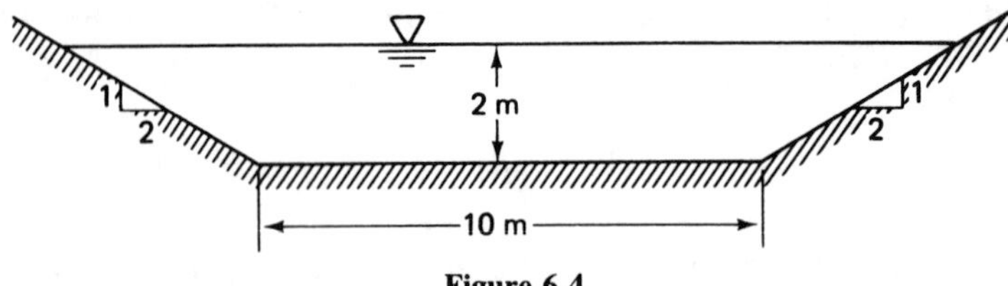

Figure 6.4

If the channel is laid on a slope of 0.0001 and carries a uniform flow at the depth of 2 m, determine the discharge.

Solution

The area of the trapezoidal section is

$$A = b \cdot d + 2d \cdot \frac{zd}{2} = (b + zd)d$$

where

$$z = \frac{1}{\text{side slope}} = \frac{1}{\frac{1}{2}} = 2$$

$$A = (10 + 2 \cdot 2) \cdot 2 = 28 \text{ m}^2$$

The wetted perimeter is

$$P = b + 2(d^2 + z^2 d^2)^{1/2} = b + 2\sqrt{1 + z^2}\,d = 18.94 \text{ m}$$

Then, the hydraulic radius is

$$R_h = \frac{A}{P} = \frac{28}{18.94} = 1.478 \text{ m}$$

Substituting the above value into Equation (6.5), we get

$$Q = \frac{1}{n} A R_h^{2/3} S_e^{1/2} = \frac{1}{0.011}(28)(1.478)^{2/3}(0.0001)^{1/2}$$

$$Q = 33.03 \text{ m}^3/\text{sec}$$

Example 6.3

If the discharge in the channel in Example 6.1 is increased to 40 m^3/sec, what is the normal depth of the flow?

Solution

The geometric parameters are

$$\text{Area:} \quad A = bd = 3d$$

$$\text{Wetted Perimeter:} \quad P = b + 2d = 3 + 2d$$

$$\text{Hydraulic Radius:} \quad R_h = \frac{A}{P} = \frac{3d}{3 + 2d}$$

Substituting these values in Equation (6.5), we get

$$Q = \frac{1}{n} A R_h^{2/3} S_e^{1/2}$$

$$40 = \frac{1}{0.022}(3d)\left(\frac{3d}{3 + 2d}\right)^{2/3}(0.041)^{1/2}$$

or

$$AR_h^{2/3} = (3d)\left(\frac{3d}{3 + 2d}\right)^{2/3} = \frac{(0.022)(40)}{(0.041)^{1/2}} = 4.346$$

Solving by trial, we find that

$$d = 1.69 \text{ m}$$

PROBLEMS

6.2.1 Determine the normal depth for the channel in Example 6.1 if the discharge were 50 m^3/sec.

6.2.2 A triangular channel with side slopes of 45° paved with brick in cement mortar should deliver 4 m^3/sec. Determine the depth of the flow if the slope is 0.0016.

6.2.3 A rectangular channel with slope 0.0004 is cut in rock. The width will be twice the depth. Determine its dimensions for a discharge of 50 m^3/sec.

6.2.4 Determine the discharge if the channel in Problem 6.2.3 is paved with smooth concrete.

6.2.5 A rectangular channel 3 m wide carries a discharge of 6.3 m^3/sec at the normal depth of 0.8 m. The channel is laid on a slope of $S_0 = 0.01$. Determine the roughness coefficient n.

6.2.6 A rectangular chute is designed to transport lumber from a high mountain region to a lower lake. The channel must remain a uniform depth of 1 m with a discharge rate of 10 m^3/sec. The channel is to be built of a material with $n = 0.016$. Determine the slope of the channel if its width is 3 m.

6.2.7 A triangular highway gutter (see Figure P6.2.7) is designed to carry a discharge of 52 m^3/min on a channel slope of 0.0016. The gutter is 0.8 m deep with one side vertical and one side sloped at 1 on z. The channel surface is clean earth excavation. Determine the side slope z.

Figure P6.2.7

6.2.8 A trapezoidal channel with 6.2-m-wide bottom and side slope $z = 2$ carries a discharge of 10 m^3/sec. The channel is built with glazed brick surface on a 0.04 channel slope. Using Chezy's formula, determine the uniform flow velocity.

6.2.9 100 m^3/sec are flowing in a rectangular channel 12 m wide. If the depth of the flow is 3 m, what will be the depth when the channel contracts to 8 m wide? Neglect losses.

6.2.10 Research in the library to find five different uniform flow formulas. List the author(s) and limitations of each formula.

6.3 HYDRAULIC EFFICIENCY OF OPEN CHANNEL SECTIONS

The uniform flow formula [Equation (6.5)] shows that for the same cross-sectional area, A, and channel slope, S, the channel section with a larger hydraulic radius, R_h, delivers a larger discharge. It is a section of higher *hydraulic efficiency*. Since the hydraulic radius is equal to the water cross-sectional area divided by the wetted perimeter, for a given cross-sectional area, the channel section with the least wetted perimeter is the *best hydraulic section*.

Among all shapes of cross-sectional area, the semicircle has the least perimeter with the given area; hence, it is the most hydraulically efficient of all sections. A channel with a semicircular cross section, however, has sides that are curved and that are almost vertical at the water surface level, which makes the channel expensive in initial construction and difficult to maintain. In practice, semicircular sections are only used in smaller flumes of prefabricated materials.

For larger channels, trapezoidal sections are most commonly used. The most efficient trapezoidal section is a half hexagon, which can be inscribed to a semicircle with its center at the free water surface and 60° angle on the sides. The next commonly used channel section is the rectangular section. The most efficient rectangular section is the half-square section, which can also be inscribed to a semicircle with the center of the circle at the free water surface. The hydraulically efficient semicircular, half-hexagon, and half-square sections are shown in Figure 6.5.

The concept of hydraulic efficient sections is only valid when the channel is lined with stablized, nonerodible materials. Ideally, a channel should be

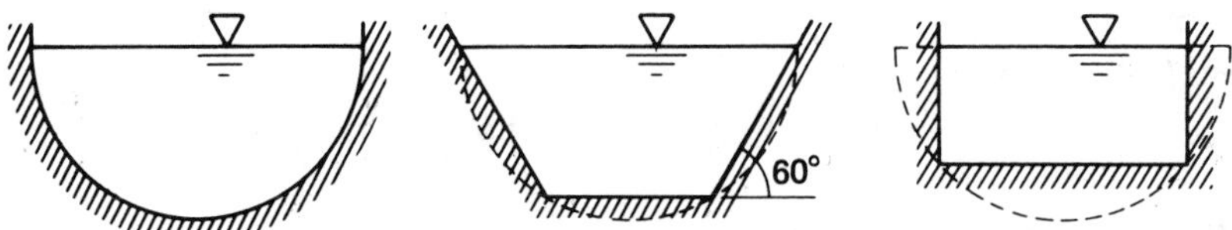

Figure 6.5 Hydraulically efficient sections.

designed for the best hydraulic efficiency, but it should be modified for practicability and construction cost. It should be noted that although the best hydraulic section offers the least water area for a given discharge, it does not necessarily have the lowest excavation construction cost. A half-hexagon section, for example, is a best hydraulic section only when the water surface reaches the level of the bank top. This section is not suitable for general applications because a sufficient distance above the water surface must be provided to prevent waves or fluctuations of water surface from overflowing the sides. The vertical distance from the designed water surface to the top of the channel banks is known as the *freeboard* of the channel.

The height of the freeboard may vary from 5% to 30% of the designed water depth, depending on the channel design and purposes. There is no accepted rule for the determination of the freeboard, since the wave actions and the fluctuations in water surface elevations may vary greatly from one channel to another. In general, the freeboard is considered to be related to the designed discharge of the channel. The U.S. Bureau of Reclamation provides a chart that can be used as a guide for average freeboard and bank heights for lined-channel design, as shown in Figure 6.6.

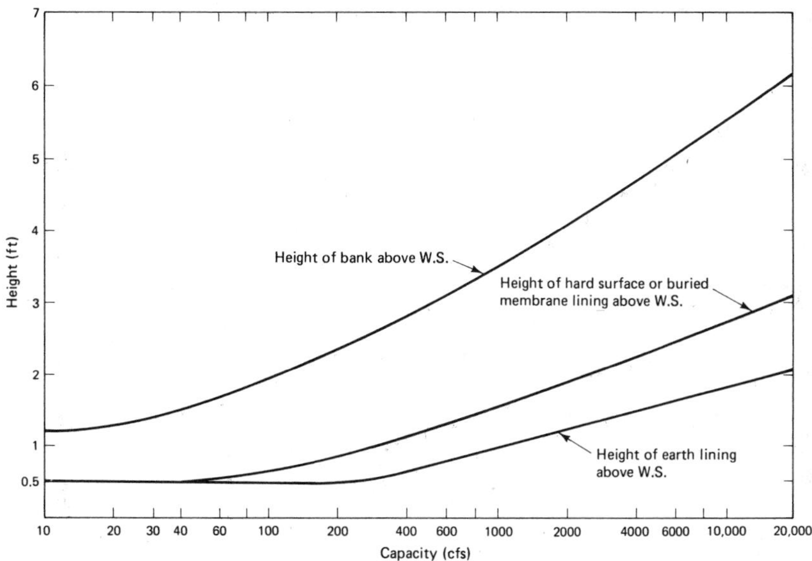

Figure 6.6 Recommended freeboard and height of banks in lined channels. (From *Linings for Irrigation Canals*, U.S. Bureau of Reclamation, 1976)

Example 6.4

Show that the best hydraulic trapezoidal section is a half hexagon.

Solution

The water cross-sectional area, A, and the wetted perimeter, P, of a trapezoidal section are

$$A = bd + zd^2 \tag{1}$$

and

$$P = b + 2d\sqrt{1 + z^2} \tag{2}$$

From Equation (1), $b = A/d - zd$. This relationship is substituted into Equation (2)

$$P = \frac{A}{d} - zd + 2d\sqrt{1 + z^2}$$

First, consider both A and z constant, and let the first derivative of P with respect to d equal zero for the minimum value of P.

$$\frac{dP}{dd} = -\frac{A}{d^2} - z + 2\sqrt{1 + z^2} = 0$$

Substituting for A from Equation (1), we get

$$\frac{bd + zd^2}{d^2} = 2\sqrt{1 + z^2} - z$$

or

$$b = 2d(\sqrt{1 + z^2} - z) \tag{3}$$

By definition, the hydraulic radius, R_h, may be expressed as

$$R_h = \frac{A}{P} = \frac{bd + zd^2}{b + 2d\sqrt{1 + z^2}}$$

Substituting the value of b from Equation (3) into the above equation and simplifying, we have

$$R_h = \frac{d}{2}$$

It shows that the best hydraulic trapezoidal section has a hydraulic radius equal to one half of the water depth. Substituting Equation (3) into Equation (2) and solving for P, we have

$$P = 2d(2\sqrt{1 + z^2} - z) \tag{4}$$

In order to determine the value of z that makes P the least, the first derivative of P is taken with respect to z. Equating it to zero and simplifying, we have

$$z = \frac{\sqrt{3}}{3} = \cot 60°$$

and thus,

$$b = 2d\left(\sqrt{1 + \frac{1}{3}} - \frac{\sqrt{3}}{3}\right) = 2\frac{\sqrt{3}}{3}d, \qquad \text{or} \qquad d = \frac{\sqrt{3}}{2}b = b \sin 60°$$

This means that the section is a half hexagon (see Figure 6.7).

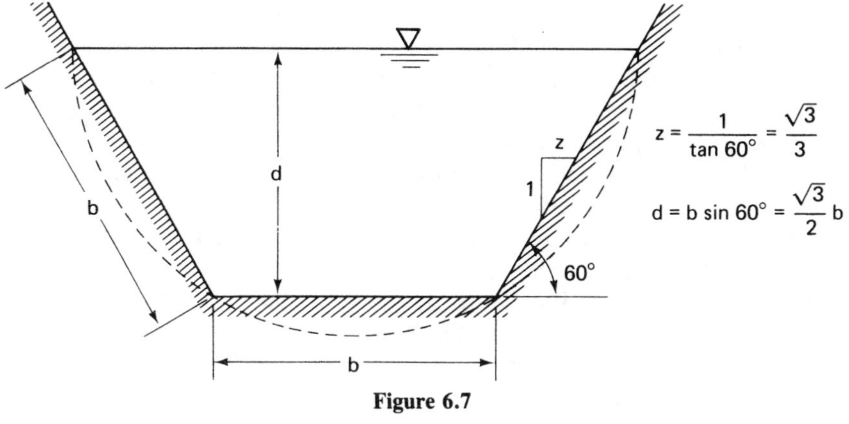

Figure 6.7

PROBLEMS

6.3.1 What is the depth-to-width ratio for the most hydraulically efficient rectangular section?

6.3.2 What would be the minimum excavated volume for a 100-m channel cut in rock (rectangular section)? It should deliver 25 m^3/sec in a slope of 0.003 m/m.

6.3.3 Determine the depth-to-diameter ratio for the most hydraulically efficient circular section.

6.3.4 Determine the side slope of the most hydraulically efficient triangular section.

6.4 ENERGY PRINCIPLES IN OPEN CHANNEL FLOWS

The energy principles derived for water flow in pipes are generally applicable to open channel flows. The energy contained in a unit weight of water flowing in an open channel may also be measured in the three basic forms

1. kinetic energy,
2. pressure energy,
3. elevation energy above a certain energy datum line.

Kinetic energy in any section of an open channel is expressed in the form of $V^2/2g$, where V is the mean velocity defined by the discharge divided by the water area, $V = Q/A$, in the section. The actual velocity of water flowing in an open channel section may vary in different parts of the section. The velocities near the channel bed are retarded due to friction and they reach a maximum near the water surface in the center part of the channel. The distribution of velocities in a cross section results in a different value of kinetic energy for each part of the section. An average value of the energy in a section of open channel may be expressed in terms of the mean velocity as $\alpha(V^2/2g)$, where α is known as the *energy coefficient*. The value of α depends on the actual velocity distribution in a particular channel section. Its value is always greater than unity. The ordinary range of α lies between 1.05 for uniformly distributed velocities and 1.20 for highly unevenly distributed velocities in a section. In simple analysis, however, the velocity heads (kinetic energy head) in an open channel are taken as $V^2/2g$ by assuming α equal to unity as an approximation.

Since open channel flow always has a free surface that is exposed to the atmosphere, the pressure on the free surface is constant and commonly taken as a zero pressure reference. Pressure energy in open channel is usually computed with reference to the free surface. If the flow in a channel is approximately along a straight-line slope, the pressure head at any submerged point A is equal to the vertical distance (the depth) measured from the free surface to the point. For a given open channel section, the water depth at the section, d, is commonly used to represent the pressure head, $p/\gamma = d$. However, if the water is flowing over a vertical curve, such as a spillway or a weir, the centrifugal force produced by the fluid mass flowing over the curved path may cause a marked difference in pressure from that directly measured from the depth. For water flows over a convex path (Figure 6.8a), the centrifugal force acts in the direction opposite to the gravity force and the pressure energy is less than that of the water depth, by mv^2/r, where m is the mass of water column immediately above a unit area and v^2/r is the centrifugal acceleration of the water mass flowing along a path with radius of curvature, r. The resulting pressure head is

$$\frac{p}{\gamma} = d - d \cdot \frac{(v^2/r)}{g} \tag{6.6a}$$

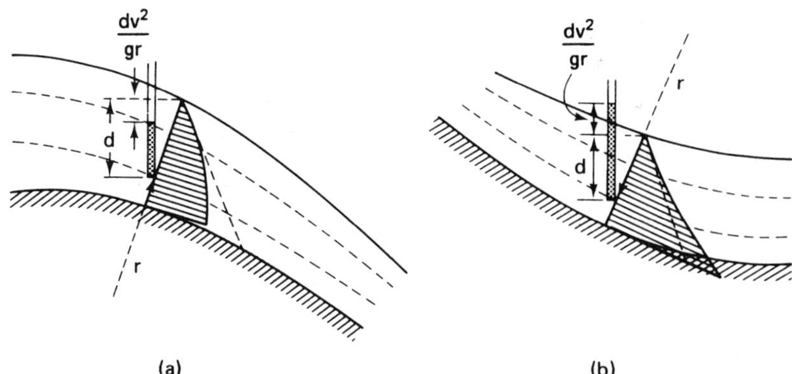

Figure 6.8 Flow over curved surfaces: (a) convex surface, (b) concave surface.

For water flows over a concave path (Figure 6.8b), the centrifugal force is in the same direction as the gravity force, and the pressure energy is greater than that represented by the water depth. The resulting pressure head is

$$\frac{p}{\gamma} = d + \frac{dv^2}{gr} \tag{6.6b}$$

where γ is the unit weight of water, d is the depth measured from the free water surface to the point of interest, v is the velocity at the point, and r is the radius of curvature of the curved flow path.

The elevation energy head of open channel flow is measured with respect to a selected horizontal datum line. The vertical distance measured from the datum to the channel bottom is commonly taken as the elevation (potential) energy head at the section.

The total energy head at any section in an open channel may therefore be generally expressed as,

$$H = \frac{V^2}{2g} + d + z \tag{6.7}$$

Specific energy in a channel section is defined as the energy head measured with respect to the channel bottom at the section. According to Equation (6.7), the specific energy at any section is

$$E = \frac{V^2}{2g} + d \tag{6.8}$$

or, the specific energy at any section in an open channel is equal to the sum of the velocity head and the water depth at the section.

Given the water area, A, and the discharge, Q, at a particular section, Equation (6.8) may be rewritten as

$$E = \frac{Q^2}{2gA^2} + d \tag{6.9}$$

Thus, for a given discharge Q, the specific energy at any section is a function of the depth of the flow only.

When the depth of the flow, d, is plotted against the specific energy for a given discharge at a given section, a *specific energy curve* is obtained (see Figure 6.9).

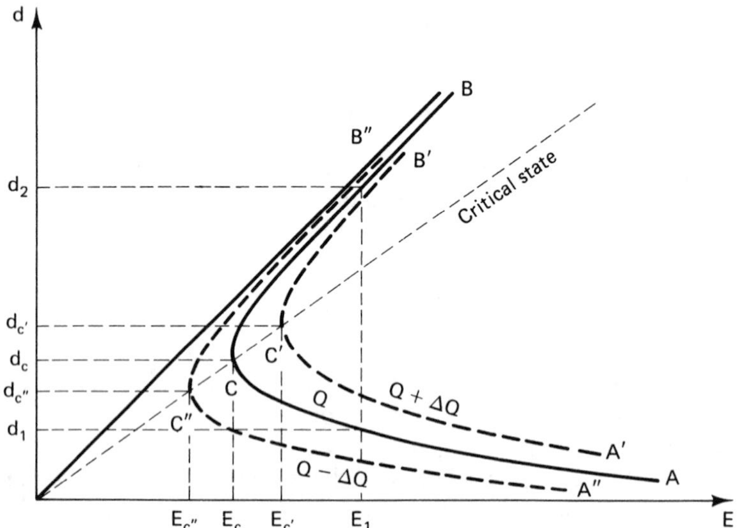

Figure 6.9 Specific energy curves of different discharges in a given open channel.

The specific energy curve has two limbs, AC and CB. The lower limb always approaches the horizontal axis toward the right and the upper limb approaches asymptotically to the 45°-line that passes through the origin.* At any point P on the specific energy curve, the ordinate represents the depth of the flow at the section and the abscissa represents the corresponding specific energy. Usually, the same scales are used for both the ordinate and the abscissa.

In general, a family of similar curves may be plotted for various values of Q's at a given section. For higher discharge, the curve moves to the right, $A'C'B'$; a lower discharge, the curve moves to the left, $A''C''B''$.

The vertex C on a specific energy curve represents the depth, d_c, at which the discharge Q may be delivered through the section at minimum energy, E_c. This depth is commonly known as the *critical depth* for the discharge Q at the given section. The corresponding flow in the section is known as the *critical flow*. At a smaller depth the same discharge can be delivered only by a higher velocity and, hence, a higher specific energy. The state of rapid and shallow flow

* For a channel of very large slope, the upper limb may approach a line with angle of inclination different from 45°.

through a section is known as the *supercritical flow* or *rapid flow*. At a larger depth the same discharge may be delivered through the section with a smaller velocity but also a higher specific energy. The tranquil, high-stage flow is known as the *subcritical flow*.

For a given value of specific energy, say E_1, the discharge may pass through the channel section at either the depth d_1 (supercritical flow) or d_2 (subcritical flow) as shown in Figure 6.9. These two depths, d_1 and d_2, are commonly known as the *alternate depths* of each other.

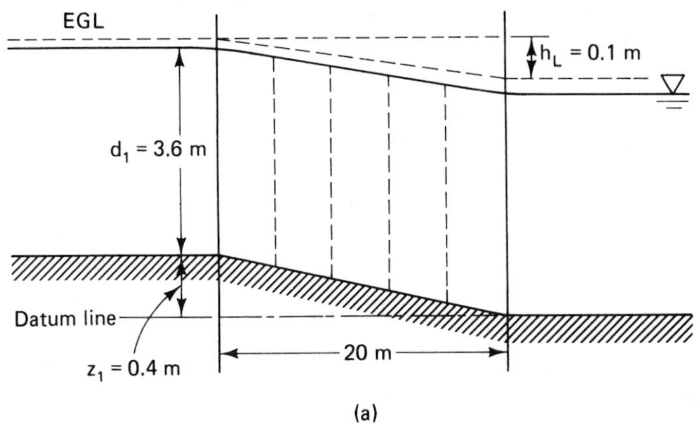

(a)

$$d_c = \sqrt[3]{\frac{5^2}{g}} = 1.37 \text{ m}$$

$$E_c = 2.05 \text{ m}$$

d (m)	E (m)
0.5	5.60
1.0	2.27
2.0	2.32
3.0	3.14
4.0	4.07

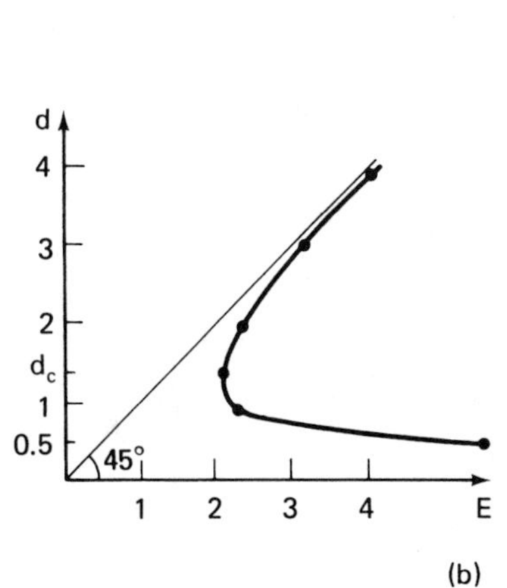

(b)

Figure 6.10

At the critical state the specific energy of the flow takes a minimum value. This value can be computed by equating the first derivative of the specific energy with respect to the water depth to zero.

$$\frac{dE}{dd} = \frac{d}{dd}\left(\frac{Q^2}{2gA^2} + d\right) = -\frac{Q^2}{gA^3}\frac{dA}{dd} + 1 = 0$$

The differential water area dA/dd near the free surface is $dA/dd = T$, where T is the top width of the channel section. Hence,

$$-\frac{Q^2T}{gA^3} + 1 = 0$$

An important parameter for open channel flow is defined by $A/T = D$, which is known as the *hydraulic depth* of the section. For rectangular cross sections, the hydraulic depth is equal to the depth of the flow. The above equation may thus be simplified to

$$\frac{dE}{dd} = 1 - \frac{Q^2}{gDA^2} = 1 - \frac{V^2}{gD} = 0 \tag{6.10}$$

or

$$\frac{V}{\sqrt{gD}} = 1 \tag{6.11}$$

The quantity $V/\sqrt{gD}$ is dimensionless. It can be derived as the ratio of the inertial force in the flow to the gravity force in the flow (see Chapter 10 for detailed discussion). This ratio may also be interpreted physically as the ratio between the mean flow velocity, V, and the speed of a small gravity (surface) wave traveling over the water surface. It is known as the *Froude number*, N_F.

$$N_F = \frac{V}{\sqrt{gD}} \tag{6.12}$$

When the Froude number is equal to unity, as indicated by Equation (6.11), $V = \sqrt{gD}$, the speed of the surface wave (disturbance wave) and that of the flow is the same. The flow is in the critical state. When the Froude number is less than unity, $V < \sqrt{gD}$, the flow velocity is smaller than the speed of a surface wave traveling on the water surface. The flow is in the subcritical state. When the Froude number is greater than unity, $V > \sqrt{gD}$, the flow is in the supercritical state.

From Equation (6.10), we may also write for critical flow,

$$\frac{Q^2}{g} = DA^2$$

In a rectangular channel, $D = d$ and $A = bd$, we have

$$\frac{Q^2}{g} = d^3b^2$$

Since this relation is derived from the critical flow conditions stated above, $d = d_c$, which is the critical depth, and

$$d_c = \sqrt[3]{\frac{Q^2}{gb^2}} = \sqrt[3]{\frac{q^2}{g}} \tag{6.13}$$

where q is the discharge per unit width of the channel. For an open channel of any sectional shape, the critical depth is always a function of the channel discharge and does not vary with the change of the channel slope.

Example 6.5

A hydraulic transition is designed to connect two rectangular channels of the same width by a sloped floor (Figure 6.10). If the channel is 3 m wide and is carrying a discharge of 15 m³/sec at 3.6 m depth, determine the water surface profile in the transition. Assume 0.1 m energy loss uniformly distributed through the transition.

Solution

The specific energy curve can be constructed based on the given discharge and the sectional geometry by using the following relationship from Equation (6.9).

$$E = \frac{Q^2}{2gA^2} + d = \frac{(15)^2}{2 \cdot 9.81 \cdot (3d)^2} + d = \frac{1.27}{d^2} + d$$

At the inlet to the transition the velocity is V_i

$$V_i = \frac{Q}{A_i} = \frac{15}{3.6 \cdot 3} = 1.39 \text{ m/sec}$$

Where A_i is the water area at the inlet and the velocity head is

$$\frac{V_i^2}{2g} = \frac{1.39^2}{2 \cdot 9.81} = 0.10 \text{ m}$$

The total energy head at the inlet as measured with respect to the datum line is

$$H_i = \frac{V_i^2}{2g} + d_i + z_i = 0.10 + 3.60 + 0.4 = 4.10 \text{ m}$$

The top horizontal line in Figure 6.10(a) shows this energy level.

At the exit of the transition the total energy available is less by 0.1 m, as indicated by the energy gradient line (EGL) in Figure 6.10a.

$$H_e = \frac{V_e^2}{2g} + d_e + z_e = H_i - 0.1 = 4.00 \text{ m}$$

E_e is the specific energy measured with respect to the channel bottom

$$E_e = H_e = 4.00 \text{ m}$$

This value is applied to the specific energy curve (Figure 6.10b) to obtain the water depth at the exit section, as shown.

Water surface elevations at four other sections 4.00 m; 8.00 m; 12.00 m; 16.00 m from the entrance section are computed by using the same method. The results for all six sections are tabulated as follows.

Section	*Inlet*	*4.00 m*	*8.00 m*	*12.00 m*	*16.00 m*	*Exit*
Specific Energy E (m)	3.70	3.76	3.82	3.88	3.94	4.00
Water Depth d (m)	3.60	3.665	3.729	3.792	3.855	3.917

PROBLEMS

6.4.1 A trapezoidal channel with a bottom width of 4 m and side slopes of $z = 1.5$ is carrying a discharge of 50 m^3/sec at a depth of 3 m. Determine the following:

(a) the alternate depth for the same specific energy;

(b) the critical depth;

(c) the uniform flow depth for a slope of 0.0004 and $n = 0.022$.

6.4.2 A 15-m-wide rectangular channel with channel slope $S = 0.0025$ and Manning's coefficient $n = 0.035$ is carrying a discharge of 62 m^3/sec. Determine the normal depth and the critical depth of the flow. Construct the specific energy curve.

6.4.3 A 3-m-wide rectangular channel carries a discharge 15 m^3/sec at a uniform depth of 1.7 m. The Manning's coefficient $n = 0.022$. Determine

(a) the channel slope,

(b) the critical depth,

(c) the Froude number.

6.4.4 At the uniform depth of 1.2 m a trapezoidal channel has a bottom width of 1.5 m and side slopes $z = 1.0$. The channel has $n = 0.025$ and slope $S = 0.004$. Determine the discharge and calculate the speed of the surface wave.

6.4.5 A flow of 100 m^3/sec occurs in a trapezoidal canal having a bottom width of 10 m, a side slope $z = 2.0$, and $n = 0.017$. Calculate the critical depth and critical slope for the flow.

6.4.6 A 12.5-m-wide rectangular channel carries 32 m^3/sec at the depth of 2 m. Is this flow subcritical or supercritical? If $n = 0.025$, what is the critical slope of this channel for this discharge? What channel slope must be provided to produce a uniform flow at the depth of 2 m?

6.4.7 Repeat Example 6.5 for an inlet depth of 1.50 m.

6.4.8 A transition is constructed to connect two trapezoidal channels with bottom slopes of 0.001 and 0.0004, respectively. The channels have the same cross-sectional shape, with a bottom width of 3 m, a side slope of $z = 2$, and Manning's coefficient $n = 0.02$. The transition is 20 m long and is designed to carry a discharge of 20 m^3/sec. Assume that the transition has the same kind of surface as that of the channels and that an energy loss of 0.02 m is uniformly distributed through the transitional length. Determine the change in the bottom elevations at the two ends of the transition.

6.4.9 A hydraulic transition 30 m long is used to connect two rectangular channels, 4 m and 2 m wide, respectively. The design discharge is 18 m^3/sec, $n = 0.013$, and the slopes of the channels are 0.0009 for both. Determine the change in the bottom elevation and the water surface profile in the transition if the energy loss in the transition is 0.5 m and is uniformly distributed throughout the transitional length. Assume uniform flows before and after the transition.

6.5 HYDRAULIC JUMPS

Hydraulic jump is a natural phenomenon in open channel. It is an abrupt reduction in flow velocity by means of a sudden increase of water depth in the downstream direction.

Through a hydraulic jump, a high-velocity supercritical flow (upstream) is changed to a low-velocity subcritical flow (downstream). Correspondingly, a low-stage supercritical depth, d_1, is changed to a high-stage subcritical depth, d_2, known, respectively, as the *initial depth* and the *sequent depth* of a hydraulic jump. In the region of the hydraulic jump the characteristic rolling of surface and boiling of water can be seen. These violent motions are accompanied by a significant loss of energy head through the jump. Given the discharge in a particular channel, the amount of energy head loss through a jump, ΔE, can be determined by simply measuring the initial and sequent depths and using the specific energy curve shown in Figure 6.11. Predicting the sequential depth by estimating the energy loss, however, is impractical because it is difficult to determine the energy loss through a jump. The relationship between the initial depth and the sequent depth in a hydraulic jump may be determined by considering the balance of forces and momentum immediately before and after the jump.

Take the free body of water volume involved in the vicinity of a jump, as shown in Figure 6.11. The balance between the hydrostatic forces and the momentum flux through section 1 and section 2, per unit width of the channel may be expressed as

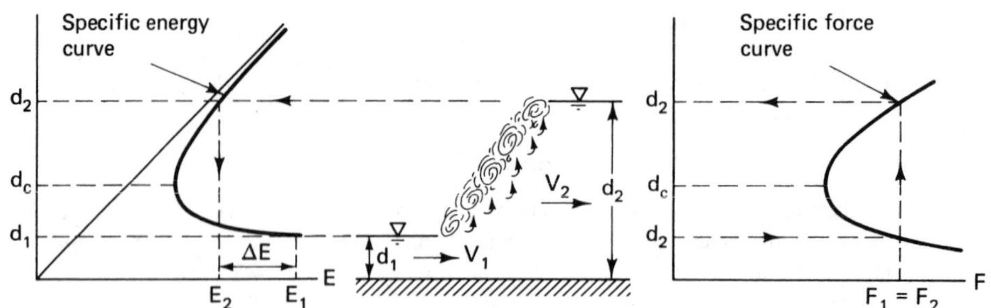

Figure 6.11 Hydraulic Jump

$$F_1 - F_2 = \rho q(V_2 - V_1) \tag{6.14}$$

where q is discharge per unit width of the channel. Substituting the following quantities

$$F_1 = \frac{\gamma}{2}d_1^2, \quad F_2 = \frac{\gamma}{2}d_2^2, \quad V_1 = \frac{q}{d_1}, \quad V_2 = \frac{q}{d_2}$$

into Equation (6.14) and simplifying, we get

$$\frac{q^2}{g} = d_1 d_2\left(\frac{d_1 + d_2}{2}\right) \tag{6.15}$$

This equation may also be rearranged into a more convenient form as follows:

$$\frac{d_2}{d_1} = \frac{1}{2}(\sqrt{1 + 8N_{F_1}^2} - 1) \tag{6.16}$$

where N_{F_1} is the Froude number of the approaching flow,

$$N_{F_1} = \frac{V_1}{\sqrt{gd_1}} \tag{6.17}$$

Example 6.6

A 3-m-wide rectangular channel carries 15 m^3/sec of water at 0.6 m depth before entering a jump. Compute the downstream water depth and the critical depth.

Solution

The discharge per unit width is

$$q = \frac{15}{3} = 5 \text{ m}^3/\text{sec/m}$$

The critical depth, by Equation (6.13), is

$$d_c = \sqrt[3]{\frac{5^2}{9.81}} = 1.60 \text{ m}$$

The approaching velocity is

$$V_1 = \frac{q}{d_1} = \frac{5}{0.6} = 8.33 \text{ m/sec}$$

The Froude number for the approaching flow can be computed by using this velocity an the initial depth $d_1 = 0.6$.

$$N_{F_1} = \frac{V_1}{\sqrt{gd_1}} = 3.43$$

Substituting this value into Equation (6.16) gives

$$\frac{d_2}{0.6} = \frac{1}{2}(\sqrt{1 + 8(3.43)^2} - 1)$$

Solve for the sequent depth d_2.

$$d_2 = 4.38 \text{ m}$$

Equation (6.14) may also be arranged to

$$F_1 + \rho q V_1 = F_2 + \rho q V_2$$

or

$$F_s = F + \rho q V \tag{6.18}$$

The quantity F_s is known as the *specific force* per unit width of the channel. For a given discharge, the specific force is a function of the water depth at a given section. When the specific force, F_s, is plotted against the water depth, it also has two limbs and a vertex that appears at the critical depth. A typical specific force curve is shown in Figure 6.11.

Hydraulic jump usually takes place in a rather short reach in a channel. Therefore, it is reasonable to assume that through a hydraulic jump the specific forces immediately before and after a jump are approximately the same. The value of F_s can be computed from the given conditions of the approaching flow. If we apply this value to the specific force curve in Figure 6.11, we can draw a vertical line that gives both the initial and sequent depths of a jump.

The energy head loss through the hydraulic jump, ΔE, may then be estimated by applying the definition

$$\Delta E = \left(\frac{V_1^2}{2g} + d_1\right) - \left(\frac{V_2^2}{2g} + d_2\right)$$

$$= \frac{1}{2g}(V_1^2 - V_2^2) + (d_1 - d_2) = \frac{q^2}{2g}\left(\frac{1}{d_1^2} - \frac{1}{d_2^2}\right) + (d_1 - d_2)$$

Substituting Equation (6.15) into the above equation and simplifying, we get

$$\Delta E = \frac{(d_2 - d_1)^3}{4d_1 d_2} \tag{6.19}$$

Example 6.7

A long, rectangular open channel 3 m wide carries a discharge of 15 m^3/sec. The channel slope is 0.004 and the Manning's coefficient is 0.01. At a certain point in the channel where the flow reaches the normal depth,

(a) Determine the state of the flow. Is it supercritical or subcritical?

(b) If a hydraulic jump takes place at this depth, what is the sequent depth at the jump?

(c) Estimate the energy head loss through the jump.

Solution

(a) The critical depth of the flow was calculated in Example 6.6, where $d_c = 1.60$ m. The normal depth of this channel can be determined by the Manning equation, Equation (6.5)

$$Q = \frac{1}{n} A_1 R_{h_1}^{2/3} S^{1/2}$$

where

$$A = d_1 b, \qquad R_h = \frac{A_1}{P_1} = \frac{d_1 b}{2d + b}, \qquad b = 3 \text{ m}$$

We have

$$15 = \frac{1}{0.01}(3d_1)\left(\frac{3d_1}{2d_1 + 3}\right)^{2/3}(0.004)^{1/2}$$

Solving the equation for d_1, we get

$$d_1 = 1.08 \text{ m}, \qquad V_1 = \frac{15}{3d_1} = 4.63 \text{ m/sec}$$

and

$$N_{F_1} = \frac{V_1}{\sqrt{gd_1}} = 1.42$$

As $N_{F_1} > 1$, the flow is supercritical.

(b) Applying Equation 6.16, we get

$$d_2 = \frac{d_1}{2}(\sqrt{1 + 8N_{F_1}^2} - 1) = 1.57d_1 = 1.70 \text{ m}$$

(c) The head loss can be estimated by using Equation (6.19).

$$\Delta E = \frac{(d_2 - d_1)^3}{4d_2 d_1} = \frac{(0.62)^3}{4 \cdot 1.7 \cdot 1.08} = 0.032 \text{ m}$$

PROBLEMS

6.5.1 A hydraulic jump occurs at an initial depth of 1.5 m in a rectangular channel 4 m wide. If the sequent depth is 2.2 m, determine the energy loss and the discharge in the channel.

6.5.2 Is the flow in the channel of Problem 6.4.3 subcritical or supercritical? Determine the sequent depth and the energy loss if a hydraulic jump takes place at a depth of 1 m.

6.5.3 Discuss how the change in discharge would affect the specific force curve.

6.5.4 Construct the specific energy and specific force curves for a circular pipe of 1 m diameter that carries 1 m^3/sec.

6.5.5 Discuss why the specific energy curve usually approaches the 45° line as a limit.

6.5.6 Plot the specific force diagram for a 10-m rectangular channel carrying 15 m^3/sec discharge. Determine the critical depth and the minimum specific energy.

6.6 GRADUALLY VARIED FLOWS

Gradually varied flows in open channels differ from the uniform flows and the rapidly varied flows (hydraulic jumps, flow through a streamlined transition, etc.), in that the changes in water depth in the channel take place very gradually with distance.

In uniform flow, the water depth remains a constant value known as the normal depth (or uniform depth). The energy gradient line is parallel to the water surface line and the channel bottom line. The velocity distribution remains unchanged throughout the reach. Thus, the computation of only one water depth is necessary for the entire reach.

In rapidly varied flows, such as hydraulic jumps, rapid changes in water depth take place in a short distance. A significant change in water velocities is associated with the rapid variation of water cross-sectional area. At this high rate of flow deceleration, the energy loss is inevitably high. The computation of water depths using the energy principles is not reliable. In this case, computations can only be carried out by applying the momentum principles [Equation (6.14)].

In gradually varied flows, the velocity changes take place very gradually with distance so that the effects of acceleration on the flow between two adjacent sections are negligible. Computation of the water surface profiles can be carried out based strictly on energy considerations.

The total energy head at any section in an open channel can be expressed as in Equation (6.7),

$$H = \frac{V^2}{2g} + d + z = \frac{Q^2}{2gA^2} + d + z$$

In order to compute the water surface profile, we must first obtain the variation of the total energy head along the channel. Differentiating H with respect to the channel distance x, we obtain the energy gradient in the direction of the flow.

$$\frac{dH}{dx} = \frac{-Q^2}{gA^3}\frac{dA}{dx} + \frac{dd}{dx} + \frac{dz}{dx} = -\frac{Q^2T}{gA^3}\frac{dd}{dx} + \frac{dd}{dx} + \frac{dz}{dx}$$

where $dA = T(dd)$. Rearranging the equation gives

$$\frac{dd}{dx} = \frac{\dfrac{dH}{dx} - \dfrac{dz}{dx}}{1 - \dfrac{Q^2T}{gA^3}} \tag{6.20}$$

The term dH/dx is the slope of the energy gradient line. It is always a negative quantity since the total energy head reduces in the direction of the flow, or $S_e = -dH/dx$. Similarly, the term dz/dx is the slope of the channel bed. It is negative when the elevation of the channel bed reduces in the direction of the flow; it is positive when the elevation of the channel bed increases in the direction of the flow. In general, we may write $S_0 = -dz/dx$.

The energy slope in gradually varied flow between two adjacent stations may also be approximated by using a uniform flow formula. For simplicity, the derivation will be demonstrated with a wide rectangular channel section where $A = bd$, $Q = bq$, and $R_h = A/P = bD/(b + 2D) \cong D = d$(in wide rectangular channels $b >> d$).

Using the Manning formula [Equation (6.5)], we get

$$S_e = -\frac{dH}{dx} = \frac{n^2Q^2}{R_h^{4/3}A^2} = \frac{n^2Q^2}{b^2d^{10/3}} \tag{6.21}$$

The slope of the channel bed may also be expressed in similar terms if uniform flow were assumed to take place in the channel. Since in uniform flows the slope of the channel bed is equal to the energy slope, the hypothetical uniform flow conditions are designated with the subscript n. We have

$$S_0 = -\frac{dz}{dx} = \left(\frac{n^2Q^2}{b^2d^{10/3}}\right)_n \tag{6.22}$$

From Equation (6.13)

$$d_c = \sqrt[3]{\frac{q^2}{g}} = \sqrt[3]{\frac{Q^2}{gb^2}}$$

or

$$Q^2 = gd_c^3b^2 = \frac{gA_c^3}{b} \tag{6.23}$$

Substituting now Equations (6.21), (6.22), and (6.23) into Equation (6.20), we have

$$\frac{dd}{dx} = \frac{S_0\left[1 - \left(\dfrac{d_n}{d}\right)^{10/3}\right]}{\left[1 - \left(\dfrac{d_c}{d}\right)^3\right]} \tag{6.24}$$

This is the general differential equation for gradually varied flows. It is also referred to as the *gradually varied flow equation*. The term dd/dx represents the slope of the water surface with respect to the bottom of the channel. For $dd/dx = 0$, the water depth remains a constant value throughout the reach. It represents the special case of a uniform flow. For $dd/dx < 0$, the water depth decreases in the direction of the flow. For $dd/dx > 0$, the water depth increases in the direction of the flow. The solutions of this equation under given conditions will provide the various water surface profiles that may occur in open channels.

6.7 CLASSIFICATION OF GRADUALLY VARIED FLOWS

In analyzing the gradually varied flows the role of critical depth, d_c, is very important. When open channel flow approaches the critical depth ($d = d_c$), the denominator of Equation (6.24) approaches zero and the value of dd/dx approaches infinity. The water surface becomes very steep. This is seen at hydraulic jumps or at a water surface entering a steep channel from a mild channel or a lake. The latter provides a unique one-to-one relationship between the discharge and the water depth in a channel and is known as the control section in open channel flows.

Depending on the channel slope, the surface conditions, the sectional geometry, and the discharge, open channels may be classified into five categories

1. steep channels,
2. critical channels,
3. mild channels,
4. horizontal channels,
5. adverse channels.

They are categorized according to the flow conditions in the channel as indicated by the relative positions of the normal depth, d_n, and the critical depth, d_c, calculated for each particular channel. The criteria are as follows:

$$\text{Steep channels: } d_n/d_c < 1.0 \quad \text{or} \quad d_n < d_c$$

$$\text{Critical channels: } d_n/d_c = 1.0 \quad \text{or} \quad d_n = d_c$$

$$\text{Mild channels: } d_n/d_c > 1.0 \quad \text{or} \quad d_n > d_c$$

$$\text{Horizontal channels: } \quad S_0 = 0$$

$$\text{Adverse channels: } \quad S_0 < 0$$

A further classification of water surface profile curves depends on the actual water depth and its relations to the critical and normal depths. The ratios of d/d_c and d/d_n may be used in the analysis, where d is the actual water depth at any section of interest in the channel.

If both d/d_c and d/d_n are greater than 1.0, the water surface profile curve is above both the critical depth line and the normal depth line in the channel (see Figure 6.12). The curve is designated as type-1 curve. There are S-1, C-1, and M-1 curves for steep, critical, and mild channels, respectively.

If the water depth d is between the normal depth and the critical depth, the curves are designated as type-2 curves. There are S-2, M-2, H-2, and A-2 curves. The type-2 curve does not exist in critical channels. In critical channels the normal depth is the critical depth, and no distance is allowed between the normal depth line and the critical depth line.

If the water depth d is less than both d_c and d_n, the water surface profile curves are type 3. There are S-3, C-3, M-3, H-3, and A-3 curves. Each of these water surface profile curves is listed and shown schematically in Figure 6.12. Examples of physical occurrence in open channels are also given.

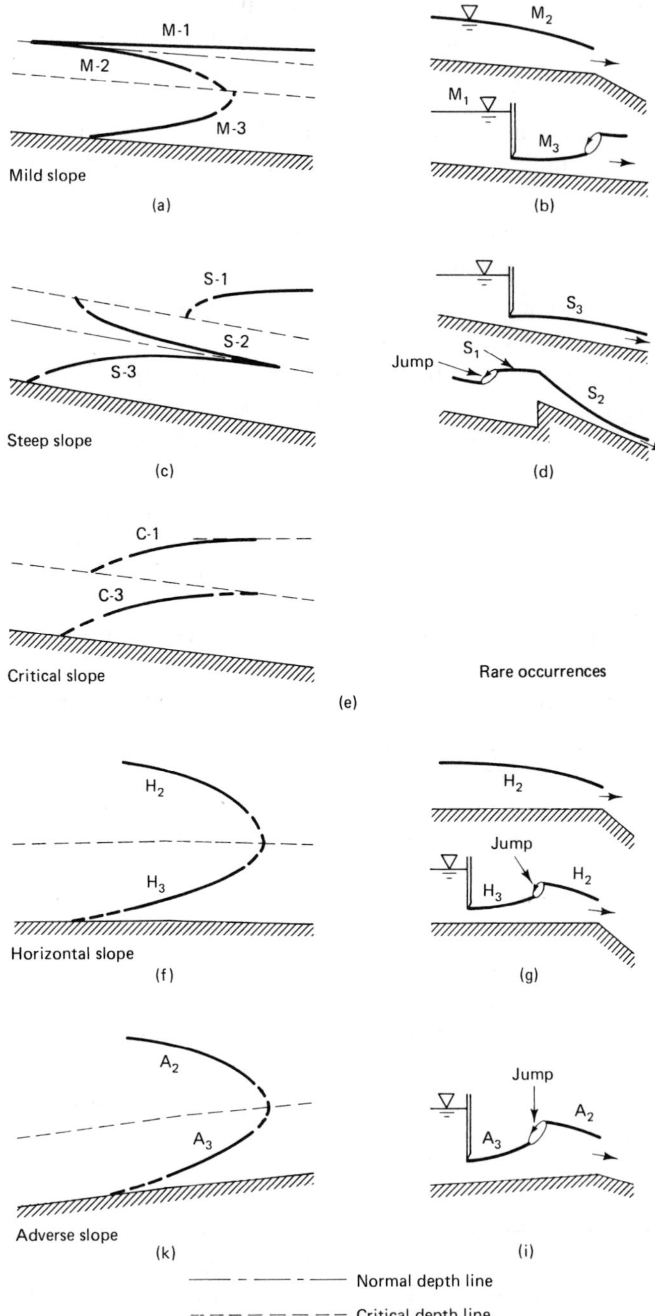

Figure 6.12 Types of gradually varied flows.

Certain important characteristics of the water surface profile curves can be demonstrated from direct analysis of the gradually varied flow equation [Equation (6.24)].

1. For type-1 curves $d/d_c > 1$ and $d/d_n > 1$, the value of dd/dx is positive, indicating that water depth increases in the direction of the flow.
2. For type-2 curves, the value of dd/dx is negative. The water depth decreases in the direction of the flow.
3. For type-3 curves, the value of dd/dx is again positive. The water depth increases in the direction of the flow.
4. When the water depth approaches the critical depth, $d = d_c$, Equation (6.24) gives $dd/dx = \infty$, indicating that the slope of the water surface profile curve is theoretically vertical. Hydraulic jump is a typical example.
5. A few types of water surface profile curves never approach the horizontal line (S-2, S-3, M-2, M-3, C-3, H-3, and A-3). Others approach the horizontal line asymptotically, except for the C-1 curve, which is horizontal throughout the channel reach. Since in a critical channel $d_n = d_c$, Equation (6.24) gives $dd/dx = S_0$, indicating that the water depth increases at the same rate as the channel bed elevation decreases, which, theoretically, results in a horizontal water surface profile.
6. In channels where $d < d_c$, the velocity of water flow is greater than that of the disturbance wave. For this reason, the flow conditions in the downstream channel will not affect those upstream. The change of water depth due to any channel disturbance propagates only in the downstream direction. Computation of the water surface profile should be carried out in the downstream direction (M-3, S-2, S-3, C-3, H-3, and A-3).
7. In channels where $d > d_c$, the speed of wave propagation is greater than the velocity of water flow. Any disturbance in the downstream channel can travel upstream and affect the flow conditions upstream as well as downstream. Any change of water depth in the downstream channel propagates upstream and may also change the water depth in the upstream channel. Computation of the water surface profile should be carried out in the upstream direction (M-1, M-2, S-1, C-1, H-2, and A-2).
8. At the break of the channel from mild to steep slope, or a significant drop of the channel bottom, the critical depth is assumed to take place in the immediate vicinity of the brink. At this point, a definite depth-discharge relationship can be obtained, and is frequently used as a *control section* for water surface profile computations.

Table 6.2 provides a summary of the water surface profile curves.

TABLE 6.2 Characteristics of Water Surface Profile Curves

Channel	Symbol	Type	Slope	Depth	Curve
Mild	M	1	$S_0 > 0$	$d > d_n > d_c$	M-1
Mild	M	2	$S_0 > 0$	$d_n > d > d_c$	M-2
Mild	M	3	$S_0 > 0$	$d_n > d_c > d$	M-3
Critical	C	1	$S_0 > 0$	$d > d_n = d_c$	C-1
Critical	C	3	$S_0 > 0$	$d_n = d_c > d$	C-3
Steep	S	1	$S_0 > 0$	$d > d_c > d_n$	S-1
Steep	S	2	$S_0 > 0$	$d_c > d > d_n$	S-2
Steep	S	3	$S_0 > 0$	$d_c > d_n > d$	S-3
Horizontal	H	2	$S_0 = 0$	$d > d_c$	H-2
Horizontal	H	3	$S_0 = 0$	$d_c > d$	H-3
Adverse	A	2	$S_0 < 0$	$d > d_c$	A-2
Adverse	A	3	$S_0 < 0$	$d_c > d$	A-3

6.8 COMPUTATION OF WATER SURFACE PROFILES

Water surface profiles in gradually varied flows may be computed by using Equation (6.24). The computation normally begins from a section where the relationship between the water surface elevation (depth) and the discharge is known. These sections are commonly known as the *control sections*. A few examples of the control sections in open channels are demonstrated in Figure 6.13.

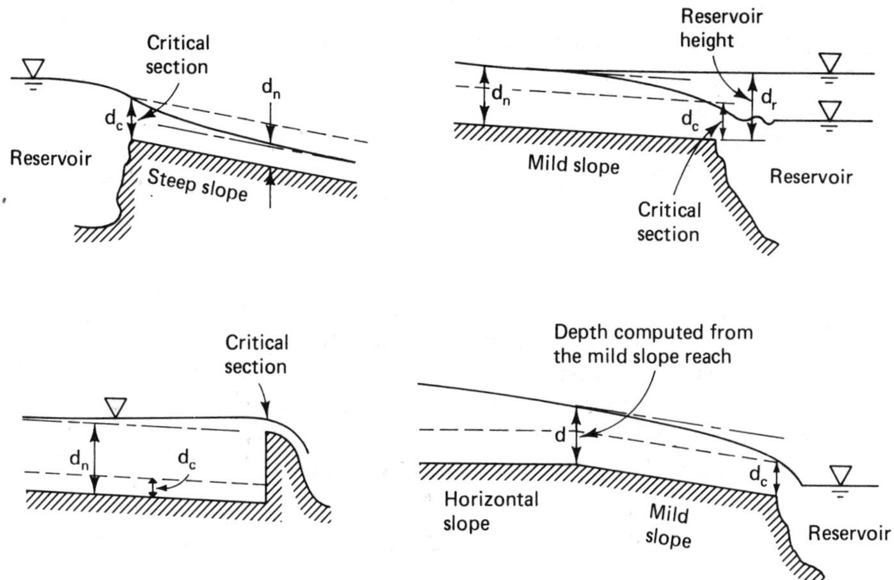

Figure 6.13 Control sections in open channels.

A successive computational procedure is used to compute the water surface elevation at the next section at a distance short enough so that the water surface between the two sections may be approximated by a straight line. A step procedure is carried out in the downstream direction for rapid (supercritical) flows and in the upstream direction for tranquil (subcritical) flows.

Two commonly used computation methods are discussed in this section

1. the *direct integration method,*
2. the *direct step method.*

These two methods are selectively presented here to demonstrate the necessary considerations and procedures involved in computing a water surface profile. Other methods are frequently used in hydraulic engineering practice, but they are not discussed here because their programming and computations are so complex. For these other methods of computation, the readers may refer to the textbook by V. T. Chow.*

Direct Integration Method For over a century, engineers have made attempts to obtain a direct integration of the gradually varied flow equation [e.g., Equation (6.24)]. In practice, a general solution is impossible because there is such a great variety of channel cross-sectional shapes. A comprehensive review of many of the existing integration methods was made by V. T. Chow in 1955.† The method described here is suggested by Chow as an outcome of his study of the various exiting methods. Chow's method can be applied directly to most of the channel shapes. In this chapter we shall discuss its application to wide rectangular channels.

The general form of the gradually varied flow equation may be rearranged and written as

$$\frac{dd}{dx} = S_0 \left[\frac{1 - \left(\dfrac{d_n}{d}\right)^N}{1 - \left(\dfrac{d_c}{d}\right)^M} \right] \tag{6.25}$$

where the exponents N and M depend on the shape of the cross section and on the flow conditions. For wide rectangular channels, $N = 10/3$ and $M = 3$, as shown in Equation (6.24). For gradually varied flow in channels of relatively simple geometries, the values of N and M may also be assumed to be constant within the range of the integration limits.

By letting $d/d_n = u$, the above equation may be written as

$$dx = \frac{d_n}{S_0}\left[1 - \frac{1}{1 - u^N} + \left(\frac{d_c}{d_n}\right)^M \left(\frac{u^{N-M}}{1 - u^N}\right)\right] du$$

* V. T. Chow, *Open Channel Hydraulics* (New York: McGraw-Hill, 1959).

† V.T. Chow, "Investigating the Equation of Gradually Varied Flow," Paper 838, *ASCE Proc.*, 81 (1955), 1–32. Discussions by C. J. Keifer, et al. Closing discussion by V. T. Chow, Paper 1177, *ASCE Proc.*, 83, HYl (1957), 9–22.

Hence,

$$x = \frac{d_n}{S_0}\left[u - \int_0^u \frac{du}{1 - u^N} + \left(\frac{d_c}{d_n}\right)^M \int_0^u \frac{u^{N-M}}{1 - u^N}\, du\right] + \text{constant} \quad (6.26)$$

The first integral term of the right-hand side may be represented as a function of u and N, or

$$F(u,N) = \int_0^u \frac{du}{1 - u^N} \quad (6.27)$$

which is known as the *varied flow function.*

The second integral term may also be expressed in similar form by letting $v = u^{N/J}$ and

$$J = \frac{N}{N - M + 1}$$

Substituting the new variables and transforming the integral into

$$\int_0^u \frac{u^{N-M}}{1 - u^N}\, du = \frac{J}{N}\int_0^v \frac{dv}{1 - v^J} = \frac{J}{N}F(v,J) \quad (6.28)$$

where

$$F(v,J) = \int_0^v \frac{dv}{1 - v^J} \quad (6.29)$$

The similarity between Equations (6.27) and (6.29) is obvious, except that the variables u and N are replaced by v and J in Equation (6.29).

Substituting Equations (6.27) and (6.29) into Equation (6.26), we may write

$$x = \frac{d_n}{S_0}\left[u - F(u,N) + \left(\frac{d_c}{d_n}\right)^M \left(\frac{J}{N}\right) F(v,J)\right] + \text{constant} \quad (6.30)$$

For wide rectangular channels, $N = 10/3$, $M = 3$, $J = N/(N - M + 1) = 5/2$, and $v = u^{N/J} = u^{4/3}$. Equation (6.30) may be simplified to

$$x = A\left[u - F\left(u, \frac{10}{3}\right) + BF\left(v, \frac{5}{2}\right)\right] + \text{constant} \quad (6.31)$$

where

$$A = \frac{d_n}{S_0}, \qquad B = \left(\frac{d_c}{d_n}\right)^M \left(\frac{J}{N}\right) = \left(\frac{d_c}{d_n}\right)^3 \left(\frac{3}{4}\right)$$

are constant when the discharge and the channel slope are given. $F(u, 10/3)$ and $F(v, 5/2)$, or simply $F(u, 3.33)$ and $F(v, 2.5)$, are the varied-flow functions for wide rectangular channels.

By Equation (6.31), the length of flow profile between two consecutive sections 1 and 2 is equal to

$$\begin{aligned} L &= x_2 - x_1 \\ &= A\{(u_2 - u_1) - [F(u_2, 3.33) - F(u_1, 3.33)] \\ &\quad + B[F(v_2, 2.5) - F(v_1, 2.5)]\} \end{aligned} \tag{6.32}$$

The numerical values for the varied-flow functions $F(u, 3.33)$ and $F(v, 2.50)$ for both positive channel slopes ($S_0 > 0$) and negative channel slopes ($S_0 < 0$) are provided in Table 6.3.

TABLE 6.3 The Varied-flow Functions for Wide Rectangular Channels

$$F(x, y) = \int_0^x \frac{dx}{1 - x^y}$$

Positive Slope			Negative (Adverse) Slope		
x \ y	2.50	3.33	x \ y	2.50	3.33
0.00	0.000	0.000	0.00	0.000	0.000
0.02	0.020	0.020	0.02	0.020	0.020
0.04	0.040	0.040	0.04	0.040	0.040
0.06	0.060	0.060	0.06	0.060	0.060
0.08	0.080	0.080	0.08	0.080	0.080
0.10	0.100	0.100	0.10	0.100	0.100
0.12	0.120	0.120	0.12	0.120	0.120
0.14	0.140	0.140	0.14	0.140	0.140
0.16	0.160	0.160	0.16	0.160	0.160
0.18	0.181	0.180	0.18	0.180	0.180
0.20	0.201	0.200	0.20	0.199	0.200
0.22	0.221	0.220	0.22	0.219	0.220
0.24	0.242	0.240	0.24	0.238	0.240
0.26	0.262	0.261	0.26	0.257	0.260
0.28	0.283	0.281	0.28	0.276	0.279
0.30	0.304	0.301	0.30	0.295	0.298
0.32	0.325	0.322	0.32	0.314	0.318
0.34	0.347	0.342	0.34	0.333	0.338
0.36	0.368	0.363	0.36	0.352	0.357
0.38	0.390	0.383	0.38	0.370	0.376
0.40	0.412	0.404	0.40	0.388	0.395
0.42	0.435	0.425	0.42	0.406	0.413
0.44	0.458	0.447	0.44	0.425	0.433
0.46	0.480	0.469	0.46	0.442	0.452
0.48	0.504	0.490	0.48	0.460	0.470

TABLE 6.3 The Varied-flow Functions for Wide Rectangular Channels (Continued)

$$F(x, y) = \int_0^x \frac{dx}{1 - x^y}$$

Positive Slope			Negative (Adverse) Slope		
x \ y	2.50	3.33	x \ y	2.50	3.33
0.50	0.528	0.513	0.50	0.477	0.488
0.52	0.553	0.535	0.52	0.492	0.506
0.54	0.578	0.558	0.54	0.510	0.524
0.56	0.604	0.680	0.56	0.525	0.542
0.58	0.630	0.605	0.58	0.542	0.560
0.60	0.658	0.629	0.60	0.558	0.577
0.61	0.672	0.642	0.61	0.566	0.585
0.62	0.686	0.656	0.62	0.575	0.593
0.63	0.699	0.666	0.63	0.582	0.602
0.64	0.715	0.680	0.64	0.589	0.619
0.65	0.729	0.692	0.65	0.595	0.617
0.66	0.746	0.705	0.66	0.603	0.624
0.67	0.761	0.719	0.67	0.610	0.633
0.68	0.777	0.733	0.68	0.618	0.641
0.69	0.743	0.746	0.69	0.625	0.659
0.70	0.800	0.760	0.70	0.633	0.657
0.71	0.828	0.776	0.71	0.640	0.664
0.72	0.845	0.791	0.72	0.647	0.672
0.73	0.863	0.806	0.73	0.654	0.680
0.74	0.882	0.822	0.74	0.661	0.687
0.75	0.901	0.838	0.75	0.668	0.694
0.76	0.921	0.855	0.76	0.675	0.702
0.77	0.941	0.872	0.77	0.681	0.708
0.78	0.963	0.889	0.78	0.687	0.715
0.79	0.984	0.907	0.79	0.693	0.722
0.80	1.008	0.926	0.80	0.699	0.732
0.81	1.032	0.946	0.81	0.705	0.736
0.82	1.056	0.966	0.82	0.712	0.743
0.83	1.083	0.987	0.83	0.717	0.749
0.84	1.110	1.010	0.84	0.724	0.755
0.85	1.140	1.033	0.85	0.730	0.767
0.86	1.170	1.058	0.86	0.736	0.768
0.87	1.203	1.084	0.87	0.742	0.774
0.88	1.238	1.112	0.88	0.748	0.780
0.89	1.277	1.143	0.89	0.753	0.786
0.90	1.318	1.176	0.90	0.759	0.797
0.91	1.364	1.211	0.91	0.765	0.798
0.92	1.414	1.251	0.92	0.771	0.804
0.93	1.470	1.295	0.93	0.777	0.810
0.94	1.535	1.345	0.94	0.782	0.816

TABLE 6.3 The Varied-flow Functions for Wide Rectangular Channels (Continued)

$$F(x, y) = \int_0^x \frac{dx}{1 - x^y}$$

Positive Slope			Negative (Adverse) Slope		
x \ y	2.50	3.33	x \ y	2.50	3.33
0.950	1.612	1.404	0.950	0.788	0.821
0.960	1.700	1.475	0.960	0.793	0.826
0.970	1.822	1.566	0.970	0.798	0.830
0.975	1.897	1.623	0.975	0.800	0.832
0.980	1.988	1.693	0.980	0.803	0.834
0.985	2.110	1.782	0.985	0.806	0.837
0.990	2.272	1.906	0.990	0.809	0.840
0.995	2.549	2.119	0.995	0.811	0.844
0.999	3.194	2.608	1.000	0.813	0.846
1.000	∞	∞	1.005	0.815	0.849
1.001	2.767	1.932	1.010	0.817	0.851
1.005	2.146	1.445	1.015	0.819	0.854
1.010	1.870	1.236	1.020	0.823	0.857
1.015	1.709	1.115	1.03	0.827	0.861
1.020	1.595	1.030	1.04	0.833	0.866
1.03	1.436	0.910	1.05	0.836	0.871
1.04	1.322	0.826	1.06	0.841	0.875
1.05	1.273	0.762	1.07	0.846	0.880
1.06	1.167	0.710	1.08	0.852	0.884
1.07	1.108	0.666	1.09	0.856	0.889
1.08	1.057	0.630	1.10	0.860	0.894
1.09	1.013	0.597	1.11	0.865	0.897
1.10	0.973	0.568	1.12	0.868	0.900
1.11	0.938	0.543	1.13	0.873	0.905
1.12	0.906	0.519	1.14	0.876	0.909
1.13	0.877	0.498	1.15	0.881	0.913
1.14	0.850	0.479	1.16	0.885	0.917
1.15	0.825	0.461	1.17	0.888	0.920
1.16	0.802	0.445	1.18	0.892	0.924
1.17	0.780	0.430	1.19	0.896	0.928
1.18	0.760	0.415	1.20	0.900	0.932
1.19	0.741	0.402	1.22	0.904	0.935
1.20	0.723	0.390	1.24	0.912	0.943
1.22	0.691	0.366	1.26	0.923	0.952
1.24	0.661	0.346	1.28	0.930	0.957
1.26	0.634	0.327	1.30	0.937	0.964
1.28	0.610	0.311	1.32	0.944	0.970
1.30	0.587	0.296	1.34	0.951	0.976
1.32	0.566	0.287	1.36	0.957	0.981
1.34	0.547	0.269	1.38	0.963	0.987

TABLE 6.3 The Varied-flow Functions for Wide Rectangular Channels (Continued)

$$F(x, y) = \int_0^x \frac{dx}{1 - x^y}$$

Positive Slope			Negative (Adverse) Slope		
x \ y	2.50	3.33	x \ y	2.50	3.33
1.36	0.529	0.257	1.40	0.969	0.991
1.38	0.512	0.246	1.42	0.975	0.997
1.40	0.496	0.236	1.44	0.982	1.002
1.42	0.482	0.227	1.46	0.986	1.006
1.44	0.468	0.217	1.48	0.992	1.010
1.46	0.455	0.209	1.50	0.998	1.014
1.48	0.443	0.201	1.55	1.010	1.023
1.50	0.431	0.194	1.60	1.022	1.033
1.55	0.404	0.171	1.65	1.034	1.041
1.60	0.380	0.163	1.70	1.045	1.049
1.65	0.359	0.150	1.75	1.055	1.056
1.70	0.340	0.139	1.80	1.064	1.062
1.75	0.323	0.129	1.85	1.073	1.067
1.80	0.307	0.120	1.90	1.082	1.075
1.85	0.293	0.112	1.95	1.090	1.080
1.90	0.279	0.105	2.00	1.098	1.083
1.95	0.267	0.099	2.10	1.112	1.091
2.00	0.256	0.093	2.20	1.125	1.098
2.10	0.237	0.082	2.3	1.138	1.106
2.20	0.218	0.074	2.4	1.147	1.112
2.3	0.204	0.066	2.5	1.156	1.117
2.4	0.181	0.060	2.6	1.165	1.121
2.5	0.178	0.054	2.7	1.173	1.124
2.6	0.167	0.050	2.8	1.181	1.126
2.7	0.157	0.045	2.9	1.187	1.130
2.8	0.148	0.042	3.0	1.193	1.133
2.9	0.141	0.038	3.5	1.221	1.144
3.0	0.134	0.035	4.0	1.237	1.149
3.5	0.105	0.025	4.5	1.252	1.153
4.0	0.086	0.018	5.0	1.264	1.156
4.5	0.072	0.014	6.0	1.272	1.160
5.0	0.062	0.011	7.0	1.282	1.162
6.0	0.047	0.007	8.0	1.290	1.163
7.0	0.037	0.005	9.0	1.294	1.165
8.0	0.031	0.004	10.0	1.299	1.165
9.0	0.026	0.003			
10.0	0.022	0.002			
20.0	0.008	0.000			

Example 6.8

A wide rectangular channel carries a discharge per unit width of 2.5 m^3/sec/m on a 0.001 slope and $n = 0.025$. Compute the backwater curve created by a low dam that has water at a depth of 2 m immediately behind the dam. The upstream computation may be carried out to a depth 1% greater than the normal depth.

Solution

The normal and critical depth for this channel can be calculated by using Equations (6.5) and (6.11), respectively. As for a wide rectangular channel $R_h \approx d$, and $D = d$, we may write, rearranging the Manning formula, Equation (6.5),

$$AR_h^{2/3} = bd_n(d_n)^{2/3} = \frac{Q \cdot n}{S_0^{1/2}}$$

or

$$d_n^{5/3} = \frac{Q}{b}\frac{n}{S_0^{1/2}} = \frac{qn}{S_0^{1/2}} = \frac{2.5 \cdot 0.025}{(0.001)^{1/2}} = 1.98$$

The normal depth is

$$d_n = (1.98)^{3/5} = 1.50 \text{ m}$$

and applying the continuity condition,

$$V_c^2 = \frac{Q^2}{A_c^2} = \frac{Q^2}{b^2 d_c^2} = \frac{q^2}{d_c^2} \tag{a}$$

For critical flow in a wide rectangular channel, we also have

$$N_F = \frac{V_c}{\sqrt{gd_c}} = 1.0, \qquad \text{or} \qquad V_c^2 = gd_c \tag{b}$$

Comparing Equations (a) and (b), we have $d_c = \sqrt[3]{\dfrac{q^2}{g}}$.

Hence,

$$d_c = \left(\frac{q^2}{g}\right)^{1/3} = \left(\frac{2.5^2}{9.81}\right)^{1/3} = 0.86 \text{ m}$$

To determine the distance between any two sections, we can apply Equation (6.32), and Table 6.3 as follows:

$$L = x_2 - x_1 = A\left\{(u_2 - u_1) - \left[F\left(u_2, \frac{10}{3}\right) - F\left(u_1, \frac{10}{3}\right)\right] + \left[BF\left(v_2, \frac{5}{2}\right) - BF\left(v_1, \frac{5}{2}\right)\right]\right\}$$

where

$$A = \frac{d_n}{S_0} = \frac{1.50}{0.001} = 1500; \qquad B = \left(\frac{d_c}{d_n}\right)^M \frac{J}{N} = \left(\frac{0.86}{1.50}\right)^3 \frac{\left(\frac{5}{2}\right)}{\left(\frac{10}{3}\right)} = 0.141$$

Water surface profile computations are carried out in the upstream direction with five assumed water depths between $d = 2.00$ m immediately upstream from the dam, and $d = 1.52$ m, which is 1% greater than the normal depth. The computation procedure is shown in Table 6.4.

TABLE 6.4

d	u	v	$F\left(u, \frac{10}{3}\right)$	$F\left(v, \frac{5}{2}\right)$	$L(m)$	$\Sigma L(m)$
2.00	1.33	1.47	0.277	0.449		
					−192	192
1.88	1.25	1.35	0.334	0.537		
					−237	429
1.76	1.17	1.24	0.430	0.661		
					−317	746
1.64	1.09	1.13	0.597	0.877		
					−931	1677
1.52	1.01	1.02	1.236	1.595		

Figure 6.14 shows the backwater surface profile that results from the above computations.

If we were only interested in the total distance between the dam and the last section where $d = 1.52$ m, one step computation alone would be enough. Applying Equation (6.32), and the corresponding values from the above table, we have

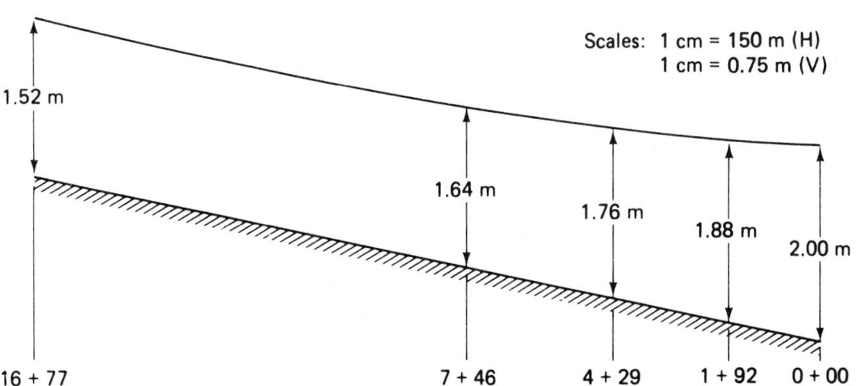

Figure 6.14 Backwater surface profile.

$$L = A\{(u_2 - u_1) - [F(u_2, 3.33) - F(u_1, 3.33)] + B[F(v_2, 2.50) - F(v_1, 2.50)]\}$$

$$= 1500\{(1.01 - 1.33) - (1.236 - 0.277) + 0.141(1.595 - 0.449)\}$$

$$= 1676 \text{ m}$$

Direct Step Method The direct step method is derived directly from the energy balance between two neighboring sections, 1 and 2, which are separated by a sufficiently short distance so that the water surface line can be approximated by a straight line. The energy relation between the two sections may be written as

$$\left(\frac{V_1^2}{2g} + d_1\right) + \Delta z = \left(\frac{V_2^2}{2g} + d_2\right) + h_L \tag{6.33a}$$

where Δz is the difference in elevation of the channel and h_L is the energy head loss between the two sections. This relationship is also illustrated in Figure 6.15.

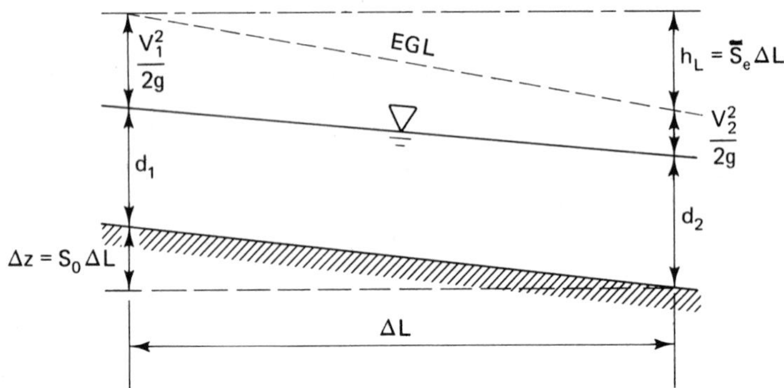

Figure 6.15 Definition for computation of water surface profile.

Equation (6.33a) may be rewritten in terms defined in Figure 6.15.

$$\left(\frac{V_1^2}{2g} + d_1\right) + S_0\Delta L = \left(\frac{V_2^2}{2g} + d_2\right) + \bar{S}_e\Delta L$$

or

$$E_1 + S_0\Delta L = E_2 + \bar{S}_e\Delta L$$

where E is the specific energy, and

$$\Delta L = \frac{E_2 - E_1}{S_0 - \bar{S}_e} \tag{6.33b}$$

The computation procedure is to determine the distance ΔL between the section with the known water depth d_1 and the section with an assumed water depth d_2.

The same procedure is applied to the next section in the same direction with d_2 as the known depth. A tabulated computation is particularly recommended as illustrated in Example 6.9. The energy slope, S_e, can be computed by applying the Manning formula

$$S_e = \frac{n^2 V^2}{R_h^{4/3}} \tag{6.34}$$

Example 6.9

Compute the backwater curve on the channel given in Example 6.8 by using the direct step method.

Solution

For a wide rectangular channel, $D \cong R_h \cong d$, and

$$V = \frac{Q}{A} = \frac{q}{d}$$

The conditions given include $S_0 = 0.001$ and $n = 0.025$. From the results of Example 6.8, we have $d_c = 0.86$ m and $d_n = 1.50$ m. From the Manning formula [Equation (6.34)], $S_e = (n^2 q^2)/d^{10/3}$, which is used to compute the energy slope in Table 6.5.

Example 6.10

A concrete-paved trapezoidal channel ($n = 0.022$) with 3.5-m bottom width, side slope $z = 2$, and bed slope of 0.012, discharges 40 m^3/sec of fresh water from a reservoir. Determine the water surface profile to within 2% of normal depth.

Solution

The normal and critical depths should be calculated first,

$$AR_h^{2/3} = \frac{[(zd_n + b)d_n]^{5/3}}{(b + 2\sqrt{1 + z^2}\, d_n)^{2/3}} = \frac{Q \cdot n}{S_0^{1/2}}$$

or

$$\frac{[(2d_n + 3.5)d_n]^{5/3}}{(3.5 + 2\sqrt{5}\, d_n)^{2/3}} = \frac{40 \cdot 0.022}{(0.012)^{1/2}} = 8.033$$

by trial and error then,

$$d_n = 1.38 \text{ m}$$

TABLE 6.5 Backwater Curve Computation—Direct Step Method (Example 6.9)

(1)	(2)	(3)	(4)	(5)	(6)	(7)	(8)	(9)	(10)	(11)	(12)
d	A	R_h	V	$\frac{V^2}{2g}$	E	ΔE	S_e	$\bar{S}_e$	$S_0 - \bar{S}_e$	Δx	L
2.00	—	2.00	1.25	0.080	2.080	—	0.000388	—	—	—	—
1.94	—	1.94	1.29	0.085	2.025	0.055	0.000429	0.000409	0.000591	93	93
1.88	—	1.88	1.33	0.090	1.970	0.055	0.000476	0.000453	0.000547	101	194
1.82	—	1.82	1.37	0.096	1.916	0.054	0.000531	0.000504	0.000496	108	302
1.76	—	1.76	1.42	0.103	1.863	0.053	0.000593	0.000592	0.000408	130	432
1.70	—	1.70	1.47	0.110	1.810	0.053	0.000666	0.000630	0.000370	143	575
1.64	—	1.64	1.52	0.118	1.758	0.052	0.000751	0.000709	0.000291	179	754
1.58	—	1.58	1.58	0.128	1.708	0.050	0.000850	0.000801	0.000199	251	1005
1.52	—	1.52	1.64	0.138	1.658	0.050	0.000967	0.000909	0.000091	550	1555

Column (1) Depth of flow (in meters) arbitrarily assigned between 2.00 m and the depth, 1.52 m (1% greater than the normal depth).
Column (2) Water cross-sectional area (in square meters) corresponding to the depth.
Column (3) Hydraulic radius (in meters) corresponding to the depth.
Column (4) Mean velocity (in meters per second) obtained by dividing the discharge by the area in column 2.
Column (5) Velocity head (in meters).
Column (6) Specific energy (in meters) obtained by adding the values from columns 1 and 5.
Column (7) Change in specific energy between two adjacent sections (in meters) obtained by subtracting the E value in column 6 and that of the previous step.
Column (8) Energy slope, or friction slope, obtained from Manning's formula [Eq. (6.34)] by using the values of V in column 4 and R_h from column 3.
Column (9) Average energy slope between the steps, equal to the averaged value of S_e just computed in column 8 and that of the previous step.
Column (10) Difference between the channel bottom slope ($S_0 = 0.001$) and that in column 9.
Column (11) Length (in meters) of the reach computed between the two adjacent sections.
Column (12) Distance from the first section where the water depth is 2.00 m.

TABLE 6.6 (Example 6.10)

d	A	R_h	V	$\frac{V^2}{2g}$	E	ΔE	S_e	$\bar{S}_e$	$S_0 - \bar{S}_e$	Δx	L
1.72	11.9	1.07	3.35	0.57256	2.29256	—	0.00499	—	—	—	—
1.68	11.5	1.05	3.47	0.61423	2.29423	0.00167	0.00549	0.00524	0.00676	0.25	0.25
1.64	11.1	1.03	3.60	0.65986	2.29986	0.00563	0.00605	0.00577	0.00623	0.90	1.15
1.60	10.7	1.01	3.73	0.70992	2.30992	0.01006	0.00668	0.00636	0.00564	1.78	2.93
1.56	10.3	0.99	3.87	0.76495	2.32495	0.01503	0.00740	0.00704	0.00496	3.03	5.96
1.52	9.9	0.97	4.02	0.82557	2.34557	0.02062	0.00821	0.00781	0.00419	4.92	10.88
1.48	9.6	0.94	4.18	0.89250	2.37250	0.02693	0.00914	0.00868	0.00332	8.11	18.99
1.44	9.2	0.92	4.35	0.96657	2.40657	0.03407	0.01019	0.00966	0.00234	14.56	33.55
1.40	8.8	0.90	4.54	1.04872	2.44872	0.04215	0.01140	0.01080	0.00120	35.13	68.68

To determine critical depth, $N_F = 1$

$$\frac{Q^2T}{gA^3} = \frac{Q^2(2zd_c + b)}{g[(zd_c + b)d_c]^3} = 1$$

by trial and error, we obtain

$$d_c = 1.72 \text{ m}$$

Thus, the channel is a steep channel, and an S-2 profile should be computed in the downstream direction starting from the critical depth at the entrance section (see Table 6.6).

PROBLEMS

6.8.1 A wide rectangular channel discharges 1.5 m^3/sec per unit width. The slope is 0.0009 and Manning's coefficient is 0.015. At a certain point the water depth is 0.75 m. How far downstream will it be 0.73 m? Use the direct integration method.

6.8.2 A trapezoidal channel with a 5-m bottom width and side slope $z = 1.0$ discharges 35 m^3/sec. The slope is 0.004 and it is paved with smooth concrete. Determine the distance between the two sections with depths of 1.69 m and 1.65 m, respectively. Use the direct step method.

6.8.3 A trapezoidal channel paved with smooth concrete has a 2.5-m bottom width and a $z = 1.0$ side slope. It discharges 18 m^3/sec on a 0.0004 channel slope. Determine the distance between the two sections with flow depths of 2.16 m and 2.20 m, respectively.

6.8.4 If the channel in Problem 6.8.3 is discharging 10 m^3/sec, determine the distance between the sections of depth 2.16 m and 2.20 m.

6.8.5 In Figure P6.8.5 a wide rectangular channel carries 1.6 m^3/sec per unit width of the channel that has $n = 0.011$ and $S_0 = 0.0016$. If a 5-m height dam is placed across the channel, determine the water surface profile upstream from the dam. Use the direct integration method with 0.30-m depth increments.

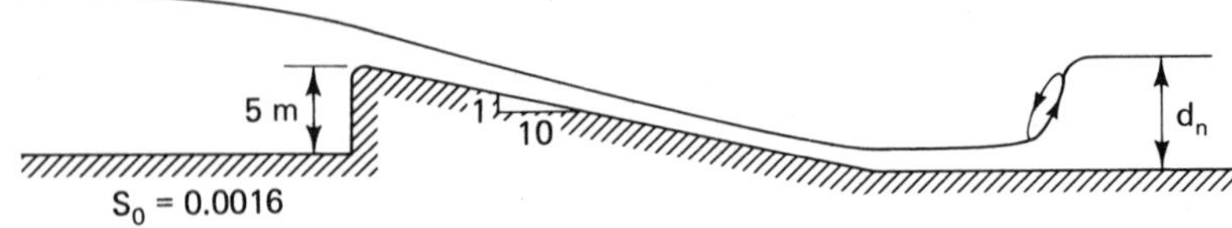

Figure P6.8.5

6.8.6 The downstream face of the dam in Problem 6.8.5 has a slope of 1/10 until it reaches the channel bottom again (see Figure P6.8.5). Determine the water surface profile over the downstream face of the dam. Use the direct integration method with depth increments of your choice.

6.8.7 Determine the water surface profile in the downstream channel immediately after the dam in Problem 6.8.6. Assume that normal depth will follow the jump.

6.8.8 Compute the water surface profile if a dam is placed across the channel of Problem 6.8.2 to raise the water level to a height of 3.4 m above the channel bottom immediately behind the dam.

6.8.9 A long trapezoidal channel with a roughness coefficient of 0.015, bottom width measuring 3.6 m, and $z = 2.0$ discharges 44 m^3/sec. If the channel slope is 0.0001, compute the water surface profile upstream from a dam that is 5 m in height.

6.9 TRANSITIONAL PHENOMENA IN SUPERCRITICAL FLOWS

Physically, the basic difference between supercritical and subcritical flows is the way surface waves are transmitted in a channel. In supercritical flow a small surface wave created by any disturbance is washed downstream because the velocity of flow in the channel is greater than the speed of the wave. In subcritical flow the velocity of the flow is less than the speed of the wave; the wave will propagate upstream as well as downstream. The patterns of surface wave propagation in stationary water ($V = 0$), subcritical flow ($V < \sqrt{gD}$), critical flow ($V = \sqrt{gD}$), and supercritical flow ($V > \sqrt{gD}$) are schematically illustrated in Figure 6.16.

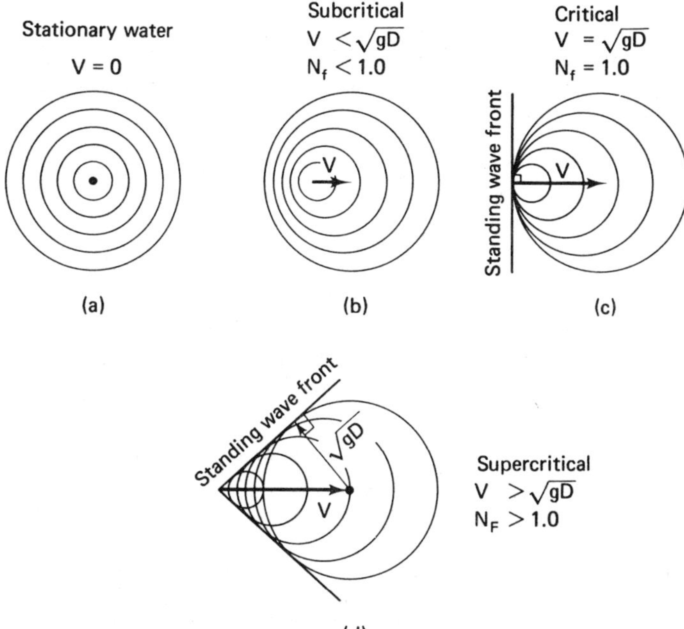

Figure 6.16 Patterns of surface wave propagation in open channel.

In Section 6.4 we said that the speed of a small surface wave traveling on water is equal to $\sqrt{gD}$, where D is the local water depth and g is the gravitational acceleration. If the velocity of water is V, then the absolute velocity of a surface wave traveling in the upstream direction is

$$\sqrt{gD} - V$$

In the case of critical flow, $V = \sqrt{gD}$, a standing wave front is created, as shown in Figure 6.16(c). The wave front in critical flow is normal to the direction of the flow.

In supercritical flows the speed of water flow V exceeds $\sqrt{gD}$. The standing wave front forms at an angle β, as shown in Figure 6.16(d). The angle β obviously decreases as the flow speed increases. If a point P is taken on the wave front, it is easily seen that the location of P after a unit time can be determined by the longitudinal displacement O–O' and the radial displacement O'–P.

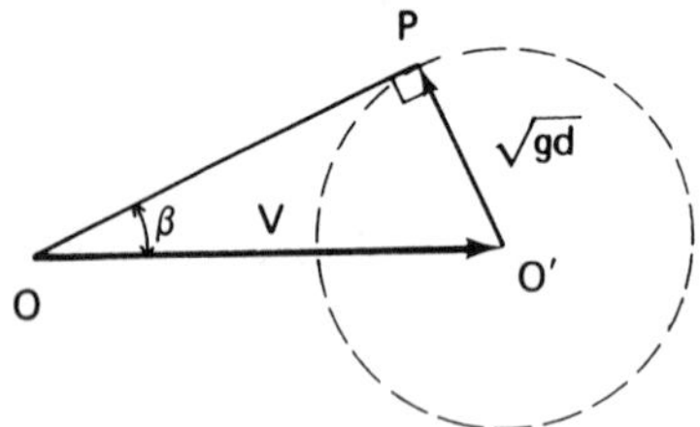

Figure 6.17 Wave angle in supercritical flow.

During a unit time period the center of disturbance O is displaced by a distance equal to the water velocity V (represented by O–O'); and the point P is carried away from the center by a radial distance equal to the wave speed $\sqrt{gD}$ (represented by O'–P), as shown in Figure 6.17. From the geometry, the magnitude of the wave angle can be determined as

$$\sin \beta = \frac{\sqrt{gD}}{V} = \frac{1}{\dfrac{V}{\sqrt{gD}}} = \frac{1}{N_F} \tag{6.35}$$

This relationship is analogous to the sound wave phenomenon in supersonic gas flows. The oblique standing wave front in supercritical open channel flow is analogous to the shock wave created at the leading front of a supersonic airplane.

Deflection of a channel wall creates a definite disturbance to the supercritical flow. The standing wave generated by a channel wall deflection may generally be classified into two categories

1. a positive deflection if the wall deflects into the flow,
2. a negative deflection if the wall deflects away from the flow.

The deflection wave front propagates across the channel, as shown in Figure 6.18. Water passing through a deflection wave front increases its water depth if

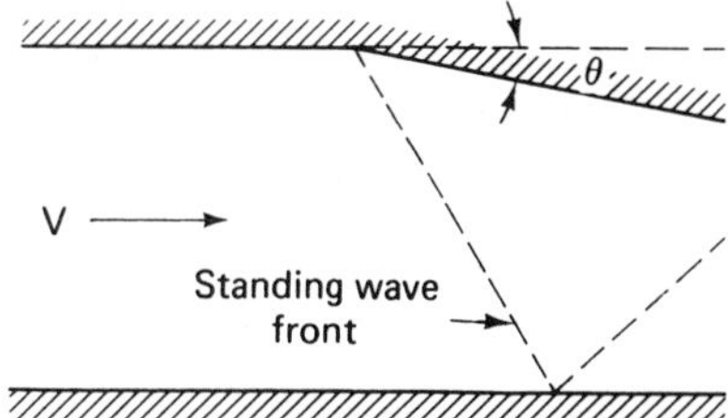

Figure 6.18 Deflection standing waves in supercritical channel.

the deflection is positive and decreases its water depth if the deflection is negative. Given the angle of wall deflection, θ, the wave angle β may be determined by considering the following factors.

Assuming flow depth d_1 and velocity V_1 for the approaching flow and d_2 and V_2 for that after the flow transverses the wave front, we may resolve the velocities V_1 and V_2 into components parallel and perpendicular to the wave front, as shown in Figure 6.19(a). The components parallel to the wave front must be equal to each other since no obvious shear should exist at the front. Figure 6.19(a) shows this relationship as

$$V_1 \cos \beta = V_2 \cos (\beta - \theta) \tag{6.36}$$

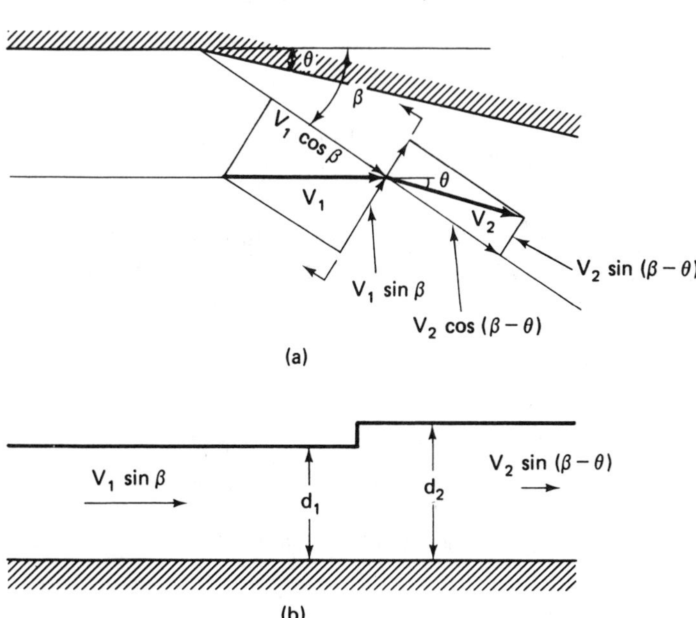

Figure 6.19 Description of a standing wave front at wall deflection: (a) top view, (b) section view.

According to the continuity equation, we may write another relationship between the upstream section and the downstream section of the wave front as

$$d_1 V_1 \sin \beta = d_2 V_2 \sin (\beta - \theta) \tag{6.37}$$

In the meantime, the momentum equation per unit width of the channel upstream and downstream from the wave front requires that

$$F_1 + \rho q(V_1 \sin \beta) = F_2 + \rho q V_2 \sin (\beta - \theta)$$

Substituting

$$F_1 = \frac{\gamma d_1^2}{2}, \qquad F_2 = \frac{\gamma d_2^2}{2}, \qquad q = d_1 V_1 \sin \beta = d_2 V_2 \sin (\beta - \theta)$$

into the above equation, we have

$$\frac{\gamma d_1^2}{2} + \rho d_1 V_1^2 \sin^2 \beta = \frac{\gamma d_2^2}{2} + \rho d_2 V_2^2 \sin^2 (\beta - \theta) \tag{6.38}$$

From Equation (6.37) we have

$$V_2 = \frac{d_1 V_1 \sin \beta}{d_2 \sin (\beta - \theta)}$$

This relationship may be substituted into Equation (6.38) and simplified for the solution of the wave angle β,

$$\sin \beta = \frac{1}{N_{F_1}} \sqrt{\left(\frac{1}{2}\right)\left(\frac{d_2}{d_1}\right)\left(1 + \frac{d_2}{d_1}\right)} \tag{6.39}$$

where $N_{F_1} = V_1/\sqrt{gD_1}$ is the Froude number of the approaching flow. The ratio d_2/d_1 is greater than unity for a positive deflection wave front and less than unity for a negative deflection wave front. For very small disturbances, the ratio d_2/d_1 is near unity and Equation (6.39) simplifies to Equation (6.35).

Combining Equations (6.36) and (6.37) the depth ratio may also be expressed

$$\frac{d_2}{d_1} = \frac{V_1 \sin \beta}{V_2 \sin (\beta - \theta)} = \frac{\tan \beta}{\tan (\beta - \theta)} \tag{6.40}$$

Substituting the above expression into Equation (6.39) and rearranging the terms, we have

$$\frac{\tan \beta}{\tan (\beta - \theta)} = \left(\frac{1}{2}\right)\left[-1 + \sqrt{1 + (8 \sin^2 \beta) N_{F_1}^2}\right] \tag{6.41}$$

The above transcendental equation expresses the relationship between the wall deflection angle θ and the wave angle β. It may be solved for β if the condition of the approaching flow, N_{F_1}, is given.

The general relations among the deflection angle, θ, the Froude number of the approaching flow, N_{F_1}, the depth ratio, d_2/d_1, and the wave angle, β, have been presented in the form of a set of charts by A. T. Ippen as shown in Figure 6.20.

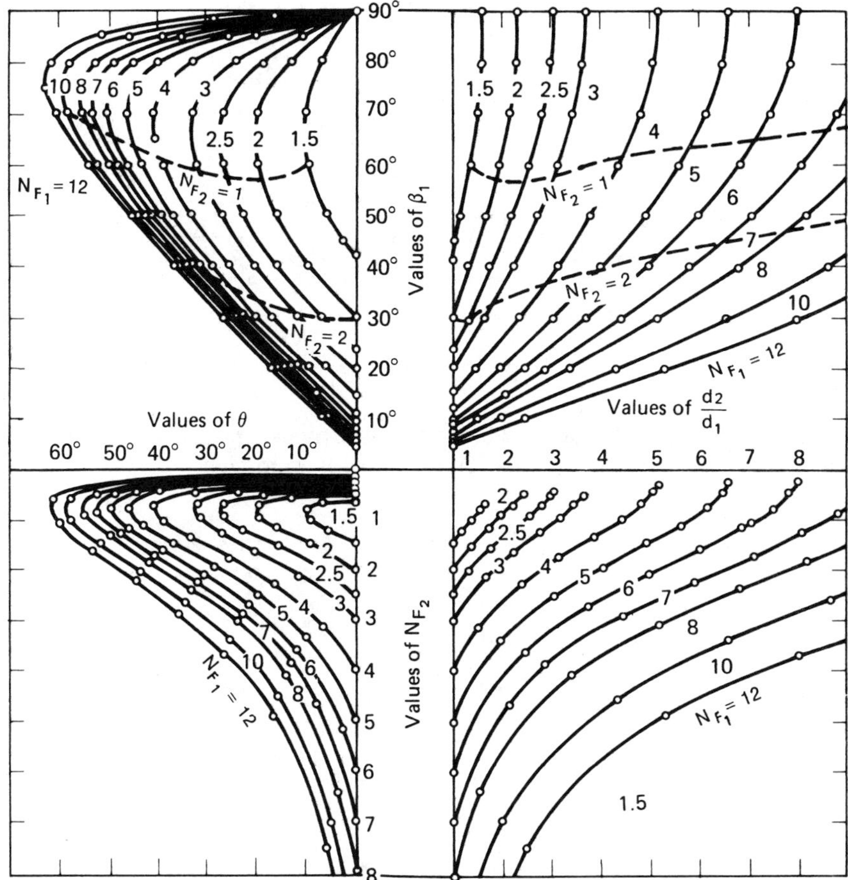

Figure 6.20 Solution chart for wave reflection angles and wave heights. (After A. T. Ippen, "Mechanics of supercritical flow, 1st paper of high-velocity flow in open channels: A symposium," *Trans. ASCE* 116 (1956), 678–94.)

Solution charts such as Figure 6.20 are useful in providing a qualitative assessment among the parameters involved and a first estimation of the problem. The accuracy in the solution would come from calculations using Equation (6.41).

Example 6.11

A rectangular channel 3.3 m wide carries a discharge of 4.95 m^3/sec. The flow encounters a positive wall deflection of $\theta = 5°$. The depth upstream from the wave is 0.5 m. Determine the standing wave angle and the water depth downstream of the wave front.

Solution

Substituting the given conditions $\theta = 5°$, $d_1 = 0.5$ m, and

$$V_1 = \frac{Q}{bd_1} = \frac{4.95}{3.3 \cdot 0.5} = 3 \text{ m/sec}$$

into Equation (6.41) gives

$$\frac{\tan \beta}{\tan (\beta - 5°)} = \frac{1}{2}\left[-1 + \sqrt{1 + 8(\sin^2 \beta)\left(\frac{3^2}{g \cdot 0.5}\right)}\right]$$

Solution may be obtained by trial. We find

$$\beta = 58°27'(58.45°)$$

The downstream depth can then be determined by using Equation (6.39). Multiplying both sides of the equation by N_{F_1} and squaring give

$$2N_{F_1}^2 \sin^2 \beta = 2\left(\frac{V_1^2}{gd_1}\right) \sin^2 \beta = \frac{d_2}{d_1}\left(1 + \frac{d_2}{d_1}\right) = 2.665$$

Hence,

$$\frac{d_2}{d_1} = \frac{-1 \pm \sqrt{1 + 10.66}}{2} = -0.5 \pm 1.707$$

Here only the positive value is taken since a negative water depth has no physical meaning. Therefore,

$$d_2 = 1.21d_1 = 0.605 \text{ m}$$

PROBLEMS

6.9.1 Repeat Example 6.11 if the initial depth and the wall deflection were, respectively, 0.45 m and 8°.

6.9.2 A wide rectangular channel discharging 3 m^3/sec per unit width encounters a positive deflection of 15°. If the wave angle is 45°, determine the upstream and downstream depths.

6.9.3 A 5-m rectangular channel that carries 15 m^3/sec encounters a wall deflection. The water depths upstream and downstream are measured, respectively, as 0.6 m and 0.7 m. Determine the wave angle and the angle of wall deflection.

6.9.4 Water flows in an open channel at 4 m/sec with a depth of 0.4 m. Determine the wave angle when the channel encounters a wall deflection of 10°.

6.10 WAVE INTERFERENCE AND DESIGN OF SUPERCRITICAL TRANSITIONS

Deflection wave fronts may be generated from more than one location in a supercritical flow structure. The standing wave fronts, positive, negative or both, meet in the flow field and form new regions of deflections. The complex surface phenomena caused by wave interference may be determined by applying the wave equations stated previously.

Figure 6.21 schematically illustrates the wave interference in a supercritical flow channel. Clearly, a standing wave pattern of this kind is not satisfactory for channel operation. With a properly chosen deflection angle, a transi-

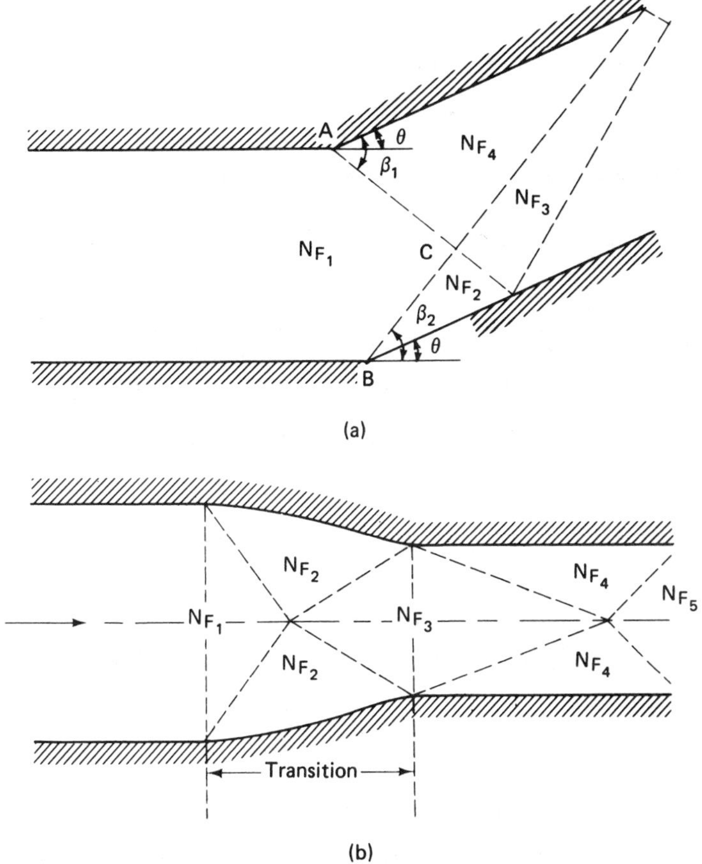

Figure 6.21 Wave interference in supercritical flow channel: (a) bend, (b) contraction.

tion may be designed to minimize the wave interference and produce a uniform flow condition downstream from the transition.

The most effective design of a *contraction in supercritical channel* is a straight-wall contraction, as shown in Figure 6.22. If the deflection angle, θ, and the transition length, L, are properly chosen, the wave front reflected from the first meeting point B of the two initial waves AB and CB will arrive at the side walls exactly at the end of the transition, E and E'. This requirement is met if the following geometric condition is satisfied:

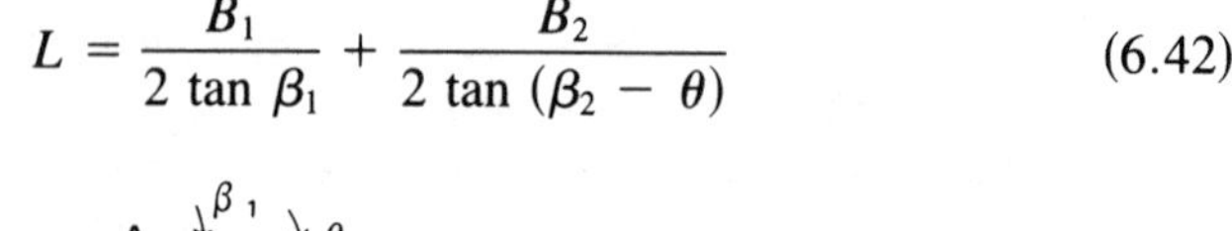

$$L = \frac{B_1}{2 \tan \beta_1} + \frac{B_2}{2 \tan (\beta_2 - \theta)} \tag{6.42}$$

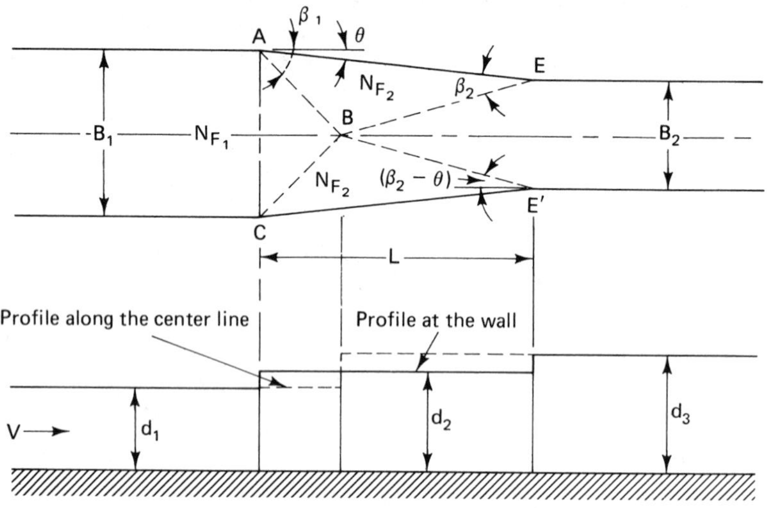

Figure 6.22 Design of a contraction transition in supercritical flow channel.

The wave angle β_1 and the transition Froude number N_{F_2}, are determined by the approaching flow conditions

$$N_{F_1} = \frac{V_1}{\sqrt{gd_1}}$$

and the angle of wall deflection, θ. The wave angle β_2 in turn depends on θ and N_{F_2}. On the other hand, the simple geometry of the transition also requires that

$$L = \frac{B_1 - B_2}{2 \tan \theta} \tag{6.43}$$

The proper value of θ may thus be determined by satisfying Equations (6.42) and (6.43) or by solving the value of θ from the following combined relationship:

$$\frac{B_1 - B_2}{2 \tan \theta} = \frac{B_1}{2 \tan \beta_1} + \frac{B_2}{2 \tan (\beta_2 - \theta)} \tag{6.44}$$

An improperly designed *expansion in a supercritical channel* usually encounters serious wave problems at the downstream end where the expansion terminates. The connection of the expansion to the walls of the exit channel sets up a positive wave front (Figure 6.23) and may become a serious problem in the downstream operation.

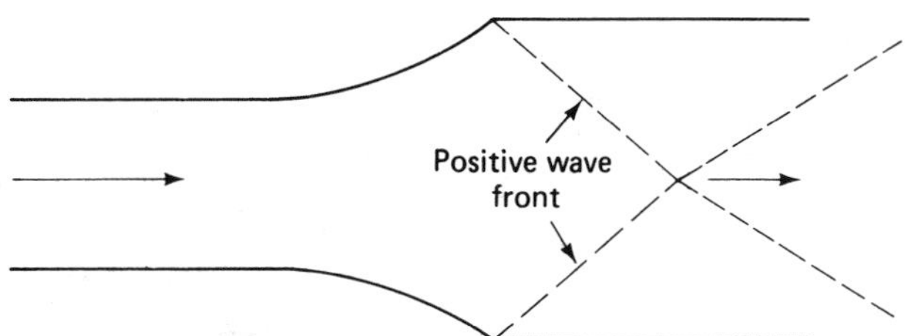

Figure 6.23 Positive waves in supercritical exit channel.

Although it is possible to design an expansion based strictly on the theory of wave mechanics to avoid the downstream wave interference, it has been found that the theoretical design usually requires an impractically long transition. Using a series of analytical and experimental studies as a basis, H. Rouse and his colleagues* developed a family of generalized expansion wall curves that give the geometry of the expansion walls covering a wide range of width ratios, b_1/b_2, for a variety of approaching flow conditions. This family of curves has been successfully used in the design of expansions that give an approximately rectilinear flow at the outlet and allow the downstream channel to be free from gross surface disturbances. The expansion curves are shown in Figure 6.24.

In a bend of a supercritical flow channel, abrupt angle changes such as the one shown in Figure 6.22 should be avoided in order to minimize the standing wave interference. A curved channel bend with a simple smooth curve causes superelevation of the water surface near the outer wall at the bend. The standing wave interference pattern, with alternating peaks and troughs, is superimposed on the elevated water surface in a rapid flow, as shown in Figure 6.25.

Two different methods may be used to eliminate the excessively high rise in water surface elevation at a supercritical channel bend. The first method is known as *water surface banking* and is accomplished by physically dropping the inner channel bottom elevation at the bend and leaving the outer bottom at the same level. In this arrangement, a transitional length must be provided both before and after the main curve in consideration of the secondary current developed in the bend. The recommended transitional length for a rectangular channel is

*H. Rouse, B. V. Bhoota, and E. Y. Hsu, "Design of Channel Expansions," *Trans. ASCE.*, 116 (1951), 347.

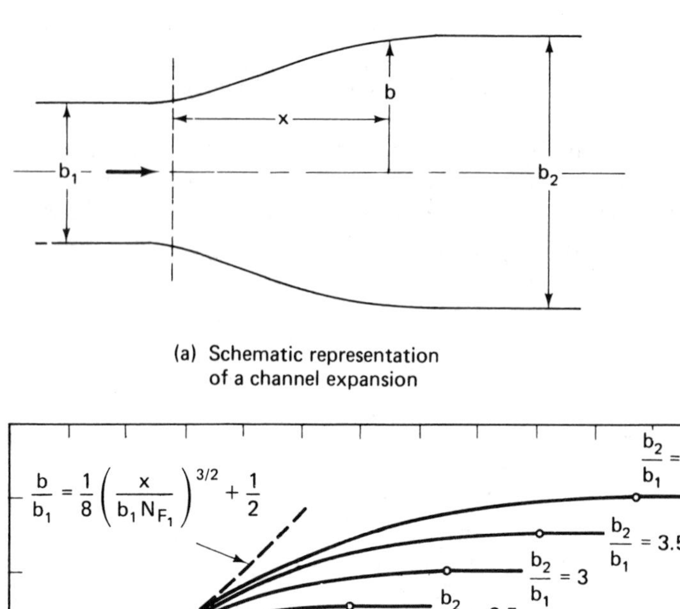

(a) Schematic representation of a channel expansion

(b) Expansion curves

Figure 6.24 General curves for supercritical channel expansion. (After Rouse, Bhoota, and Hsu, "Design of channel expansions," *Trans. ASCE*, 116 (1951), 347.)

$$L = 15\left(\frac{V^2B}{gR}\right) \tag{6.45}$$

where V is the mean velocity, B is the channel width, and R is the centerline radius at the bend.

The second method uses a compound curve for the bend, as shown in Figure 6.26. The compound curve usually includes a series of three curves, a main curve and two *transition curves*, each at one end of the main curve. With different radii of curvature and central angle from the main curve, the transition curves produce a disturbance wave pattern that offsets the wave pattern created by the main curve at a fixed phase angle. The geometry of the transition curves can be selected so that a maximum depth occurs just at the point where the disturbance due to the main curve would begin. With this arrangement, the

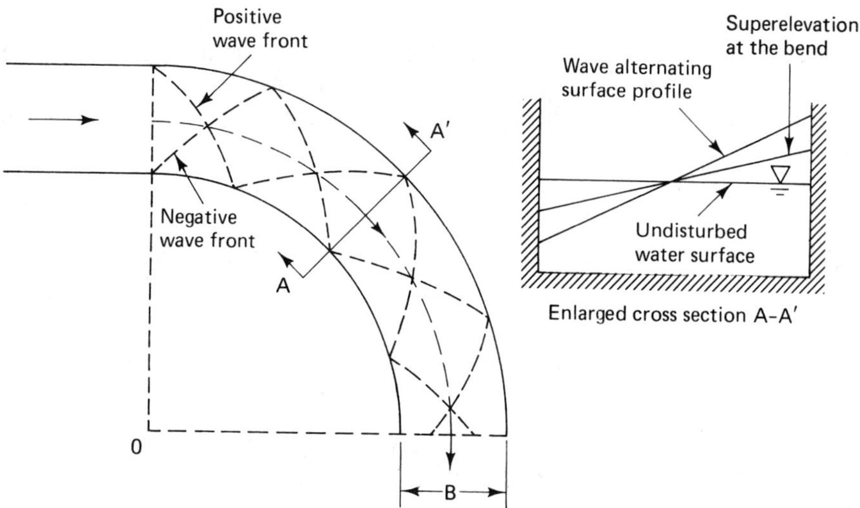

Figure 6.25 Supercritical flow at a channel bend.

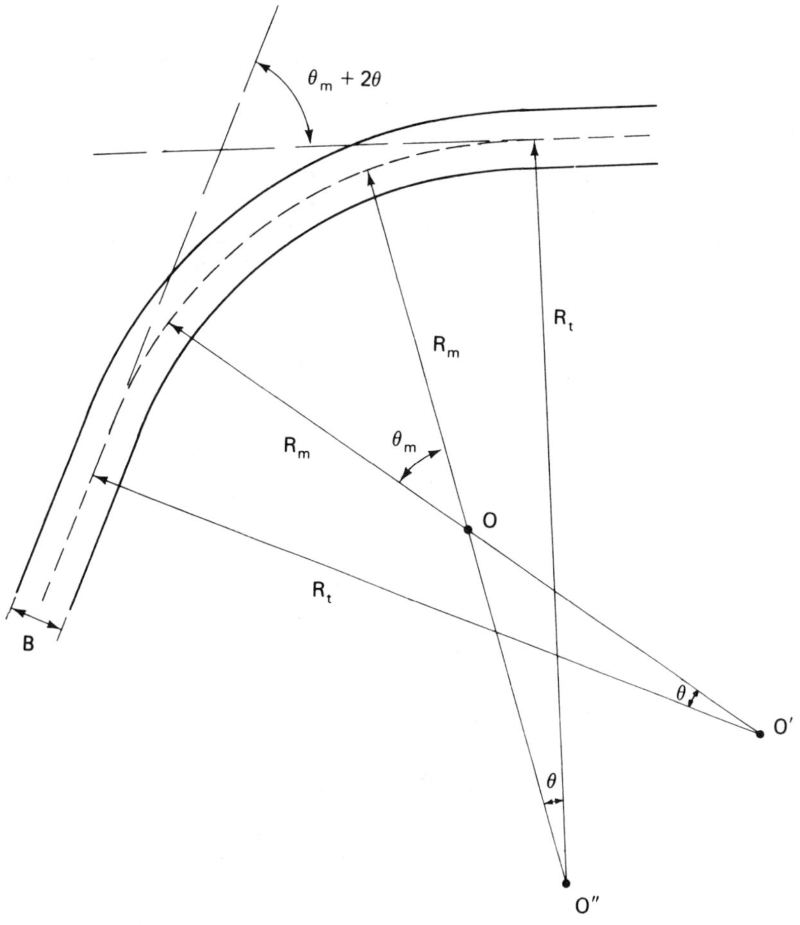

negative disturbance from the beginning of the inner transition curve reaches the outer wall just in time to cancel the positive disturbance that would be generated there, and, therefore, no disturbance is generated.

The transition curves should be made with radius equal to twice that of the main curve,

$$R_t = 2R_m \tag{6.46}$$

and the central angle of the transition curves may be approximated by

$$\theta = \tan^{-1}\left[\frac{\left(\dfrac{B}{\tan \beta_m}\right)}{R_t + 0.5B}\right] \tag{6.47}$$

where β_m is the wave angle of the main curve when subjected to the undisturbed flow conditions. The wave angle may be approximated by the following formula.

$$\beta_m = \sin^{-1}\left(\frac{\sqrt{gD}}{V}\right) \tag{6.48}$$

where D and V are the undisturbed water depth and velocity, respectively.

Example 6.12

A rectangular channel carrying a discharge of 20 m^3/sec at normal depth is required to contract from a width of 3.5 m to a width of 2.8 m. If the channel slope is 0.016 and Manning's coefficient is 0.011, design the most effective transition for the contraction so that the wave interference can be prevented downstream.

Solution

The depth of the approaching flow can be calculated using the Manning formula [Equation (6.5)], and,

$$d_1 = 0.757 \text{ m}$$

The velocity is

$$V_1 = \frac{Q}{b_1 d_1} = \frac{20}{3.5 \cdot 0.757} = 7.55 \text{ m/sec}$$

The corresponding Froude number is

$$N_{F_1} = \frac{V_1}{\sqrt{gd_1}} = 2.77$$

If a value is assumed for θ, the length of the transition can be computed by Equation (6.43)

$$L = \frac{B_1 - B_2}{2 \tan \theta}$$

and the wave angle by Equation (5.41)

$$\frac{\tan \beta_1}{\tan (\beta_1 - \theta)} = \frac{1}{2}[-1 + \sqrt{1 + 8(\sin \beta_1 N_{F_1})^2}]$$

Figure 6.20 gives the value for a first approximation. The velocity and the Froude number in the transition region can be determined through Equations (6.36) and (6.40)

$$V_2 = V_1 \cdot \frac{\cos \beta_1}{\cos (\beta_1 - \theta)}$$

and

$$d_2 = d_1 \frac{V_1}{V_2} \frac{\sin \beta_1}{\sin (\beta_1 - \theta)} = d_1 \frac{\tan \beta_1}{\tan (\beta_1 - \theta)}$$

so that

$$N_{F_2} = \frac{V_2}{\sqrt{g d_2}} = \frac{V_2}{\sqrt{g d_1 \dfrac{\tan \beta_1}{\tan (\beta_1 - \theta)}}}$$

The second wave angle is determined similarly.

$$\frac{\tan \beta_2}{\tan (\beta_2 - \theta)} = \frac{1}{2}\left[-1 + \sqrt{1 + 8(\sin \beta_2 N_{F_2})^2}\right]$$

Equation (6.42) is now used to check the length

$$L = \frac{B_1}{2 \tan \beta_1} + \frac{B_2}{2 \tan (\beta_2 - \theta)}$$

A new wall deflection, θ, is chosen and the procedure repeated until the length checks fairly well. Table 6.7 was developed to follow this procedure, where $A = \frac{1}{2}\left[-1 + \sqrt{1 + 8(\sin \beta N_F)^2}\right]$. The wall deflection should be 2.74° and the transition length is 7.31 m.

TABLE 6.7

θ	$L(m)$	β_1	$\frac{\tan \beta_1}{\tan (\beta_1 - \theta)}$	A_1	V_2	N_{F_2}	β_2	$\frac{\tan \beta_2}{\tan (\beta_2 - \theta)}$	A_2	$L(m)$
5°	4.00	25°	1.2812	1.2300						
		26°	1.2706	1.2892						
		25.75°	1.2731	1.2744	7.27	2.365	31.00°	1.2319	1.2932	
							30.00°	1.2381	1.2455	
							29.85°	1.2391	1.2382	6.65
2.7°	7.42	23°	1.1475	1.1108						
		23.5°	1.1447	1.1407						
		23.56°	1.1443	1.1442	7.406	2.541	27.65°	1.1261	1.1845	
							25.70°	1.1338	1.1366	
							25.65°	1.1340	1.1339	7.32
2.78°	7.21	23.64°	1.1487	1.1490	7.402	2.534	25.8°	1.1378	1.1379	7.29
2.74°	7.31	23.6°	1.1465	1.1466	7.404	2.537	25.73°	1.1359	1.1359	7.31

PROBLEMS

6.10.1 Design a proper expansion for the channel in Example 6.12 in which the width will change from 2.8 m to 5.6 m.

6.10.2 A rectangular channel, 4 m wide carries a discharge of 18 m^3/sec at a normal depth of 0.6 m. The channel must be contracted to a width of 3 m. Design a transition that will prevent wave interference in the downstream channel.

6.10.3 A 4-m-wide supercritical channel has a rectangular cross section and carries 40 m^3/sec discharge. The flow approaches a transition with the depth of 0.6 m. Design the transition to narrow the width of the channel by 50 percent. Determine the water velocity and the depth at the end of the transition.

7 HYDRAULICS OF WELLS AND SEEPAGE

Ground water occurs in permeable, water-bearing geologic formations known as *aquifers*. There are basically two types of aquifers:

1. The *confined aquifer* is a relatively high-permeable, water-bearing formation (e.g., sand or gravels) confined below a layer of very low permeability (e.g., clay).
2. The *unconfined aquifer* is a water-bearing formation with a free *water table* below which the soil is saturated and above which there is air under atmospheric pressure.

Movement of ground water occurs under hydraulic gradient or gravitational slope in the same way as water movement occurs in pipes or open channels.

Pressure gradient in an aquifer may occur naturally in courses (water table gradient) or it may occur as a result of artificial means (such as well pumps).

Figure 7.1 schematically shows several examples of ground water occurrence in both confined and unconfined aquifer formations.

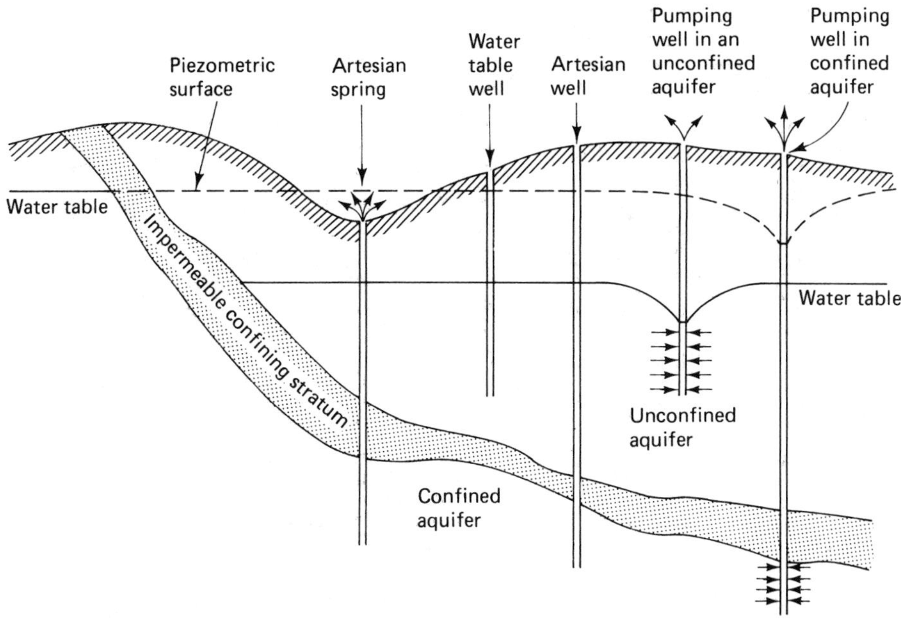

Figure 7.1 Ground water occurrence in confined and unconfined aquifers.

The pressure level (head) in a confined aquifer is represented by the piezometric surface that usually originates from a distant source such as the water table of a recharging area. An artesian spring is formed if the impermeable confining stratum is perforated at a location where the ground surface falls below the piezometric surface. It becomes an artesian well if the ground surface rises above the piezometric surface. The water table of an unconfined aquifer is usually unrelated to the piezometric surface of a confined aquifer in the same region, as shown in Figure 7.1. This is because the confined and unconfined aquifers are hydraulically separated by the impermeable confining stratum.

The capacity of a ground water reservoir in a particular geological formation strictly depends on the percentage of void space in the formation and on how the interstitial spaces are connected. Figure 7.2 schematically shows several types of rock formation and their relation to the interstitial space.

The percentage ratio of void space to the total volume of the formation is known as the *porosity* of the formation, defined as

$$\alpha = \frac{100\omega}{Vol} \tag{7.1}$$

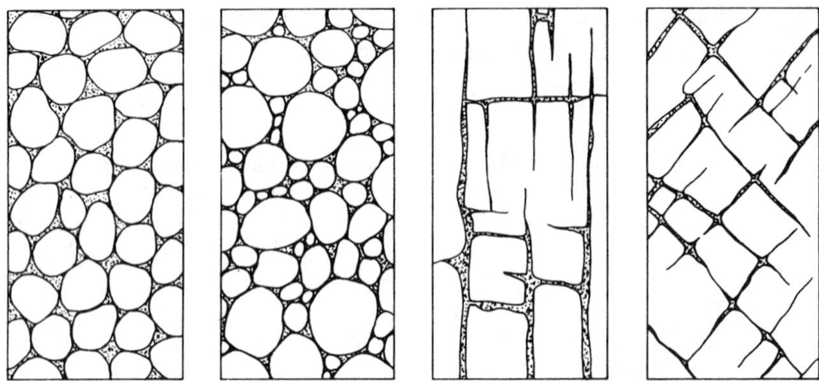

Figure 7.2 Examples of rock texture and interstices.

where ω is the volume of the void space, and *Vol* is the total volume of the formation. Table 7.1 lists the porosity range for common water-bearing formations.

TABLE 7.1 Porosity Range in Common Water-bearing Formations

Material	$\alpha(\%)$
Clay	45–55
Silt	40–50
Medium to coarse mixed sand	35–40
Fine to medium mixed sand	30–35
Uniform sand	30–40
Gravel	30–40
Gravel and sand mixed	20–35
Sandstone	10–20
Limestone	1–10

7.1 MOVEMENT OF GROUND WATER

Movement of ground water is proportional to the hydraulic gradient in the direction of the flow. The *apparent velocity* of ground water movement in a porous medium is governed by *Darcy's law**

$$V = K\frac{dh}{dL} \tag{7.2}$$

where dh/dL is the hydraulic gradient in the direction of the flow path (dL) and K is a proportionality constant known as the *coefficient of permeability*. The

* H. Darcy, *Les Fontaines Publiques de la Ville de Dijon* (Paris: V. Dalmont, 1856).

apparent velocity is defined by the quotient of discharge divided by the cross-sectional area of the medium through which it flows.

The coefficient of permeability depends not only on the medium of the formation but also on the properties of the fluid. By dimensional analysis, we may write

$$K = \frac{Cd^2\gamma}{\mu} \tag{7.3}$$

where Cd^2 is a property of the medium formation only, γ is the specific gravity, and μ is the viscosity of the fluid, respectively. The constant C represents the various properties of the medium formation that affect the flow other than d, which is a representative dimension proportional to the size of the interstitial space in the medium formation. The coefficient of permeability can be effectively determined by laboratory experiments or field tests applying Darcy's equation [Equation (7.2)], as will be discussed later.

Physically, the *apparent velocity* is the speed at which the bulk of ground water migrates between two stations ΔL apart in the porous medium.

$$V = \frac{\Delta L}{\Delta t}$$

This is not the actual speed that an individual water particle travels in the pores. The actual distance that water particles travel between any two stations ΔL apart in the porous medium cannot be a straight line; hence it must be longer than ΔL. In this chapter the movement of ground water is treated from a hydraulic engineer's point of view. We are only interested in the movement of water in quantity. The fluid mechanical aspect of water particle movement in pores will not be included in this discussion.

For an area A perpendicular to the direction of the water movement in an aquifer, the discharge may be expressed as

$$Q = AV = KA\frac{dh}{dL} \tag{7.4}$$

Laboratory measurement of the coefficient of permeability can be demonstrated by the following example.

Example 7.1

A small sample of an aquifer is packed in a test cylinder (see Figure 7.3) to form a column 30 cm long and 4 cm in diameter. At the outlet of the test cylinder, 21.3 cm^3 of water is collected in 2 min. During the testing period a constant

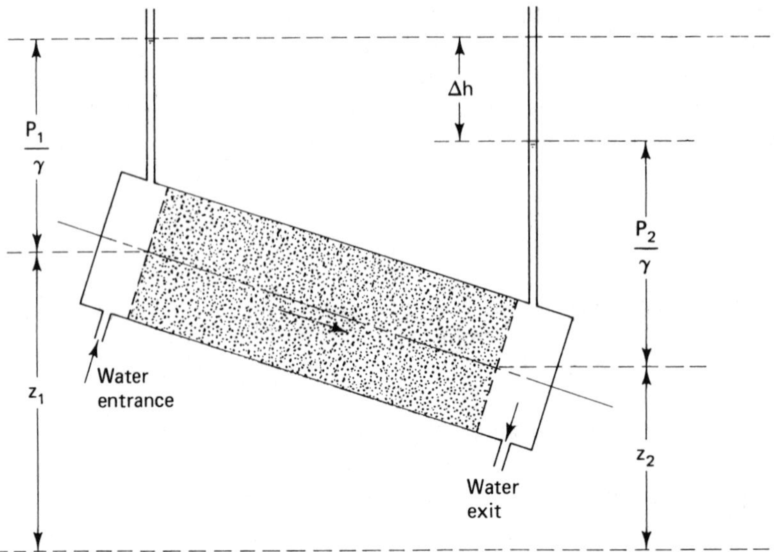

Figure 7.3 Laboratory determination of coefficient of permeability.

piezometric head difference of $\Delta h = 14.1$ cm is observed. Determine the coefficient of permeability of the aquifer sample.

Solution

The cross-sectional area of the sample volume is

$$A = \frac{\pi}{4}(4)^2 = 12.56 \text{ cm}^2$$

The hydraulic gradient is equal to the change of piezometric head per unit length of the aquifer measured in the direction of the flow

$$\frac{dh}{dL} = \frac{14.1}{30} = 0.47$$

The discharge rate

$$Q = 21.3 \text{ cm}^3/2 \text{ min} = 10.65 \text{ cm}^3/\text{min}$$

Applying Darcy's law [Equation (7.4)], we may calculate

$$K = \frac{Q}{A}\frac{1}{\left(\dfrac{dh}{dL}\right)} = \frac{10.65}{12.56} \cdot \frac{1}{0.47} = 1.80 \text{ cm}^3/\text{min}/\text{cm}^2$$

Based on Darcy's law, several different types of permeameters have been developed for laboratory measurements of permeability in small samples of aquifers.* Although laboratory testing is conducted under controlled conditions, such measurements may not represent the field permeability. When unconsolidated samples are taken from the field and repacked in laboratory permeameters, the texture, porosity, grain orientation, and packing may be markedly disturbed and changed; consequently, permeabilities are modified. For better reliability, field permeability of aquifer can be determined by well pumping tests in the field. This method will be discussed in Section 7.3

PROBLEMS

7.1.1 In a laboratory experiment a sand sample is packed into a cylindrical sample space 4 cm in diameter and 20 cm long. Under a steady head of 24 cm, 100 cm^3 of water is collected in 5 min. What is the apparent velocity of water flowing through the sample? Determine the coefficient of permeability of the sample.

7.1.2 The experiment in Problem 7.1.1 was done at room temperature (20°C). Apply Equation (7.3) and determine the discharge in 5 min if the same experiment had been performed at a temperature of 5°C. What is the permeability coefficient of the sample at this temperature?

7.1.3 The sand filter shown in Figure P7.1.3 is 1 m deep and has a surface area of 4 m^2. If the permeability is 0.65 cm/sec, determine the discharge through the filter for a differential head of 0.8 m.

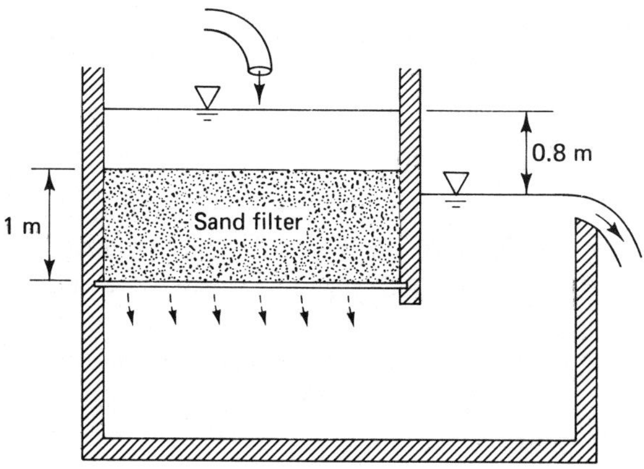

Figure P7.1.3

*L. K. Wenzel, *Methods for Determining Permeability of Water-bearing Materials with Special Reference to Discharging-Well Methods,* U.S.G.S. Water Supply Paper #887 (1942).

7.2 RADIAL FLOW TO A WELL

Removal of water from the aquifer surrounding a well by pumping results in radial flow to the well, because the pumping action lowers the water table (or the piezometric surface) at the well and forms a region of pressure depression that surrounds the well. At any given distance from the well the *drawdown* of the water table (or piezometric surface) is defined by the vertical distance measured from the original to the lowered water table (or piezometric surface). Figure 7.4(a) shows the drawdown curve of the water table in an unconfined aquifer; Figure 7.4(b) shows the drawdown curve of the piezometric surface in a confined aquifer.

In a uniform aquifer formation the axisymmetric drawdown curve describes a conic-shaped geometry commonly known as the *cone of depression*. The outer limit of the cone of depression defines the *area of influence* of the well.

Darcy's law may be directly applied to derive the *radial flow equation* that relates the discharge to the drawdown of the water table at a well in the unconfined aquifer after a steady state of equilibrium is reached. By using plane polar coordinates with the well as the origin, the discharge to the well at any distance r from the well may be expressed as

$$Q = AV = 2\pi rh\left(K\frac{dh}{dr}\right) \tag{7.5}$$

Integrating between the limits of the well (where $r = r_w$, $h = h_w$) and the undisturbed water tabl‐ (where $r = r_0$, $h = h_0$) yields

$$Q = \pi K\frac{h_0^2 - h_w^2}{\left(\ln\frac{r_0}{r_w}\right)} \tag{7.6}$$

The selection of the radius of influence, r_0, is arbitrary. The variation of Q is rather small for the wide range of r_0, because the influence on the well by the water table at far distances is small. In practice, approximate values of r_0 may be taken between 100 and 500 m, depending on the nature of the aquifer and the operation of the pump.

In confined aquifers, water delivery to a well is confined to the depth of well penetration into the aquifer, b. Applying Darcy's law, we find that the discharge flowing through the cylindrical boundary with radius r from the center of the well equals

$$Q = AV = 2\pi rb\left(K\frac{dh}{dr}\right) \tag{7.7}$$

Integrating between the boundary conditions at the well ($r = r_w$, $h = h_w$) and

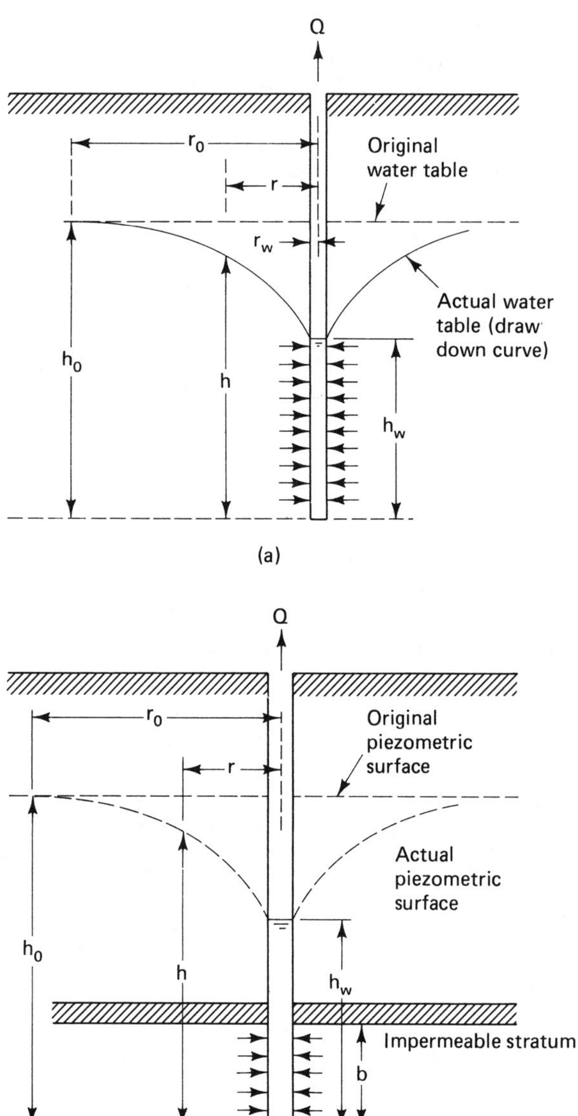

Figure 7.4 Radial flow to a pumping well from (a) an unconfined aquifer, and (b) a confined aquifer.

at the radius of influence ($r = r_0$, $h = h_0$) yields

$$Q = 2\pi Kb \frac{h_0 - h_w}{\ln\left(\dfrac{r_0}{r_w}\right)} \tag{7.8}$$

The above equations apply only to steady flow where the discharge remains a constant value at any distance from the well. A more general equation for the discharge may be written for any distance, r

$$Q = 2\pi Kb \frac{h - h_w}{\ln\left(\frac{r}{r_w}\right)} \tag{7.9}$$

Eliminating Q between Equations (7.8) and (7.9) results in

$$h - h_w = (h_0 - h_w)\frac{\ln\left(\frac{r}{r_w}\right)}{\ln\left(\frac{r_0}{r_w}\right)} \tag{7.10}$$

which shows that the piezometric head varies linearly with the logarithm of the distance from the well, regardless of the rate of discharge.

7.3 FIELD DETERMINATION OF PERMEABILITY COEFFICIENT

Laboratory tests of soil permeability (Example 7.1) are performed on small samples of soil. Their value in solving engineering problems depends on how well they represent the masses of materials in the field. When used with consideration of field conditions and careful handling of the samples, the laboratory test methods can be very valuable. Nevertheless, important hydraulic engineering projects often require field tests that measure the undisturbed soil permeabilities in situ.

EQUILIBRIUM TEST

The aquifer permeability coefficient, K, can be effectively measured in the field by well pumping tests.

In addition to the pumped well, the pumping test requires two observation wells to penetrate the aquifer. The observation wells are located at two arbitrary distances r_1 and r_2 from the pumped well, as schematically represented in Figure 7.5. After pumping the well at a constant discharge Q for a long period, the water levels in the observation wells, h_1 and h_2, will reach final equilibrium values. The equilibrium water levels in the observation wells are measured to calculate the aquifer's permeability coefficient.

For unconfined aquifers, the coefficient can be calculated by integrating Equation (7.6) between the limits of the two observation wells

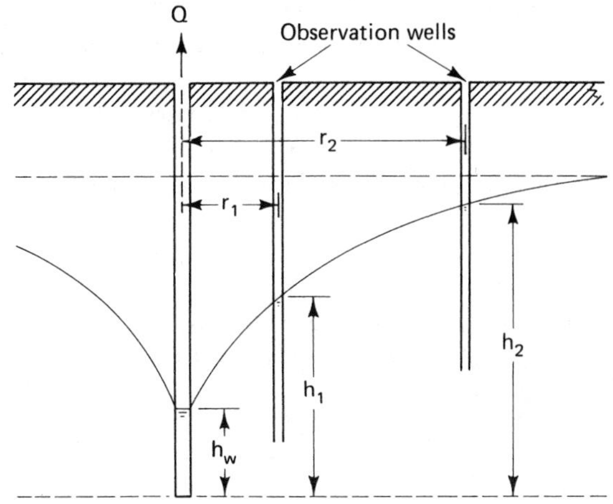

Figure 7.5 Field determination of permeability coefficient in unconfined aquifers.

$$K = \frac{Q}{\pi(h_2^2 - h_1^2)} \ln\left(\frac{r_2}{r_1}\right) \tag{7.11}$$

For confined aquifers, the coefficient can be calculated by rearranging Equation (7.8) as follows:

$$K = \frac{Q}{2\pi b(h_2 - h_1)} \ln\left(\frac{r_2}{r_1}\right) \tag{7.12}$$

where b is the thickness of the confined aquifer, as shown in Figure 7.6.

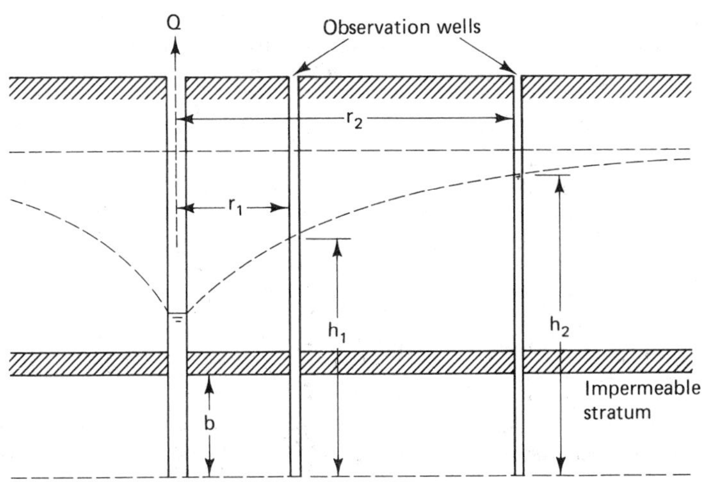

Figure 7.6 Field determination of permeability coefficient in confined aquifers.

The derivation of Equations (7.11) and (7.12) assumes that the aquifer is homogeneous and isotropic and that it extends over an area larger than the area that the radius of influence can reach. In practice, however, the equations have been widely used to provide good estimations of permeability coefficients for a variety of aquifer conditions, in spite of these restrictive conditions.

Table 7.2 gives an indication of the numerical values for the coefficient of permeability for some natural soil formations.

TABLE 7.2 Magnitude of Coefficient of Permeability of Some Natural Soil Formations

Soils	*K(m/sec)*
Clays	$< 10^{-9}$
Sandy clays	$10^{-9} - 10^{-8}$
Peat	$10^{-9} - 10^{-7}$
Silt	$10^{-8} - 10^{-7}$
Very fine sands	$10^{-6} - 10^{-5}$
Fine sands	$10^{-5} - 10^{-4}$
Coarse sands	$10^{-4} - 10^{-2}$
Sand with gravels	$10^{-3} - 10^{-2}$
Gravels	$> 10^{-2}$

Example 7.2

A well 20 cm in diameter penetrates 30 m deep into the undisturbed water table of an unconfined aquifer. After a long period of pumping at the constant rate of 0.1 m^3/sec, the drawdown at distances of 20 m and 50 m from the well are observed to be 4 m and 2.5 m, respectively. Determine the coefficient of permeability of the aquifer. What is the drawdown at the pumped well?

Solution

Conditions given are $Q = 0.1$ m^3/sec, $r_1 = 20$ m, $r_2 = 50$ m; hence, $h_1 =$ 30.0 m − 4 m = 26 m, and $h_2 =$ 30.0 m − 2.5 m = 27.5 m, in reference to Figures 7.4 and 7.5. Substituting these values into Equation (7.11), we have

$$K = \frac{0.1}{\pi(27.5^2 - 26^2)} \ln\left(\frac{50}{20}\right) = \frac{0.00114}{\pi} = 0.000363 \text{ m/sec}$$

The drawdown at the pumped well can be calculated by using the same equation with the calculated value of the coefficient of permeability and the well diameter.

At the well

$$r = r_0 = \frac{d_0}{2} = \frac{0.2}{2} = 0.1 \text{ m}$$

and

$$h = h_0$$

Therefore,

$$0.000363 = \frac{Q}{\pi(h_1^2 - h_0^2)} \ln\left(\frac{r_1}{r_0}\right) = \frac{0.1}{\pi(26^2 - h_0^2)} \ln\left(\frac{20}{0.1}\right)$$

From which we have

$$h_0 = 14.53 \text{ m}$$

The drawdown at the well is (30 − 14.53) = 15.47 m.

PROBLEMS

7.3.1 A confined aquifer of uniform thickness (18 m) is completely penetrated by a pumping well. After a long period of pumping at the constant rate of 0.3 m^3/sec, the water elevations in the observation wells (r_1 = 20 m, r_2 = 65 m) are stabilized. The drawdown measured at the observation wells are, respectively, 16.25 m and 3.42 m. Determine the coefficient of permeability.

7.3.2 A sample of an aquifer is taken from a circular island in a lake to a laboratory where the coefficient of permeability is determined to be 0.000142 m/sec. The island (see Figure P7.3.2) is approximately 800 m in diameter, and a 30-cm-diameter well is installed at the center of the island. Determine the minimum depth that the well must penetrate below the surface elevation of the lake in order to yield a constant discharge of 0.2 m^3/sec.

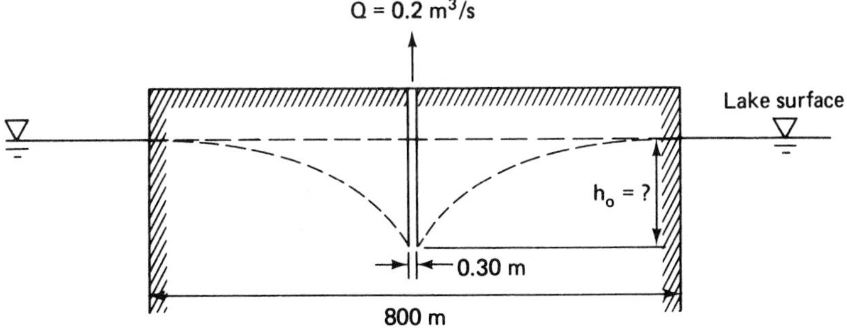

Figure P7.3.2

7.3.3 Obtain the free surface of the water table for the well in Figure P7.3.3. If the well yields 1.5 m^3/hr, what is the permeability coefficient?

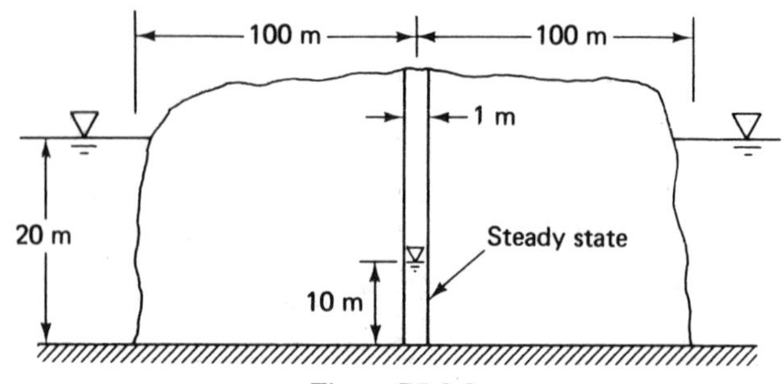

Figure P7.3.3

7.3.4 If water is taken from the well in Problem 7.3.3 at a rate of 1.85 m^3/hr, what would be the water surface elevation in the well as measured with respect to the impervious floor? Assume $K = 0.0087$ m/hr.

NONEQUILIBRIUM TEST

Field determination of permeability K using the equilibrium theory assumed that a new well must be pumped at a constant flow rate until an equilibrium condition is established. In reality, however, many years may be required for true equilibrium to develop in many confined aquifers. It is usually not possible to reach even approximate equilibrium when a new well is pumped for a reasonable time.

C.V. Theis* first presented an analysis of well flow that took into consideration the effect of time and the storage characteristic of the confined aquifer. The Theis equation is

$$h_0 - h = \frac{Q}{4\pi T} \int_u^\infty \frac{e^{-u}}{u} \, du \tag{7.13}$$

where $h_0 - h$ is the drawdown in an observation well at a distance r from the pumping well and u is a dimensionless parameter given by

$$u = \frac{r^2 S_c}{4Tt} \tag{7.14}$$

In Equation (7.14) t is the time since the pumping began, and S_c is the *storage constant* of the aquifer, a dimensionless number defined as the water yield from a column of aquifer of unit area by lowering the water table or piezometric surface by one unit height. T is the *transmissibility* of the aquifer, defined as

*C.V. Theis, "The relation between the lowering of the piezometric surface and the rate and duration of discharge of a well using ground water storage," *Trans. Am. Geophys. Union,* 16, (1935), 519–524.

$$T = \text{(aquifer thickness)(permeability)} = bK$$

The integral in Equation (7.13) is commonly written as $W(u)$, known as the well function of u. It is not directly integratable but may be evaluated by the series

$$W(u) = -0.5772 - \ln u + u - \frac{u^2}{2 \cdot 2!} + \frac{u^3}{3 \cdot 3!} \cdots \quad (7.15)$$

For a reasonably large value of t, and a small value of r, u can be made small enough that the terms following $\ln u$ become small and may be neglected. When $u < 0.01$, the Theis equation may be modified to*

$$s = h_0 - h = \frac{Q}{4\pi T}\left[-0.5772 - \ln \frac{r^2 S_c}{4Tt}\right]$$

$$= \frac{-2.30Q}{4\pi T} \log_{10} \frac{0.445 r^2 S_c}{Tt} \quad (7.16)$$

The observed drawdown, $h_0 - h$, may be plotted against the pumping time t on a semilogarithmic scale as shown in Figure 7.7.

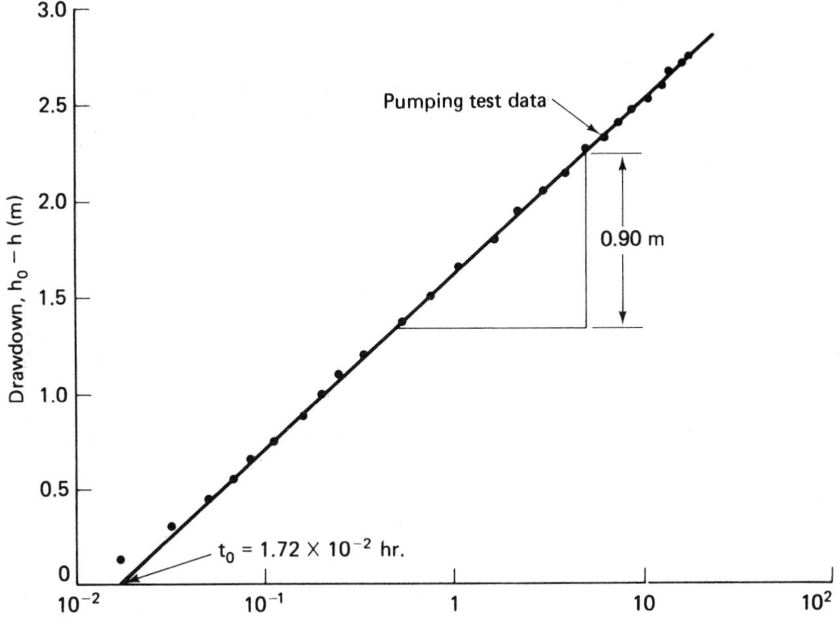

Figure 7.7 Jacob's method for solution of the Theis equation, Equation (7.13).

*C.E. Jacob, "Drawdown test to determine effective radius of artesian well," *Trans. ASCE*, 112 (1947), 1047–70.

Taking from the plotted data the value of $\Delta(h_0 - h)$ as the change in drawdown occurring over one log cycle of time, such as from 10 hr to 100 hr after pumping begins, the value of T can be calculated as follows:

$$\Delta(h_0 - h) = (h_0 - h_1) - (h_0 - h_2) = (h_2 - h_1)$$
$$= \frac{-2.30Q}{4\pi T}\left[\log_{10}\left(\frac{0.455r^2S_c}{Tt_1}\right) - \log_{10}\left(\frac{0.445r^2S_c}{Tt_2}\right)\right]$$

or

$$T = \frac{2.30Q}{4\pi(h_1 - h_2)}\log_{10}\left(\frac{t_2}{t_1}\right) \tag{7.17}$$

In the semilogarithmic scale, Equation (7.16) is represented by a straight line that intersects the horizontal axis at a point designated as t_0. At this point, $(h_0 - h) = 0$ and we may write

$$\log_{10}\frac{0.445r^2S_c}{Tt_0} = 0$$

or,

$$\frac{0.445r^2S_c}{Tt_0} = 1$$

Hence,

$$S_c = \frac{2.246Tt_0}{r^2} \tag{7.18}$$

Example 7.3

Using the Jacob method described above, determine the transmissibility, T, the permeability K, and the storage constant S_c of the aquifer based on the data in Figure 7.7. The well pumps at a constant rate of 8.5 m^3/hr, in an aquifer estimated to be 40 m thick, and the observation well is located 20 m from the pumping well.

Solution

The drawdown observed during one cycle of time, $t_1 = 0.5$ hr and $t_2 = 5.0$ hr, is 0.9 m. Applying Equation (7.17), the transmissibility T is

$$T = \frac{2.3\cdot 8.5}{4\pi\cdot 0.9}\cdot\log_{10}(5.0/0.5) = 1.73 \text{ m}^3/\text{hr/m}$$

The permeability K is

$$K = T/b = 1.73/40 = 0.043 \text{ m/hr}$$

The storage constant S_c can be estimated by using Equation (7.18).

$$S_c = \frac{2.246 \cdot 1.73 \cdot 1.72 \cdot 10^{-2}}{(20)^2} = 1.67 \cdot 10^{-4}$$

7.4 AQUIFER BOUNDARIES

When a well is located close to an aquifer boundary, the shape of the drawdown curve may be significantly modified, which, in turn, affects the discharge rate as predicted by the radial flow equations.

The solution of aquifer boundary problems can often be simplified by applying the *method of images*. A hydraulic image is an imaginary source or sink, with the same strength as the original well, placed on the opposite side of the boundary to represent the effect of the boundary. Figure 7.8 shows the effect of an impermeable boundary on a well located at a close distance from it. Figure 7.9 shows the application of the method of images by placing an imaginary well of the same strength, Q, at the same distance on the opposite side of the boundary. The original boundary conditions are thus hydraulically replaced by the two-well system in a hypothetically uniform aquifer of infinite extent. The two wells pumping at the same discharge rate hydraulically affect each other in such a way that the drawdown curves are linearly superimposed. The imaginary well and the real well offset one another at the boundary line causing a zero hydraulic gradient ($dh/dr = 0$); as a result, there is no flow across the boundary.

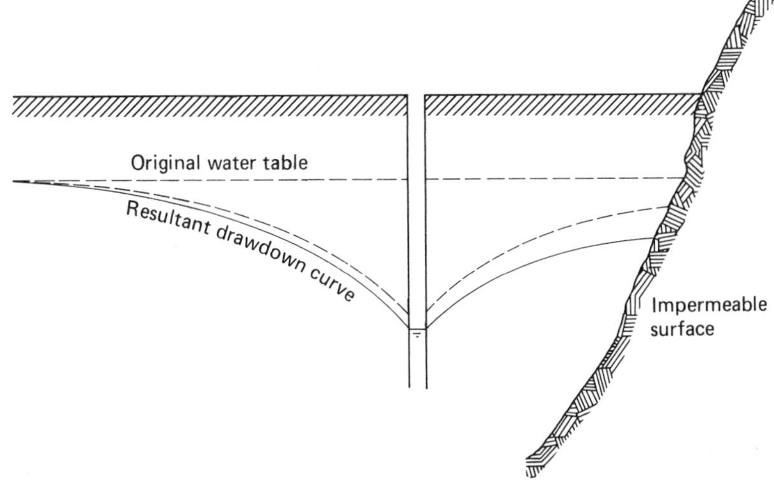

Figure 7.8 Pumping well near an impermeable boundary.

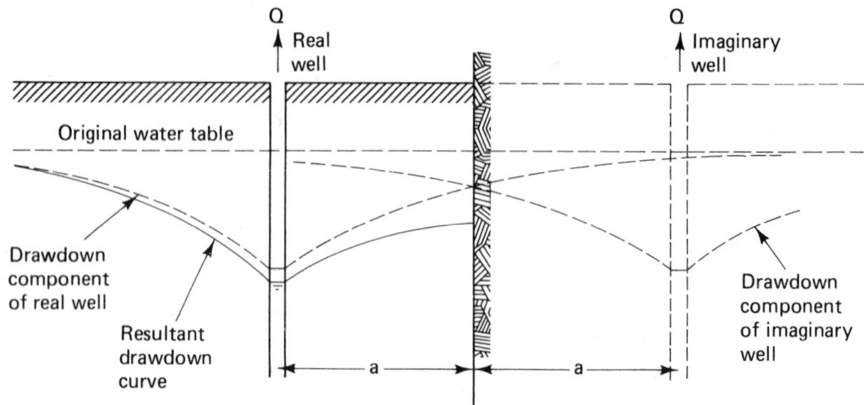

Figure 7.9 Equivalent hydraulic system with imaginary well.

The presence of a lake, river, or other large body of water in the vicinity of a well increases flow to the well. The effect of such a water body on the drawdown is exactly opposite to that of an impermeable boundary. Instead of an imaginary pumping well, the equivalent hydraulic system involves an imaginary *recharge well* placed at an equal distance across the boundary. The recharge well infuses water at a discharge rate, Q, into the aquifer under positive pressure. The resulting drawdown curve is obtained by linearly superimposing the drawdown component of the real well and the drawdown component of the imaginary recharge well that replaces the water boundary, as shown in Figures 7.10 and 7.11. The resulting drawdown curve of the real well intersects the boundary line at the elevation of the free water surface. The steeper hydraulic gradient causes more water to flow across the boundary line.

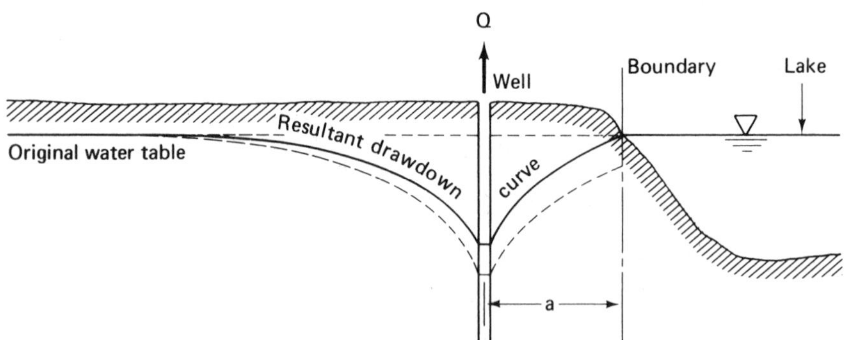

Figure 7.10 Pumping well near a perennial water body.

Replacing the actual boundaries by an equivalent hydraulic system with imaginary wells can be applied to a variety of ground water boundary conditions. Figure 7.12(a) shows a discharge well pumping water from an aquifer bounded on two sides by impermeable boundaries. A system of three imaginary wells are

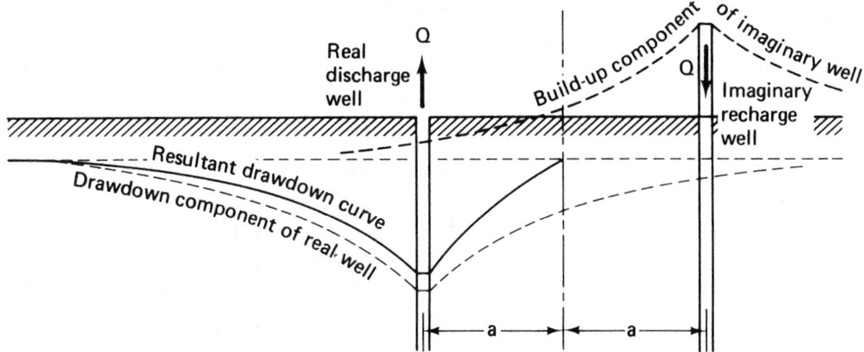

Figure 7.11 Equivalent hydraulic system with imaginary recharge well.

required to provide the equivalent flow. The imaginary discharge wells, I_1 and I_2, provide the required absence of flow across the boundaries; a third imaginary well, I_3, is necessary to balance the system. The three imaginary wells all have the same discharge rate (Q) as the real well. All wells are placed at equal distances from the physical boundaries.

Figure 7.12(b) represents the situation of a discharge well pumping water from an aquifer bounded on one side by an impermeable boundary and on the other side by a perennial stream. The equivalent hydraulic system includes three imaginary wells of equal strength Q. I_1 is a recharge well and I_2 is a discharge well. A third imaginary well, I_3, a recharge well like I_1, is necessary to balance the system.

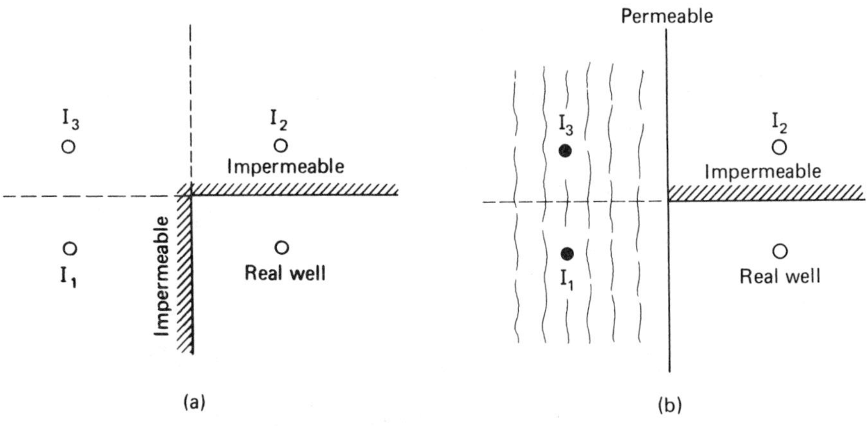

Figure 7.12 Discharging well near multiple boundaries.

Example 7.4

A factory located near a river bank wants to extract a discharge of 0.04 m^3/sec from a confined aquifer (K = 0.00012 m/sec and b = 20 m). Local authorities require that at a distance of 30 m from the bank, the ground water table may not be lower than 0.1 m from the surface elevation in the river. Determine the minimum distance from the bank that the well should be located.

Solution

As shown in Figures 7.10 and 7.13 a pumping well located near a perennial water body, such as a river, may be hydraulically replaced by an imaginary recharging well of the same strength on the opposite side and at the same distance from the boundary. The resultant drawdown curve may be obtained by superimposing the real and the imaginary wells and assuming an infinite extent of the aquifer without the boundary.

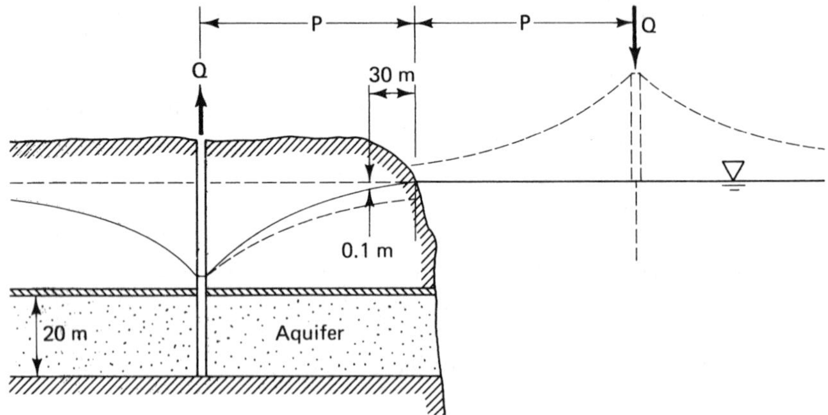

Figure 7.13

Assume P to be the distance between the pumping well and the river bank. Then the drawdown (at 30 m from the bank or $P - 30$ m from the pumping well and $P + 30$ m from the discharge well) is the sum of the piezometric surfaces at the point from the pumping well and the recharging well.

Rearranging Equation (7.12), we have

$$h_2 - h_1 = s_1 - s_2 = \frac{Q}{2\pi bK} \ln\left(\frac{r_2}{r_1}\right)$$

$$(s_1 - s_2)_d = \frac{Q_{disch}}{2\pi bK} \ln\left(\frac{P}{P - 30}\right) \quad \text{(discharge well)}$$

$$(s_1 - s_2)_r = \frac{Q_{rech}}{2\pi bK} \ln\left(\frac{P + 30}{P}\right) \quad \text{(recharge well)}$$

Superimposing the two wells, as $s_{2_d} = s_{1_r}$, we have

$$s_{1_d} - s_{2_r} = \frac{Q}{2\pi bK}\left(\ln\frac{P}{P - 30} + \ln\frac{P + 30}{P}\right) = \frac{Q}{2\pi bK} \ln\left(\frac{P + 30}{P - 30}\right)$$

$$0.1 = \frac{0.04}{2\pi(20)(0.00012)} \ln\left(\frac{P + 30}{P - 30}\right)$$

and

$$\ln\left(\frac{P + 30}{P - 30}\right) = 0.0377$$

or

$$P + 30 = 1.038(P - 30)$$

Finally,

$$P = 1592 \text{ m}$$

PROBLEMS

7.4.1 A 30-cm diameter well penetrates the thickness of a 40-m thick aquifer that has a coefficient of permeability of 0.6 m/hr. The well is pumped at a constant rate of 0.10 m^3/sec. Determine the profile of the drawdown curve. If two such wells of the same strength are placed at the distance of 20 m apart, what would be the drawdown curve?

7.4.2 If the well in Problem 7.4.1 were placed 20 m from a straight, impermeable aquifer boundary, determine the profile of the drawdown curve.

7.4.3 A 50-m thick aquifer, with permeability coefficient of 0.5 m/hr lies between an impermeable wall and a freshwater lake as shown in Figure P7.4.3. Determine the profile of the drawdown curve if the well is pumped at a constant rate of 0.1 m^3/sec, and the well radius is 20 cm.

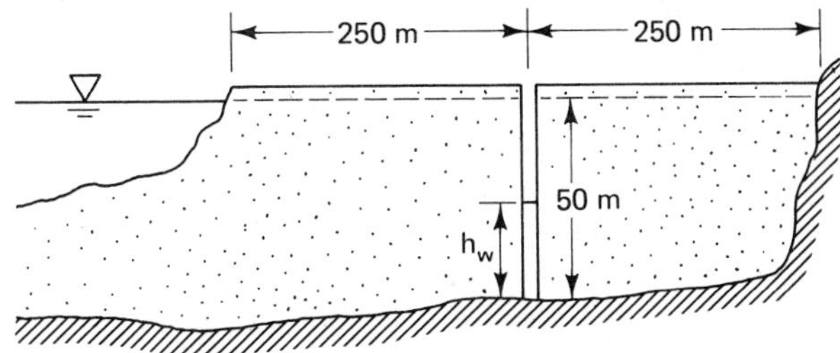

Figure P7.4.3

7.4.4 Determine the discharge flowrate from the well in Figure P7.4.3 if the equilibrium h_w is 30 m.

7.4.5 Determine the coefficient of permeability of the aquifer in Figure P7.4.3 if the well discharges 0.12 m^3/sec at equilibrium depth h_w of 28 m.

7.5 INVESTIGATION OF GROUND WATER

Locating subsurface ground water by information obtained on the earth's surface is an ancient art known as *divination*. At the turn of this century geophysical methods had been developed for petroleum and mineral explorations. A few of these methods have proved useful for locating and analyzing ground water. Information obtained by surface methods can only provide indirect indication of ground water. Correct interpretation of the data usually requires supplemental information that can only be obtained by subsurface investigations. Two of the most commonly used geophysical methods are described below.

The Electrical Resistivity Method Electrical resistivity of rock formations varies over a wide range. The measured resistivity of a particular formation depends on a variety of physical and chemical factors, such as the material and the structure of the formation, the size, shape, and distribution of pores, and the water content. The distinguishable difference between a rock formation alone and the same formation with large amounts of water filling the interstitial spaces is the key in detecting ground water.

The procedure involves measuring the electrical potential difference between two electrodes placed in the ground surface. When an electric current is applied through two other electrodes outside but along the same line with the potential electrodes, an electrical field penetrates the ground and forms a current flow network, as shown in Figure 7.14.

A deeper penetration of the electrical field will occur by increasing the spacing between the electrodes. A variation in apparent resistivity is plotted against the electrode spacing from which a smooth curve can be drawn.*

The interpretation of such a resistivity-spacing curve in terms of subsurface formation is frequently complex and often difficult. Nevertheless, with certain supplemental data from subsurface investigations to substantiate the surface measurement, correct estimations can often be made.

Seismic Wave Propagation Methods By shocking the earth's surface with a small dynamite explosion or the impact of a heavy weight, the time required for the sound or shock wave to reach a certain point at a known distance away can be measured. Seismic waves propagate through a medium the same way that light waves do. They may be refracted or reflected at the interface of any two materials of different elastic properties. A change in propagation velocity takes place at the interface. The wave speed is greatest in solid igneous rock and the least in unconsolidated formations. The water content in a particular formation will significantly alter the wave speed in the formation. As the seismic wave is

*H.M. Mooney, and W.W. Wetzel, *The Potentials About a Point Electrode and Apparent Resistivity Curves for a Two-, Three-, and Four-Layered Earth* (Minneapolis: University of Minnesota Press, 1956).

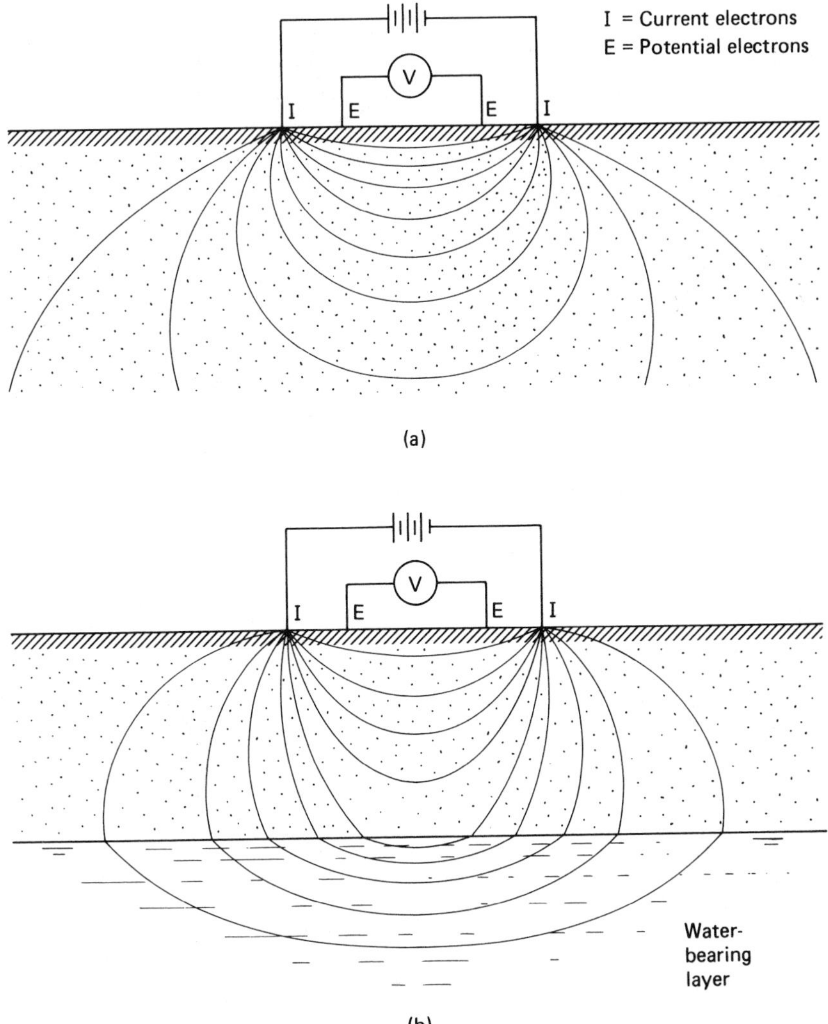

Figure 7.14 Schlumberger arrangement for resistivity determination: (a) current lines in homogeneous medium, (b) current lines distorted by the presence of a water-bearing layer.

traveling several hundred meters deep into the ground, subsurface information may be obtained by placing several seismometers at various distances from the shock point along the same line. The wave travel time is plotted against the distance, as shown in Figure 7.15. The sudden change in the slope of the time-spacing curve can be interpreted to determine the depth of a ground water table.

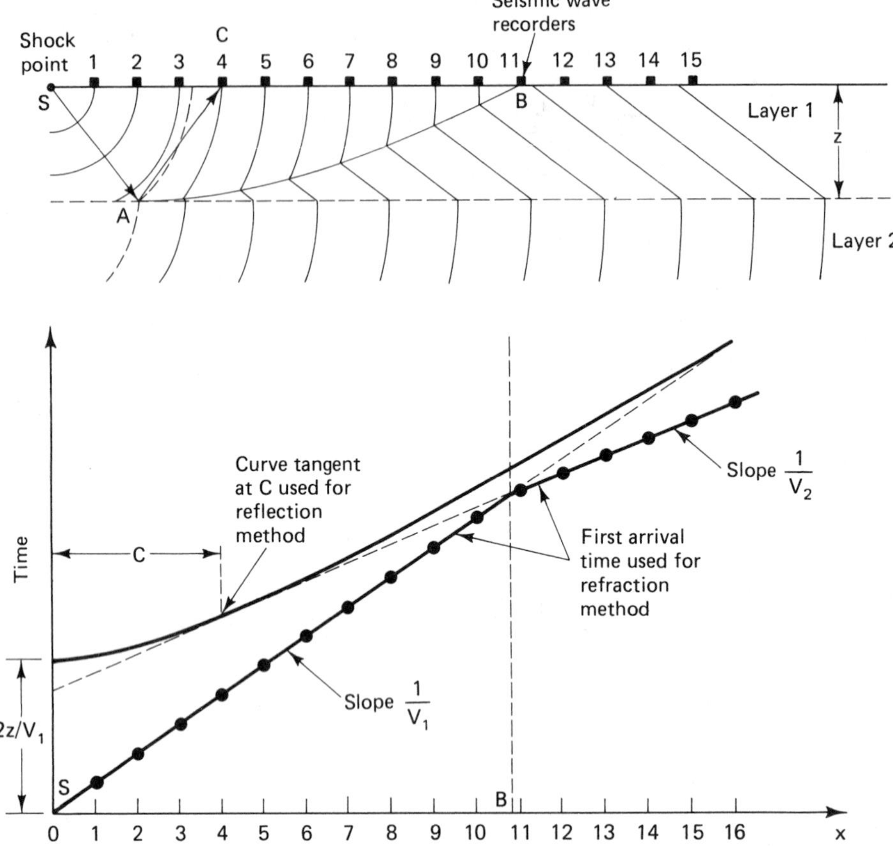

Figure 7.15 Propagation of seismic waves in a two-layer medium. Waves propagate at speed V_1 in the upper (dry) layer and at a higher speed, V_2, in the lower (water-bearing) layer. For points to the lower right of line *AB*, the wave refracted at the point *A* through the lower layer, 2, and reflected back to the surface arrives sooner than the wave propagating directly through the upper layer, 1.

7.6 SEA WATER INTRUSION IN COASTAL AREAS

Along the coastline the coastal aquifers are in contact with sea water. Under natural conditions, fresh ground water is discharged into the sea under the water table, as shown in Figure 7.16. However, with an increased demand of ground water in certain coastal regions, the seaward flow of fresh ground water has been reduced or even reversed, causing saltwater from the sea to enter and penetrate the inland aquifers. This phenomenon is commonly known as *sea water intrusion*. If the salt water travels far enough inland and enters the water supply wells, the ground water supply becomes useless. Furthermore, once the coastal aquifer is contaminated by salt, it is very difficult to remove the salt from the

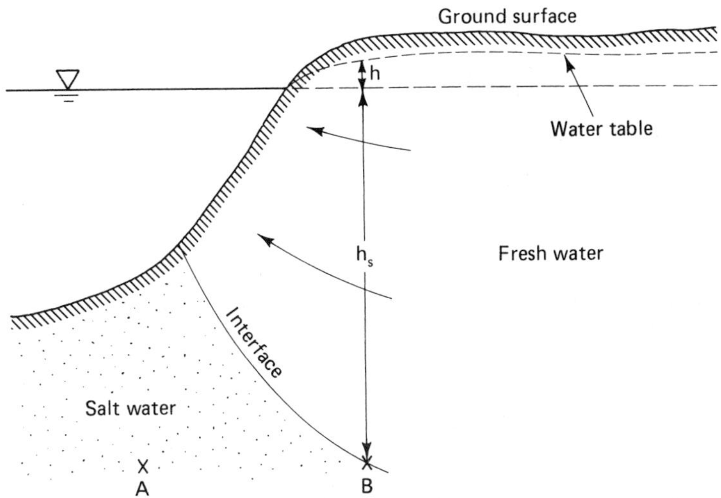

Figure 7.16 Schematic representation of freshwater and saltwater distribution in unconfined coastal aquifers.

formation and the aquifer may be permanently damaged. Engineering prevention and control of sea water intrusion will be discussed in this section.

Overdrafting of coastal aquifers results in lowering the water table in unconfined aquifers or the piezometric surface in confined aquifers. The natural gradient originally sloping toward the sea is reduced or reversed. Because of the difference in densities of salt water and fresh water, an interface is formed when the two liquids are in contact. The shapes and movements of the interface are governed by the pressure balance of the salt water on one side and fresh water on the other side of the interface.

It has been found that the interface that occurrs underground does not take place at sea level but at a depth below sea level that is approximately 40 times the height of the freshwater table above the sea level, as shown in Figure 7.16. This distribution is due to the equilibrium of hydrostatic pressure that exists between these two liquids, which have different densities.

Figure 7.16 shows the cross section of a coastal aquifer. The total hydrostatic pressure at point A at a depth h_s below the sea level is

$$P_A = \rho_s g h_s$$

where ρ_s is the density of the salt water and g is the gravitational acceleration. Similarly, the hydrostatic pressure at a point inland, B, at the same depth as A, and on the interface is

$$P_B = \rho_f g h + \rho_f g h_s$$

where ρ_f is the density of fresh water. For a stationary interface, the pressure at

A and B must be the same, and we may write

$$\rho_s g h_s = \rho_f g h + \rho_f g h_s \tag{7.19}$$

Solving Equation (7.19) for h_s yields

$$h_s = \frac{\rho_f}{\rho_s - \rho_f} h \tag{7.20}$$

By taking $\rho_s = 1.025 \text{ g/cm}^2$ and $\rho_f = 1.000 \text{ g/cm}^2$, the above relationship gives

$$h_s = \frac{1.000}{1.025 - 1.000} h = 40 \text{ h} \tag{7.21}$$

This is commonly known as the *Ghyben-Herzberg relation*.

This relation shows that a small depression in the water table near the coastline due to well pumping could cause a big rise in the interface. Similarly, a buildup on the water table near the coastline due to artificial recharge could drive the saltwater wedge deep into the ground, thus forcing it to move seaward. These phenomena are schematically demonstrated in Figure 7.17.

Obviously, artificial recharge of an overdrafted coastal aquifer is an effective method of controlling sea water intrusion. By proper management, artificial recharge of the aquifer can eliminate the overdraft and maintain the proper water table level and gradient.

In addition to the artificial recharge, several other methods have also been effectively applied for control of sea water intrusion. The most common methods are described as follows.

1. *Pumping Trough:* A pumping trough is a line of discharge wells situated along the coastline. By pumping the wells, a depression trough line is formed, as shown in Figure 7.18. While salt water is taken into the wells, a certain amount of fresh water in the aquifer is also removed. The freshwater motion is in the seaward direction toward the wells. This movement of fresh ground water can stabilize the saltwater and fresh water interface.
2. *Pressure Ridge:* A pressure ridge is a series of recharge wells installed in parallel to the coastline. Fresh water is pumped into the coastal aquifer to maintain a fresh water pressure ridge along the coastline to control the salt water intrusion. The pressure ridge must be large enough to repel the sea water and must be located far enough inland; otherwise, the salt water inland of the ridge will be driven farther inland, as demonstrated by Figure 7.19. Inevitably, a small amount of fresh water will be wasted to the sea; the remainder that moves landward can be used to supply part of the pumping draft. Reclaimed

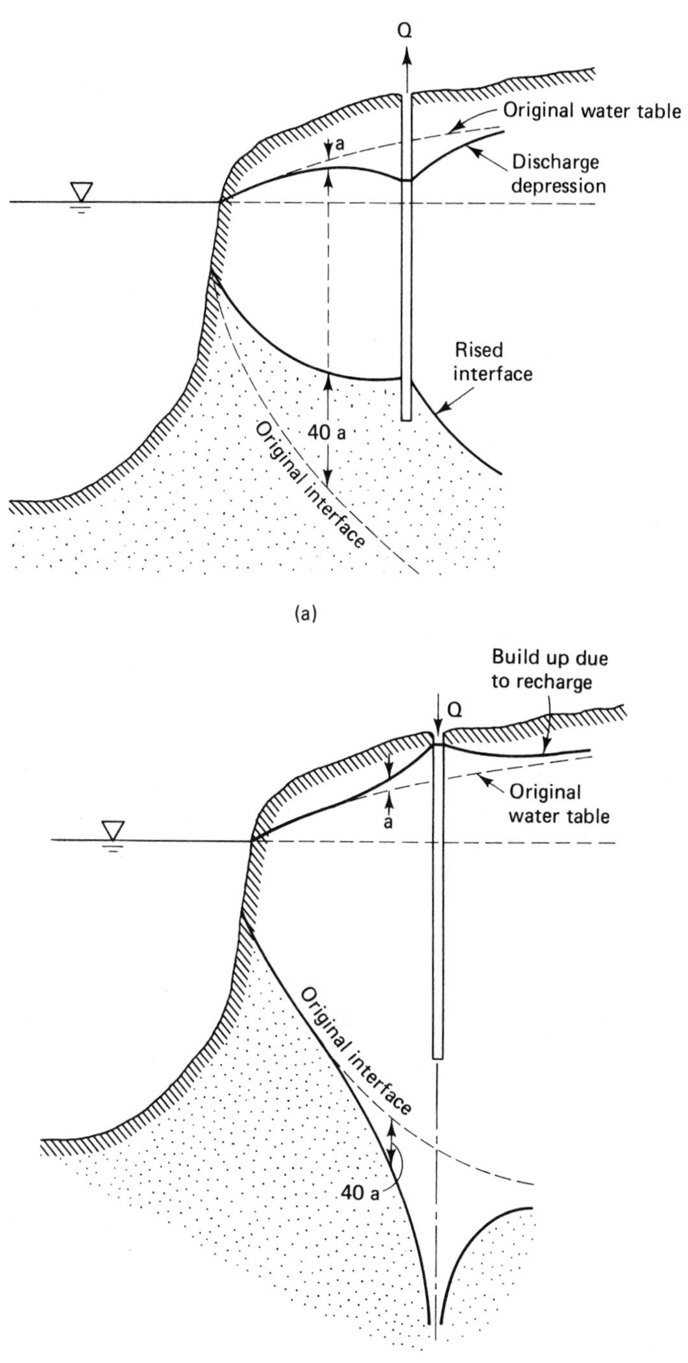

Figure 7.17 Sea water intrusion under the influence of: (a) discharge well, and (b) recharge well.

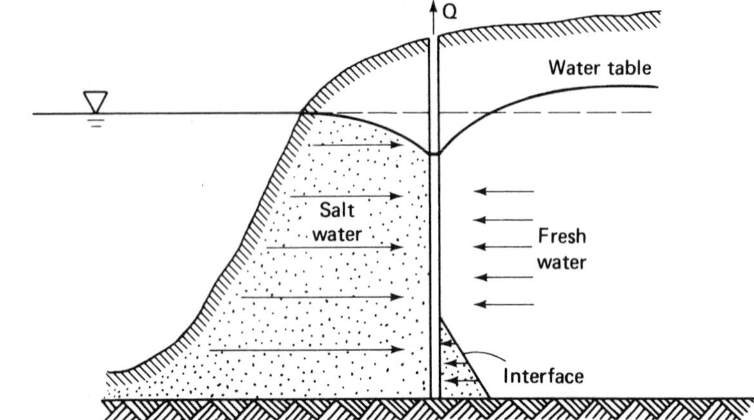

Figure 7.18 Control of sea water intrusion by pumping trough.

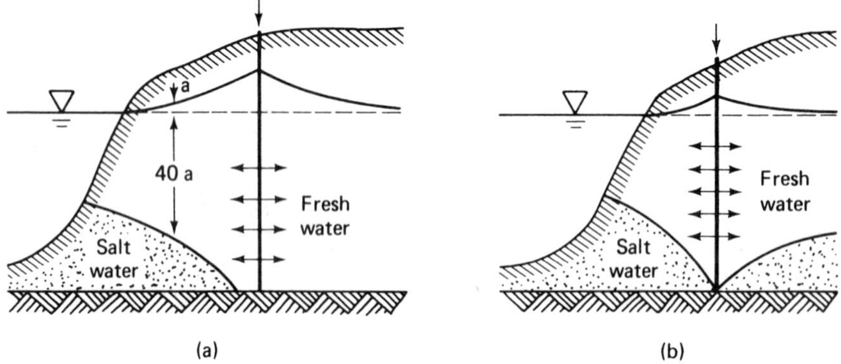

Figure 7.19 Control of sea water by pressure ridge: (a) proper recharge arrangement, (b) improper recharge arrangement.

waste water may be used to meet part of the recharge need. This method's advantage is that it doesn't deplete the usable ground water capacity, but the disadvantages of high initial and operating costs and the need for supplemental fresh water often make small-scale operations impractical.

3. *Subsurface Barriers:* Subsurface barriers may be built along the coastline to reduce the coastal aquifer permeability. In relatively shallow-layer aquifers, subsurface dikes may be constructed with sheet piling, puddled clay, or even concrete materials. An impermeable subsurface barrier may be formed by injecting flowable materials such as slurry, silicone gel, or cement grout into the aquifers through a line of holes. Subsurface barriers are best suited for locations such as narrow, alluvial canyons connected to large inland aquifers. Although initial cost of installing the barriers may be very high, there is almost no operation or maintenance expense.

7.7 SEEPAGE THROUGH DAM FOUNDATIONS

Dams constructed to store water in a reservoir may continuously lose water through seepage. Impermeable concrete dams constructed on an alluvial foundation may lose water through foundation seepage, while earth dams may also lose water through the dam body. Water movement in seepage flow is governed by Darcy's law in the same way as for ground water. Seepage flow can best be analyzed by applying the *flow net* technique.

A flow net is a graphical representation of flow patterns expressed by a family of *streamlines* and their corresponding *equipotential lines*. Streamlines are always drawn in the direction of the flow and they divide the flow field into a certain number of flow channels, each carrying the same discharge. Equipotential lines connect all points in the field that have equal velocity potential. The two sets of lines always meet at right angles and form an orthogonal net in the field. Figure 7.20 represents a portion of a flow net formed by streamlines and equipotential lines.

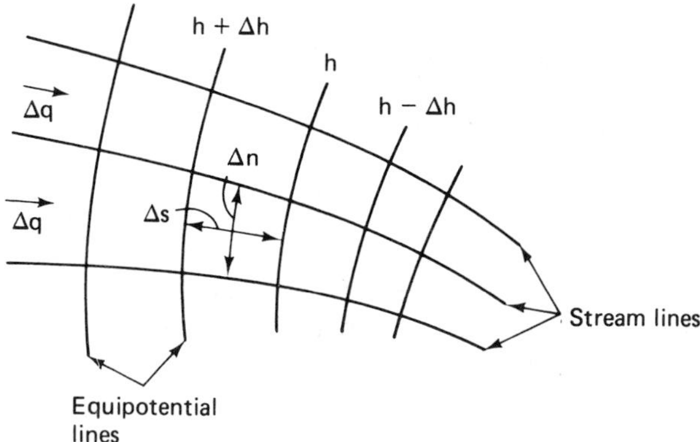

Figure 7.20 Flow net.

Flow nets are usually constructed with the distance between a pair of adjacent streamlines, Δn, and that between a pair of equipotential lines, Δs, in the same square in equal length.

$$\Delta n = \Delta s$$

The approximate velocity at any point in the field can be approximated by the following relation:

$$v = \frac{\Delta q}{\Delta n} \propto \frac{\Delta h}{\Delta s}$$

Darcy's law for steady flow at a low rate through a porous medium is

$$v = K\frac{dh}{ds} = K\frac{\Delta h}{\Delta s}$$

The rate of flow through a stream channel, per unit width of the dam, is

$$\Delta q = KA\frac{\Delta h}{\Delta s} \tag{7.22}$$

where A is the width of the square, or

$$A = \Delta n \tag{7.23}$$

Substituting Equation (7.23) into Equation (7.22), and $\Delta n = \Delta s$, we have

$$\Delta q = K(\Delta n)\frac{\Delta h}{\Delta s} = K\Delta h \tag{7.24}$$

Since Δh is a constant value of potential drop between any two adjacent equipotential lines, we may write

$$\Delta h = \frac{H}{n} \tag{7.25}$$

H is the difference between the reservoir water and tail water levels (Figure 7.21) and n is the number of squares in each stream channel of the flow net. The above equation may be rewritten as

$$\Delta q = K\frac{H}{n} \tag{7.26}$$

For m channels counted in a graphic flow net, the total flow rate per unit width of the dam is

$$q = m\Delta q = K\left(\frac{m}{n}\right)H \tag{7.27}$$

The total seepage discharge can be calculated by merely determining the ratio of m/n from a graphically constructed flow net.*

Figure 7.21 shows an example of seepage under a concrete dam with and without a cut-off wall. A *cut-off wall* is a thin layer of impermeable material or sheet piling partially penetrating the aquifer under the dam. Depending on its location relative to the base of the dam, the cut-off wall alters the seepage flow

*Methods for construction of a flow net are presented in Appendix A.

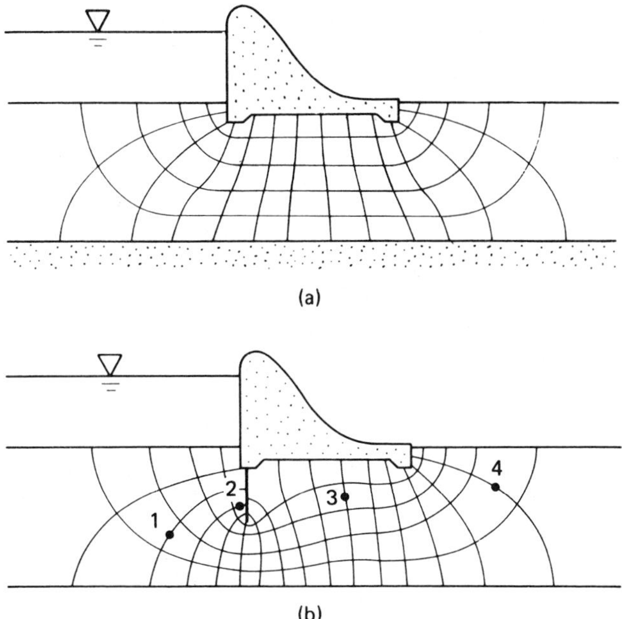

Figure 7.21 Seepage through pervious dam foundation: (a) without cut-off wall, (b) with cut-off wall.

pattern by lengthening the seepage paths, thus producing a vertical drop in the piezometric gradient. The cutoff walls effectively decrease the seepage rate and reduce the total uplift force on the base of the dam.

Example 7.5

A concrete gravity dam built on an alluvial channel bed as shown in Figure 7.21, stores water at a depth of 50 m. If the coefficient of permeability is $K = 2.14$ m/day, estimate the seepage per meter width of dam for (a) an unmodified aquifer and (b) a dam with a cut-off wall.

Solution

From Equation (7.27)

$$Q = K\left(\frac{m}{n}\right)H$$

(a) For the unmodified aquifer without the cut-off wall, we count the number of flow channels, $m = 5$, and the number of equipotential drops, $n = 13$, in Figure 7.21(a).

$$Q = 2.14 \cdot \frac{5}{13} \cdot 50 = 48.64 \text{ m}^3\text{/day/m width}$$

(b) For the modified aquifer with cut-off wall, as shown in Figure 7.21(b), we count the number of flow channels, $m = 5$, and the number of equipotential drops, $m = 16$. Therefore,

$$Q = 2.14 \cdot \frac{5}{16} \cdot 50 = 33.44 \text{ m}^3/\text{day/m width}$$

PROBLEMS

7.7.1 If the water elevation in the upstream reservoir of Example 7.5 is lowered to 40 m, determine the seepage rate per unit width of the dam foundation. Estimate the magnitude and direction of the velocity at locations 1, 2, 3, and 4.

7.7.2 Determine the required length of an impervious floor at the upstream heel of the dam, x, in Figure P7.7.2 to reduce the quantity of foundation seepage by 50%.

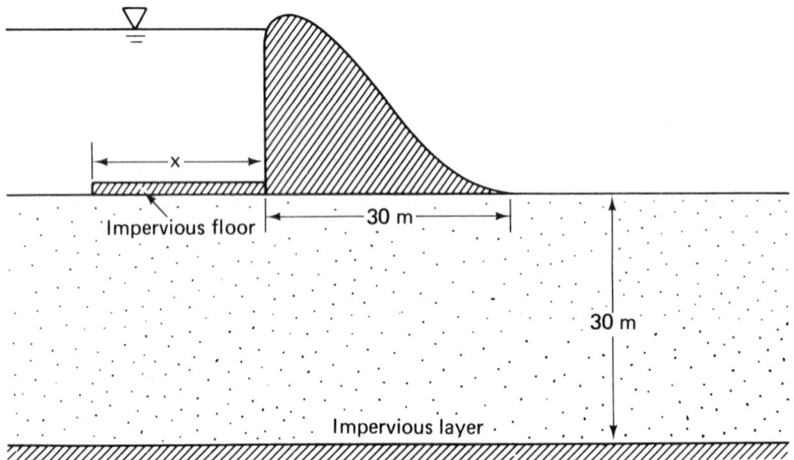

Figure P7.7.2

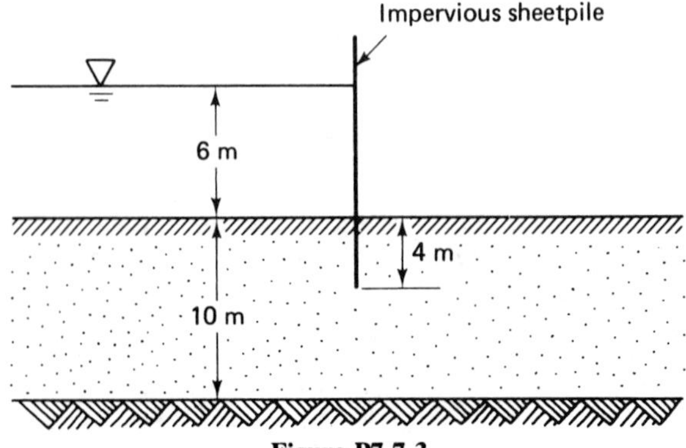

Figure P7.7.3

7.7.3 Determine the quantity of seepage per unit width of the channel bed in Figure P7.7.3. The permeability $K = 0.00051$ cm/sec.

7.8 SEEPAGE THROUGH AN EARTH DAM

Since an earth dam is built with pervious materials, it is of particular engineering concern. It may be damaged by seepage through the dam body, by piping, for example. Seepage analysis should be made for each earth dam designed by applying the method of flow nets.

Seepage through an earth dam body can be treated as flow through an unconfined porous medium. The upper surface of the flow, known as the *surface of saturation* or *phreatic surface*, is at atmospheric pressure. The typical shape of a *phreatic line* in a homogeneous earth dam is shown in Figure 7.22. The phreatic line is a streamline whose intersection with the equipotential lines is equally spaced vertically by the amount of $\Delta H = H/n$, where H is the total head available and n is the number of potential drops counted in a graphic flow net. This line, which provides the upper boundary of the flow net, must be initially located by trial. An empirical rule for locating the phreatic line was suggested by Casagrande* and is shown in Figure 7.22.

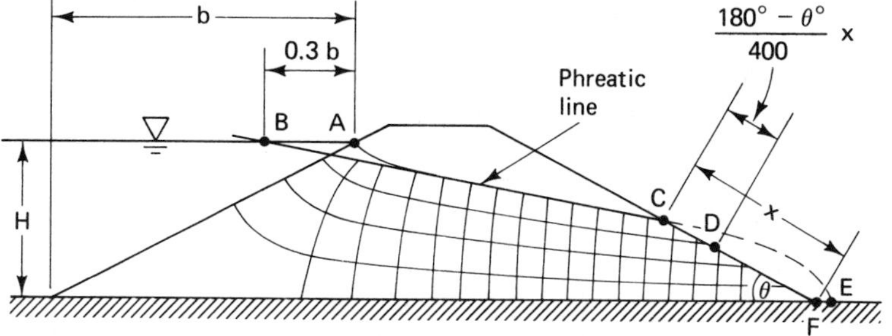

Figure 7.22 Seepage flow net through a homogeneous earth dam. (A large portion of the phreatic line AD can be approximated by the parabola BCE with F as the focus and passing through the point B. Point A on the upstream face of the dam is the intersection of the water surface with the dam. Point D is the downstream transition where the seepage is exposed to the atmosphere.)

Section DF on the lower part of the downstream dam face must be protected against soil piping, which may eventually lead to dam failure. The seepage water may be removed permanently from the downstream surface by a properly designed drainage system. For a nonstratified, homogeneous earth dam, a narrow, longitudinal drain will effectively intercept all the water seeping through the embankment. Figure 7.23 schematically shows the dimensions of a typical earth dam drainage.

*A. Casagrande, "Seepage Through Dams," *J. New Eng. Water Works Assoc.*, 51 (1937), 139.

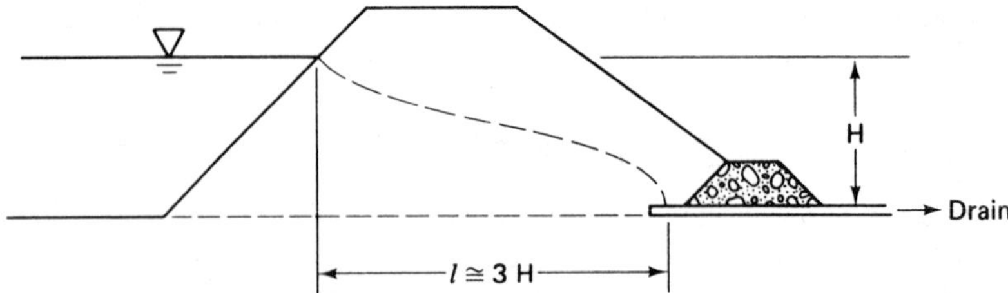

Figure 7.23 Drainage blanket in an earth dam.

The total discharge through an earth dam can be determined by the graphically constructed flow net, as discussed in Section 7.7. Equation (7.27) gives the seepage discharge for the unit width of the dam.

$$q = K\left(\frac{m}{n}\right)H$$

where m is the number of flow channels and n is the number of equipotential drops in the flow net.

PROBLEMS

7.8.1 An earth dam (see Figure P7.8.1) is 30 m high and has a 3-m freeboard and a 3-m crest width. The dam has a 1(V): 2(H) upstream face slope and a 1(V): 3(H) downstream face slope. The dam material has a permeability coefficient of 0.0001 cm/sec. Compute the seepage per unit width from the flow net based on the formula of the equivalent parabola.

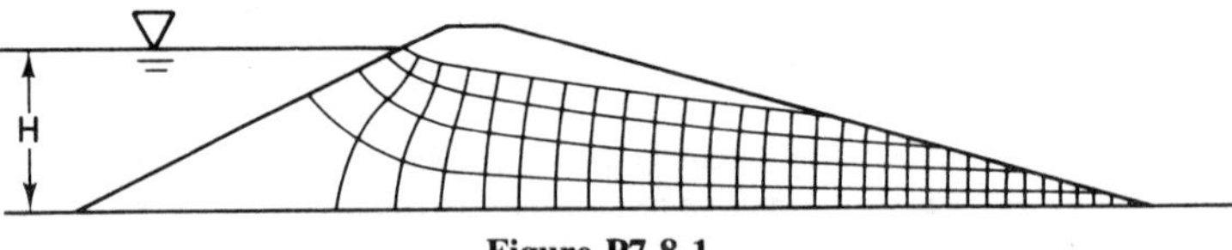

Figure P7.8.1

7.8.2 An earth dam (see Figure P7.8.2) is constructed with a uniform material having

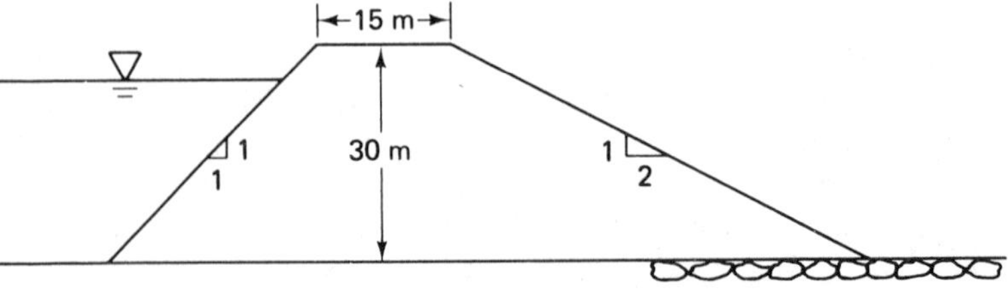

Figure P7.8.2

a coefficient of permeability $K = 2 \cdot 10^{-4}$ m/sec on a relatively impervious foundation. The dam is 30 m high and has a top width of 15 m. The upstream face of the dam has a 1(V): 1(H) slope and a downstream face of 1(V): 2(H). Compute the seepage rate in units of cubic meters per day per unit width of the dam.

7.8.3 Determine the seepage rate per unit width of the dam in Problem 7.8.2 if a drainage blanket extends back 10 m from the toe of the dam.

8

HYDRAULIC STRUCTURES

8.1 FUNCTIONS OF HYDRAULIC STRUCTURES

Water is useful to people only if it is properly controlled, conveyed, and contained. Hydraulic structures are designed and built to serve these purposes. Because of the great variety of hydraulic structures that can be built to serve one or more of these purposes, any classification of hydraulic structures will inevitably be discretionary, and there are no clearly defined general classifications. A general classification of hydraulic structures based on use is not satisfactory because many identical structures may be used to serve completely different purposes. For example, a low dam could be built across a channel as a device to measure discharge or a low dam could be built to raise the water level at the entrance to an irrigation canal to permit diversion of water into the canal. Instead of classifying the various hydraulic structures into arbitrary categories, we have listed the common functions of hydraulic structures and the basic design criteria.

1. Storage structures are designed to hold water under hydrostatic condi-

tions. A storage structure usually has a large capacity for a relatively small change in hydrostatic head (water elevation).

2. Conveyance structures are designed to transport water from one place to another. The design normally emphasizes delivery of a given discharge with a minimum consumption of energy.
3. Waterway and navigation structures are designed to ensure water transportation. The protection and maintenance of a minimum water depth throughout the entire waterway under a variety of possible conditions are emphasized in the design.
4. Measurement or control structures are used to quantify the discharge in a particular conduit. Stable performance and a one-to-one relationship between the indicator and the corresponding discharge are necessary.
5. Energy conversion structures are designed to transform hydraulic energy into mechanical or electric energy (e.g., hydraulic turbine systems) or electrical or mechanical energy into hydraulic energy (e.g., hydraulic pumps). The design emphasis is on the efficiency of the system.
6. Sedimentation or fish control structures are designed to direct or regulate the movement of the nonhydraulic elements in water. An understanding of the basic mechanisms and behaviors of the elements involved is an essential requirement for the design.
7. Energy dissipation structures are used to disperse excess water energy to prevent channel erosion.
8. Collection structures are designed to gather and admit water to a conduit or system. A typical example is a surface drainage inlet used to collect storm runoff to a sewer system.

Obviously, detailed consideration of all these functions and their design criteria are beyond the scope of this book. Only the most commonly encountered hydraulic structures are discussed here in order to demonstrate the fundamental considerations used in designing hydraulic structures.

8.2 GRAVITY DAMS AND ARCH DAMS

A dam is a barrier structure placed across a watercourse to store or slow the normal water flow. Water stored behind a dam may reach a great height depending on the height of the dam. Low dams are usually a few meters in height; high dams are those over 100 meters in height. A typical gravity dam is a massive structure (Figure 8.1). The enormous weight of the dam body provides the necessary stability against tip-over about the toe of the dam or shear failure along a horizontal plane. Arch dams are normally built on solid rock foundations that provide resistance to the load by arch actions (Figure 8.2). Generally, low dams are gravity dams and high dams (100 m and higher) are arch dams.

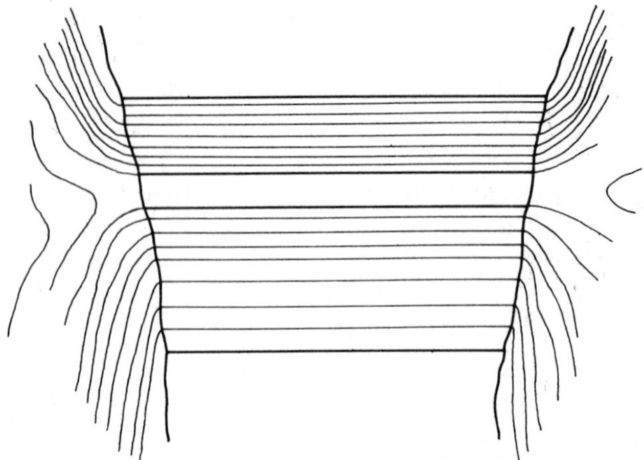

Figure 8.1 Top-view representation of a gravity dam.

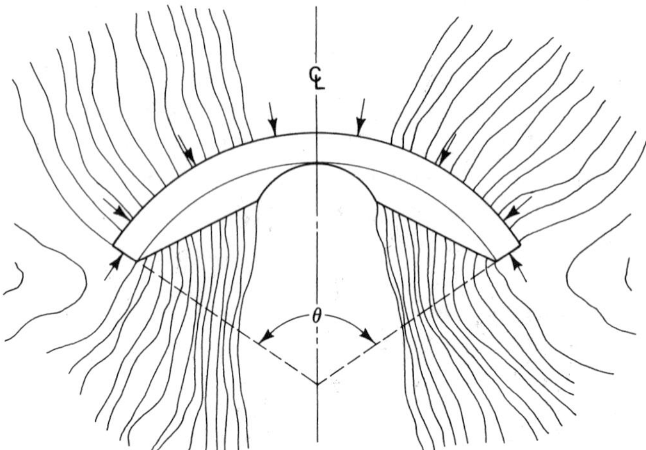

Figure 8.2 Top-view representation of an arch dam and reactions.

Combining gravity action and arch action in dam design is a common practice. A dam that has a combination design may be called an *arch-gravity dam* or a *gravity-arch dam,* depending on whether the arch or gravity action predominates.

8.3 STABILITY OF GRAVITY DAMS

The major forces acting on a gravity dam (or any other type of dam) are represented in Figure 8.3. They are

1. Hydrostatic force, F_{HS},
2. Weight of the dam, W,

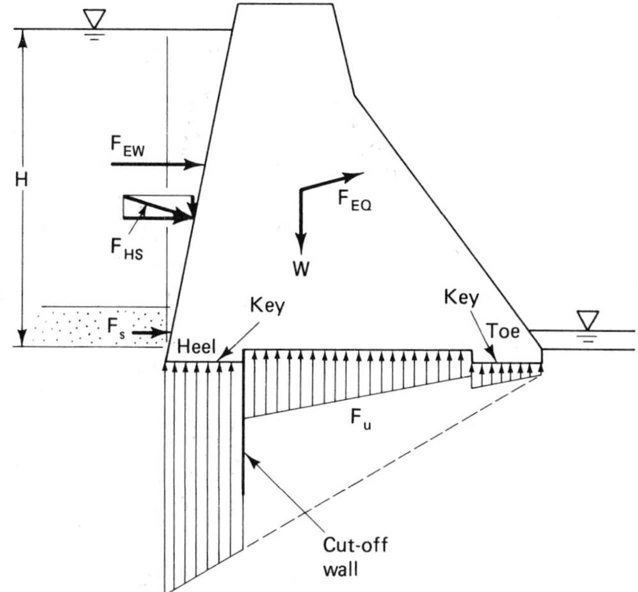

Figure 8.3 Cross-section representation of a gravity dam.

3. Uplifting force on the base of the dam, F_u,
4. Sedimentation of silt deposit pressure force, F_s,
5. Earthquake force on the dam, F_{EQ},
6. Earthquake force due to the water mass behind the dam, F_{EW}.

The straight-line structure of the gravity dam (Figure 8.1) allows force analysis per unit width of the dam. The binding forces between each unit width segment are neglected in the analysis because they can only add to the stability of the dam.

The hydrostatic force acting on the upstream face of the dam may be resolved into a horizontal component and a vertical component. The horizontal component of the hydrostatic force acts along a horizontal line $H/3$ above the base of the dam. This horizontal force forms a clockwise moment about the toe of the dam, and it is a major factor in the possible overturning of the dam. It may also cause dam failure by shear along a horizontal plane. The vertical component of the hydrostatic force is equal to the weight of the water mass directly above the upstream face of the dam. It acts along a vertical line that passes through the centroid of that mass. The vertical component of hydrostatic force always forms a counterclockwise moment about the toe. It is a major stabilizing factor in gravity dams.

The largest stability force is the weight of the dam, which depends not only on the dimensions but also on the material used. The unit weight of most masonry or solid earth materials is approximately from 2.4 to 2.6 times that of water. The importance of this single stability force of gravity explains the name gravity dam.

The uplifting force on the base of a dam can be determined by foundation seepage analysis (Chapter 7). Acting in the opposite direction of the gravity, this force should be minimized in every dam design, if possible. It weakens the foundation by uplifting and it tends to overturn the dam. If the foundation soil is porous and homogeneous in structure, the uplifting pressure on the base varies linearly from full hydrostatic head H at the heel of the dam to the full hydrostatic head of the tail water at the toe. The total resulting uplifting force can be determined by integrating the trapezoidal force distribution area represented in Figure 8.3. The magnitude of the uplifting force as well as the overturning (clockwise) momentum can be greatly reduced by installing an impermeable cut-off wall a few meters into the foundation. The cut-off wall alters the seepage course by lengthening the pathway, thus reducing the seepage and uplifting force.

The water flow velocity immediately behind the dam is very slow or near zero. The water loses its ability to carry sediments or other suspended matters. These heavier materials are deposited on the bottom of the lake near the base of the dam. The silt-water mixture is approximately 50% heavier than water (sp. gr. 1.5) and forms an excess pressure force near the heel. Normally, the thickness of the silt layer will increase slowly with time. This force may contribute to dam failure by shear along the base.

In earthquake zones, the forces generated by earthquake motion must be incorporated into dam design. Earthquake force on dams is due to acceleration associated with the earthquake motions. The magnitude of the earthquake force in the dam body, F_{EQ}, is proportional to the acceleration and the mass of the dam body. The force may be acting in any direction through the centroid of the dam body.

Earthquake force due to acceleration of the water body behind the dam is approximately equal to

$$\frac{5}{9}\left(\frac{a\gamma}{g}\right)H^2$$

where a is the earthquake acceleration, γ is the specific weight of water, and H is the hydrostatic head, or depth of water, immediately behind the dam. The earthquake force of the water body acts in a horizontal direction at distance $(4\pi/3)H$ above the base of the dam.*

In order to ensure stability, the safety factors of the dam against sliding failure or overturning must be greater than unity. In addition, the maximum pressure exerted on the foundation must not exceed the bearing strength of the foundation.

* J. I. Bustamante, "Water Pressure on Dams Subjected to Earthquakes," *J. Engr. Mech., Div., ASCE,* 92 (October, 1966), 115–27.

The *ratio of forces against sliding* is defined by the ratio of the total horizontal resistance force that the foundation can develop to the sum of all forces acting on the dam that tend to cause sliding. This ratio may be expressed as

$$\text{force ratio against sliding} = \frac{\mu(\Sigma F_v) + A_s \tau_s}{\Sigma F_H} \tag{8.1}$$

where μ is the coefficient of friction between the dam base and the foundation (ordinarily $0.4 < \mu < 0.75$), ΣF_v is the summation of all vertical force components acting on the dam, τ_s is the shear stress strength of keys, and A_s is the total shear area provided by the keys. ΣF_H is the summation of all horizontal force components acting on the dam.

The keys are the part of the dam body built into the foundation to add resistance against dam sliding. Horizontal forces are transmitted to the foundation via the shear force in the keys. The total shear force provided by the keys, $\tau_s A_s$, must be larger than the difference between the total horizontal force acting on the dam, ΣF_H, and the friction force provided by the base, $\mu(\Sigma F_v)$.

$$\tau_s A_s > [\Sigma F_H - \mu(\Sigma F_v)]$$

The *ratio of forces against overturning* is defined by the ratio of the resisting moments (counterclockwise moments about the toe) to the overturning moments (clockwise moments about the toe)

$$\text{force ratio against overturning} = \frac{W \cdot l_w + (F_{HS})_v \cdot l_v}{\Sigma F_H \cdot Y_H + F_u \cdot l_u} \tag{8.2}$$

where $(F_{HS})_v$ is the vertical component of the hydrostatic force and l_w, l_v, and l_u are the horizontal distances from the toe to the lines of action of the weight, the vertical component of hydrostatic force, and the uplifting force, respectively. Y_H is the vertical distance measured from the toe to the lines of action of each respective horizontal force component, F_H.

We may assume that the vertical pressure on the foundation is a linear distribution between the toe and heel as shown in Figure 8.4. If we let R_v represent the resultant of all vertical forces acting on the base of the dam and P_T and P_H represent the pressure at the toe and heel, respectively, we may write

$$R_v = \frac{(P_T + P_H)}{2} \cdot B$$

and

$$R_v = \frac{(P_T - P_H)}{2}\left(\frac{B}{e}\right) \cdot \left(\frac{B}{6}\right)$$

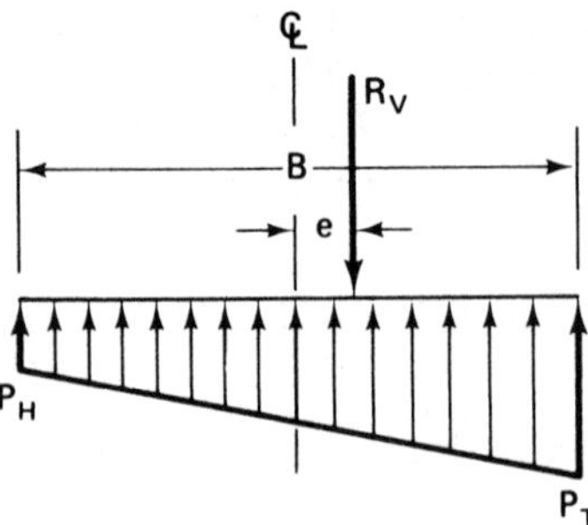

Figure 8.4 Pressure distribution on dam foundation.

Solving these two equations simultaneously, we have

$$P_T = \left(\frac{R_v}{B}\right)\left(1 + \frac{6e}{B}\right) \tag{8.3}$$

$$P_H = \left(\frac{R_v}{B}\right)\left(1 - \frac{6e}{B}\right) \tag{8.4}$$

The vertical resultant force normally acts through a point on the downstream side of the base centerline. Therefore, P_T is usually the critical pressure in design. The value of P_T must be kept less than the bearing strength of the foundation. The pressure at the heel, P_H, is less important. Nevertheless, it is desirable to keep P_H a positive value at all times in order to prevent tension cracks from developing in the heel region. Negative pressure indicates tension, and masonry materials have very low resistance to tension stress. A positive P_H value can be ensured if the vertical resultant force, R_v, is kept within the middle third of the base, or

$$e < 6B \tag{8.5}$$

8.4 STABILITY OF ARCH DAMS

The force load on an arch dam is essentially the same as on a gravity dam. To resist these forces, the dam foundaton must provide a horizontal arch reaction. The large horizontal reactions can only be provided by strong, solid rock abutments at the two ends of the arch (Figure 8.2). Arch dams are mostly high dam construction and are only built in relatively narrow rock canyon sections. The efficiency of using material strength, and the slim dam base in relation to the height of the dam, make the arch dams the best choice in many situations. Since the arch dam combines the resistance from the arch action with gravity, there is high stress in each segment of the dam and detailed stress analysis is required.

The stability analysis on an arch dam is usually carried out on each horizontal rib. Take a rib situated h meters below the designed reservoir water level. The forces acting in the direction of the dam centerline may be summed up as follows:

$$2R \sin \frac{\theta}{2} = 2r\gamma h \sin \frac{\theta}{2}$$

where R is the reaction from the abutment, θ is the central angle of the rib, r is the outer radius (extrados) of the arch, and γh is the hydrostatic pressure acting on the rib. The previous equation can be simplified for the abutment reaction

$$R = r\gamma h \tag{8.6}$$

This value is determined by considering only the arch reaction for resisting the hydrostatic load on the dam. In practice, however, several other resistive forces, as discussed in the gravity dams, should also be included. The analysis should consider the combination resistance from both the arch and gravity actions.

The volume of an arch dam is directly related to the thickness of each rib, t, the width of the rib, B, and the center angle, θ. For the minimum volume of the dam, it can be shown that $\theta = 133°34'$. Other factors, such as topographic conditions, often prevent the use of this optimal value. Values within the range of $110° < \theta < 140°$ are commonly used in arch dam design.

A simple approach to arch dam design is to keep the central angle constant while the radii vary from rib to rib, as shown in Figure 8.5. Another approach frequently used is to keep the radii of the ribs at a constant value and allow variations of the central angle, as shown in Figure 8.6.

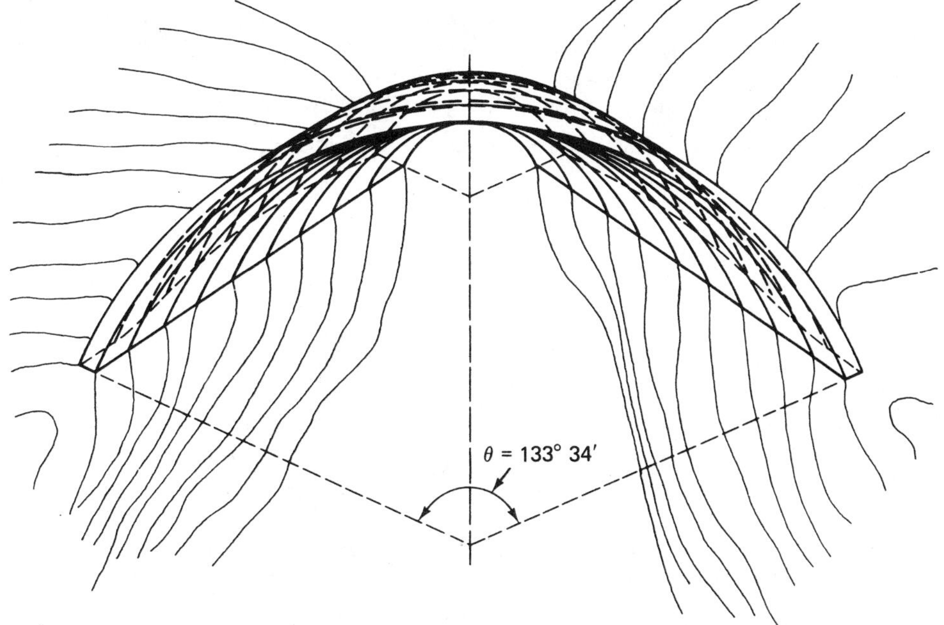

Figure 8.5 Constant-angle arch dam.

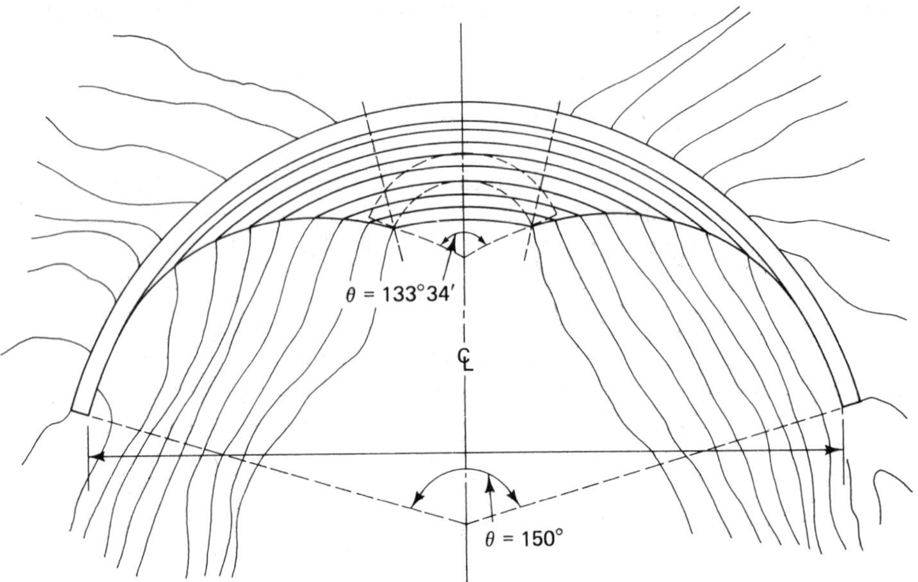

Figure 8.6 Constant-radius arch dam.

PROBLEMS

8.4.1 For the concrete gravity dam shown in Figure P8.4.1, compute the foundation bearing pressure at the heel and toe. Assume that the uplift force takes a triangular distribution with maximum magnitude one-third that of the hydrostatic pressure at the heel and zero at the toe. The reservoir is full to its designed elevation with 3 m of freeboard, and the masonry has specific gravity of 2.65.

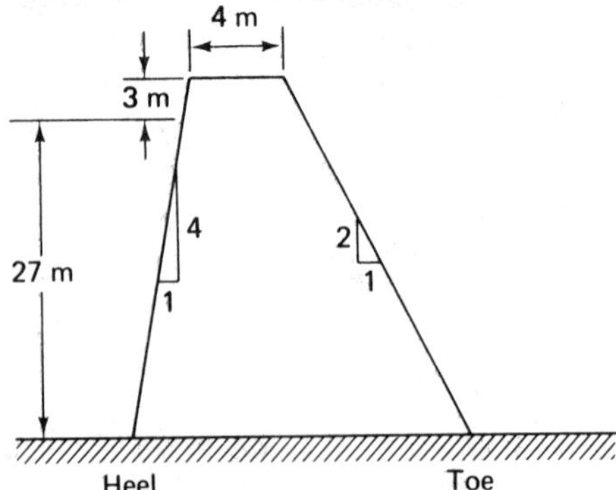

Figure P8.4.1

8.4.2 Check the dam in Problem 8.4.1 for stability against overturning and sliding if the coefficient of friction between the dam base and the foundation is 0.65.

8.4.3 The specific gravity of the dam shown in Figure P8.4.3 is 2.63, and the coefficient of friction between the dam and the foundation is 0.53. The freeboard is 2.5 m when the reservoir is filled to design capacity. Assume that the uplift force has a triangular distribution with maximum magnitude of 60% of that of the hydrostatic pressure. Check the stability of the dam and compute the foundation bearing pressure distribution.

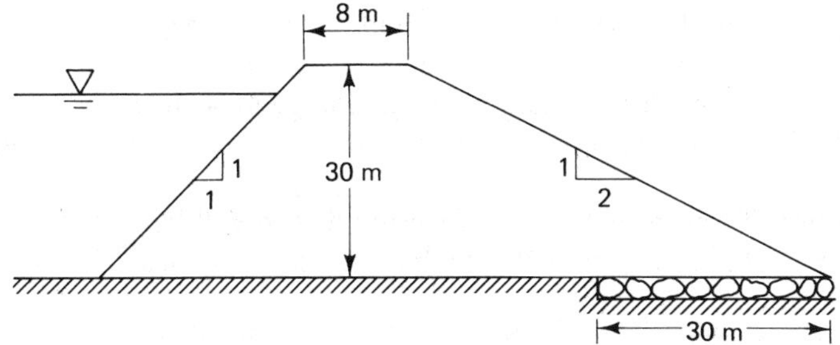

Figure P8.4.3

8.4.4 A constant angle arch dam ($\theta = 133°34'$) is designed to span a uniform canyon 150 m wide. The height of the dam is 78 m including 3 m of freeboard. If the dam has 4 m thickness at the crest and has a symmetrical cross section that increases its thickness to 11.8 m at the base, determine the masonry stress in the dam, using the cylinder method (constant radii) for approximate analysis at

(a) the crest,

(b) the midheight,

(c) the dam base.

8.5 WEIRS

Any obstruction of a streamflow over which water flows can be called a *weir*. Since weirs are usually installed in stream flows that have free surfaces, the flow behavior over a weir is governed by gravity forces. As a result of a sudden reduction of area, water over the weir is accelerated to pass the crest. This fact has been utilized in the past to keep flood currents from overflowing a bridge, as schematically shown in Figure 8.7. By placing an obstruction of adequate dimensions in an otherwise subcritical stream flow, the water level is raised upstream from the weir. With the increased available head, the flow is acceler-

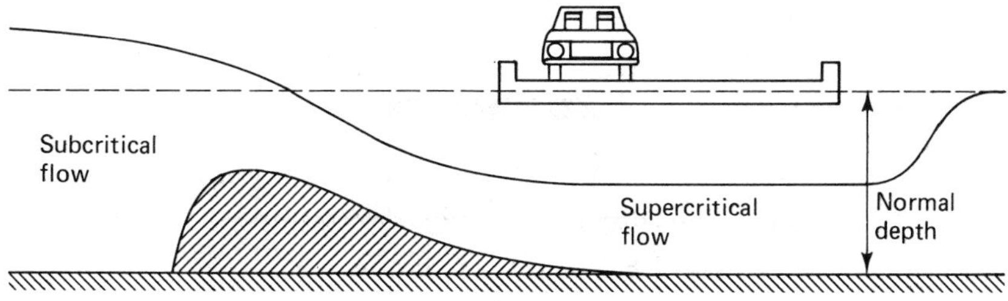

Figure 8.7 Acceleration of flow over a weir.

ated to pass over the crest. This acceleration causes the depth of water to decrease and reach the depth of supercritical flow after passing the weir. Some distance downstream from the weir the flow returns to a normal subcritical depth through a hydraulic jump. This arrangement makes the bridge structure free from the high flood water.

Flow acceleration over the weir provides a unique one-to-one relationship between the approaching water height and the discharge for each type of weir. Weirs are commonly built to measure the discharge in open channels. Weirs are also used to raise stream-flow levels in order to divert a desired amount of discharge.

The use of a weir as a flow-measuring device will be discussed in detail in Chapter 9. In this section the hydraulic characteristics of weirs will be presented.

A weir structure increases the depth immediately upstream of the weir and reduces the water cross section at the crest. The increase of water depth slows down the water velocity upstream; the sudden reduction of cross-sectional area causes the flow to speed up quickly as it passes over the crest. The occurrence of critical flow on weirs is the essential feature of weir structures.

The hydraulics of an overflow weir may be examined by using an ideal frictionless weir (see Figure 8.8). At the occurrence of the critical depth, the discharge per unit width of the weir can be determined by using the critical flow conditions

$$\frac{V_c}{\sqrt{gd_c}} = 1 \tag{6.11}$$

and

$$d_c = \sqrt[3]{\frac{q^2}{g}} \tag{6.13}$$

Rearranging Equation (6.11), we may write

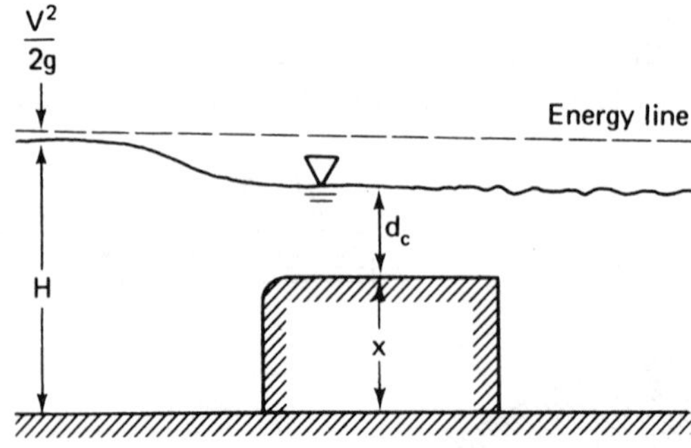

Figure 8.8 Flow over a frictionless weir.

$$\frac{V_c^2}{2g} = \frac{d_c}{2}$$

The energy at the critical section is

$$E = d_c + \frac{V_c^2}{2g} = d_c + \frac{d_c}{2} = \frac{3}{2}d_c \tag{6.8}$$

If the approaching velocity head can be neglected, the energy of the approaching flow is approximately equal to the water depth upstream of the weir, H. The same amount of energy is available at the critical section

$$E + x = H = \frac{3}{2}d_c + x \tag{8.7}$$

Combining Equations (8.7) and (6.13) and defining $H_s = H - x$, we have

$$q = \sqrt{gd_c^3} = \sqrt{g\left(\frac{2H_s}{3}\right)^3} = 1.705H_s^{3/2} \tag{8.8}$$

This is the basic form of the weir equation. The coefficient 1.705 is higher than the coefficients obtained in experiments because friction loss is neglected in the above analysis.

A *sharp-crest weir* is shown in Figure 8.9. Upstream from the weir the velocity of all water elements is nearly uniform and parallel. The water elements near the bottom of the channel rise from the bottom in the immediate neighborhood of the weir in order to pass the crest. The vertical component of the flow near the upstream face of the weir causes the lower surface of the stream to separate from the weir and form a *nappe* [see Figure 8.9(a)]. The nappe usually

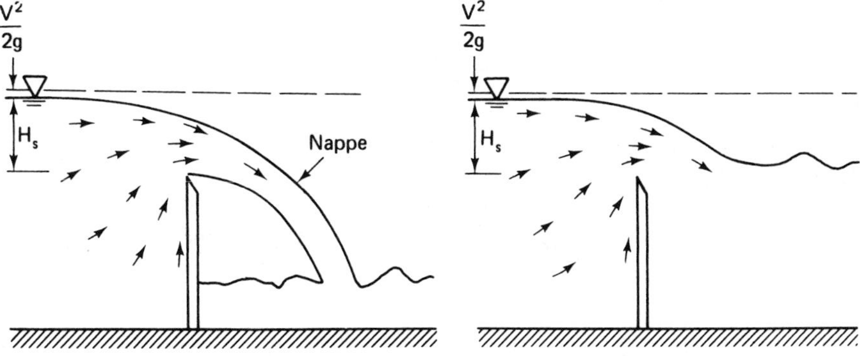

Figure 8.9 Water flows over sharp-crest weir: (a) free-falling nappe; (b) submerged flow.

traps a certain amount of air between the lower nappe surface and the downstream side of the weir. If no means of restoring air are provided, a void will appear that represents a negative pressure on the structure. The nappe will also cling intermittently to the side of the weir and cause the flow to be unstable. The dynamic effect of this unstable flow may result in added negative pressure. This negative pressure may eventually damage the structure.

When the downstream water level rises over the weir crest, the weir is said to be *submerged* [see Figure 8.9(b)]. In this case, the negative pressure no longer exists, and a new set of flow parameters may be considered in the determination of the discharge coefficient.

Example 8.1

Uniform flow occurs in a long rectangular channel 4 m wide at a depth of 2 m. The channel is laid on a slope of 0.001 and the Manning coefficient is 0.025. Determine the minimum height of a low weir that can be built on the floor of this channel (see Figure 8.10) to produce critical depth.

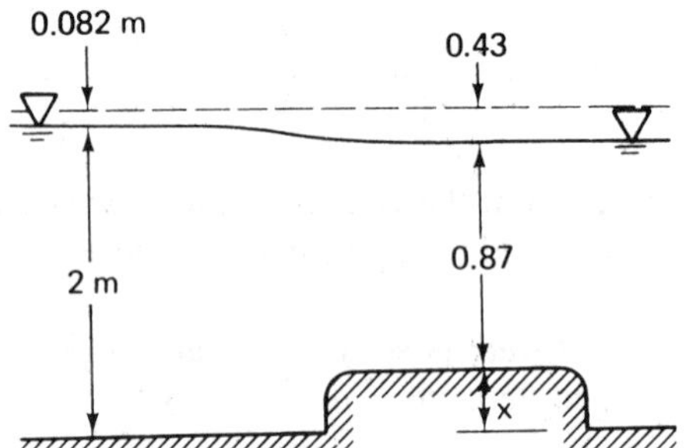

Figure 8.10

Solution

For a uniform flow condition, the Manning uniform flow formula [Equation (6.5)] may be used to determine the channel discharge, Q

$$Q = \frac{1}{n} A R_h^{2/3} S_0^{1/2}$$

Here, $A = 2\text{ m} \cdot 4\text{ m} = 8\text{ m}^2$, $P = 2 \cdot 2\text{ m} + 4\text{ m} = 8\text{ m}$, and $R_h = A/P = 1.0\text{ m}$. Hence,

$$Q = \frac{1}{0.025}(8)(1.0)^{2/3}(0.001)^{1/2} = 10.12\text{ m}^3/\text{sec}$$

and

$$V = \frac{Q}{A} = \frac{10.12}{8} = 1.26\text{ m/sec}$$

The specific energy is

$$E = d + \frac{V^2}{2g} = 2 + \frac{(1.26)^2}{2 \cdot 9.81} = 2 + 0.082 = 2.082 \text{ m}$$

For a critical flow condition to occur, the condition of Equation (8.7) must be satisfied, or

$$E_c = \frac{3}{2} d_c = \frac{3}{2} \sqrt[3]{\frac{q^2}{g}}$$

Hence, $d_c = 0.87$ m.

The corresponding velocity is

$$V_c = \frac{Q}{4d_c} = \frac{10.12}{4 \cdot 0.87} = 2.92 \text{ m/sec}$$

and the critical velocity head is

$$\frac{V_c^2}{2g} = 0.43 \text{ m}$$

If there is no energy loss at the weir, the minimum height of the weir that can be built in the floor to produce critical flow in this channel is x.

$$x = E - \left(d_c + \frac{V_c^2}{2g}\right) = 2.082 - (0.87 + 0.43) = 0.781 \text{ m}$$

PROBLEMS

8.5.1 If a rectangular channel carries a discharge 2 m^3/sec per unit channel width, and the energy line measured 2.7 m above the channel bottom and the weir crest is 1.4 m in height, what is the flow velocity over the weir? Determine the coefficient of discharge considering energy losses [in reference to Equation (8.8)].

8.5.2 The elevation of the crest of an overflow weir is 100 m above the mean sea level at a certain port. If the weir is 5 m long and discharges 10 m^3/sec, determine the water surface elevation upstream from the weir. Neglect friction loss.

8.5.3 A weir 1.0 m in height is built across the floor of a 4-m-wide rectangular channel. If the water depth on the crest measures 0.3 m, determine the discharge and the water depth upstream from the weir.

8.5.4 The depth of water upstream from a 10-m-wide frictionless weir 1 m in height is 1.5 m. Determine the discharge.

8.6 OVERFLOW SPILLWAYS

A spillway is an essential part of a large dam and provides an efficient, safe means of releasing flood water that exceeds the design capacity of the reservoir. Basically, an overflow spillway is an open channel on large slopes that allows the excess water to flow out at supercritical velocities. The ideal longitudinal profile of an overflow spillway should flow along the same curve as the underside of the free-falling water nappe (Figure 8.11) to minimize the pressure on the spillway surface. However, caution must be exercised to avoid any negative pressure on the surface. Negative pressure is caused by separation of the high-speed water flows over the spillway, resulting in a pounding action of the spillway surface that can cause significant damage to the spillway structure.

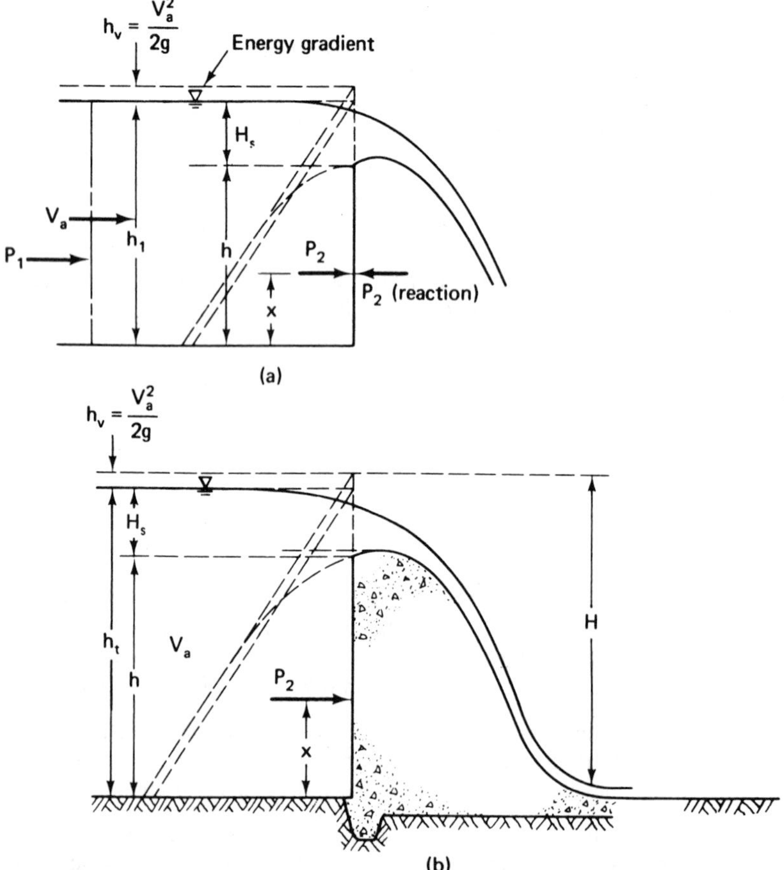

Figure 8.11 Ideal longitudinal profile of an overflow spillway: (a) water nappe over sharp-crest weir, (b) flow profile in overflow spillway.

The U.S. Waterways Experimental Station suggests a set of simple crest profiles that have been found to agree with actual prototype measurements. The geometry of the U.S. Waterways Experimental Station spillway crest profiles is shown in Figure 8.12.

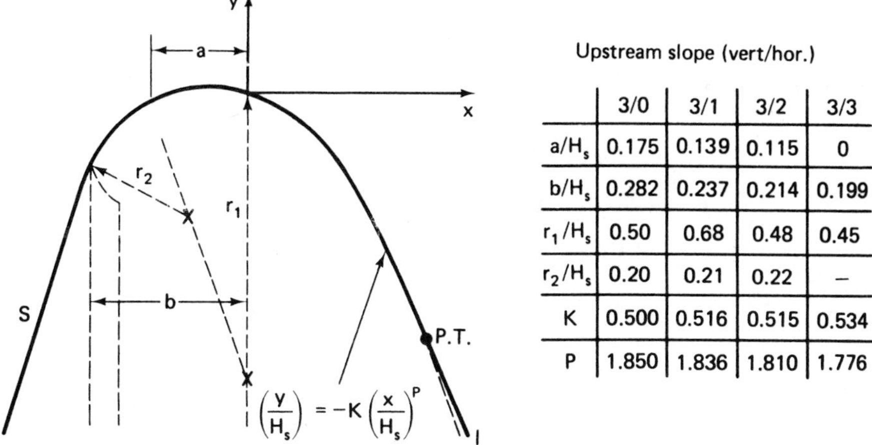

Upstream slope (vert/hor.)

	3/0	3/1	3/2	3/3
a/H_s	0.175	0.139	0.115	0
b/H_s	0.282	0.237	0.214	0.199
r_1/H_s	0.50	0.68	0.48	0.45
r_2/H_s	0.20	0.21	0.22	–
K	0.500	0.516	0.515	0.534
P	1.850	1.836	1.810	1.776

Figure 8.12 Overflow spillway profile.

The discharge of a spillway may be calculated by a formula similar to that derived for flow over a weir, [Equation (8.8)]

$$Q = CLH_s^{3/2} \tag{8.9}$$

where C is the coefficient of discharge, L is the width of the spillway crest. H_a is the sum of the static head, H_s, and the approaching velocity head, $V_a^2/2g$, at the crest,

$$H_a = H_s + \frac{V_a^2}{2g} \tag{8.10}$$

The coefficient of discharge of a particular spillway crest is often determined by scaled model tests (Chapter 10) and accounts both for the energy losses and the magnitude of the approaching velocity head. The value of the coefficient normally ranges between 1.66 to 2.26. A detailed discussion of Equation (8.9) will be presented in Chapter 9.

Example 8.2

An overflow spillway 80 m wide carries a maximum discharge of 400 m³/sec. Define the crest profile for the spillway. Consider a 3:1 upstream slope and a 2:1 downstream slope and assume C = 2.22.

Solution

Applying Equation (8.9), we get

$$Q = 2.22LH_s^{3/2}$$

and

$$H_s = \left(\frac{Q}{2.22L}\right)^{2/3} = \left(\frac{400}{2.22 \cdot 80}\right)^{2/3} = 1.72 \text{ m}$$

From the table in Figure 8.12 we have

$$a = 0.139H_s = 0.239 \text{ m}; \qquad r_1 = 0.68H_s = 1.170 \text{ m}$$

$$b = 0.237H_s = 0.408 \text{ m}; \qquad r_2 = 0.21H_s = 0.361 \text{ m}$$

$$K = 0.516; \qquad P = 1.836$$

and

$$\left(\frac{V}{H_s}\right) = -K\left(\frac{x}{H_s}\right)^P = -0.516\left(\frac{x}{H_s}\right)^{1.836}$$

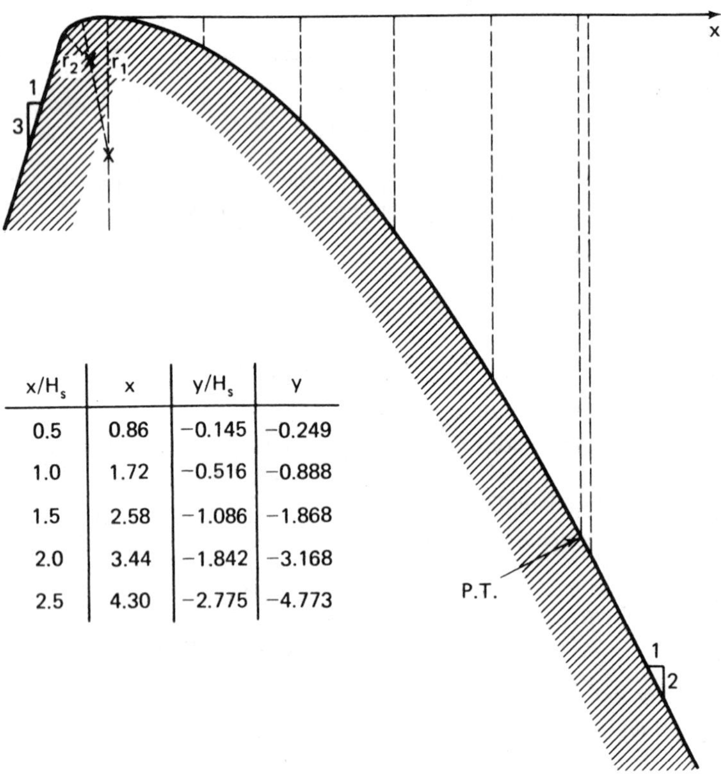

x/H_s	x	y/H_s	y
0.5	0.86	−0.145	−0.249
1.0	1.72	−0.516	−0.888
1.5	2.58	−1.086	−1.868
2.0	3.44	−1.842	−3.168
2.5	4.30	−2.775	−4.773

Figure 8.13

The downstream end of the profile curve will be matched to a straight line with slope 2 : 1. The position of the point of tangency is determined by

$$\frac{d\left(\frac{Y}{H_s}\right)}{d\left(\frac{X}{H_s}\right)} = -KP\left(\frac{X}{H_s}\right)^{P-1} = -0.947\left(\frac{X}{H_s}\right)^{0.836} = -2$$

Hence,

$$\frac{X}{H_s} = 2.45, \qquad X_{\text{P.T.}} = 4.21 \text{ m}$$

$$\frac{Y}{H_s} = 2.674, \qquad Y_{\text{P.T.}} = 4.599 \text{ m}$$

The crest profile curve of the spillway is shown in Figure 8.13.

PROBLEMS

8.6.1 An overflow spillway is designed to discharge 230 m^3/sec and has a maximum head of 2.2 m. Determine the width and the profile of the spillway crest having a vertical upstream slope and a 1.5:1 downstream slope. Assume $C = 2.18$.

8.6.2 Determine the maximum discharge for an overflow spillway 38 m wide and having an available head of 1.86 m. Define the crest profile for both the upstream and downstream slopes of 1:1. Assume $C = 2.22$.

8.6.3 A spillway needs to be designed to carry a peak flow of 50 m^3/sec with the reservoir elevation 1 m above the crest of the spillway. The elevation difference between the reservoir and the tailwater is 15 m. If an overflow spillway is used, with a crest coefficient of 2.0, determine the length of the spillway crest required to handle the discharge.

8.6.4 The coefficient of discharge for an overflow spillway is 1.89 at the head of 0.98 m. The length of the spillway is 9.50 m. Determine the increase in discharge if the head increases to 1.25 m while the discharge coefficient increases to 1.91.

8.7 SIDE-CHANNEL SPILLWAYS

A side-channel spillway carries water from an overflow crest in a channel parallel to the crest (Figure 8.14).

The discharge per unit width of the overflow crest can be determined by Equation (8.8), and the discharge through any section of the side-channel at distance x from the upstream end of the channel is

$$Q_x = xCH_s^{3/2} \tag{8.11}$$

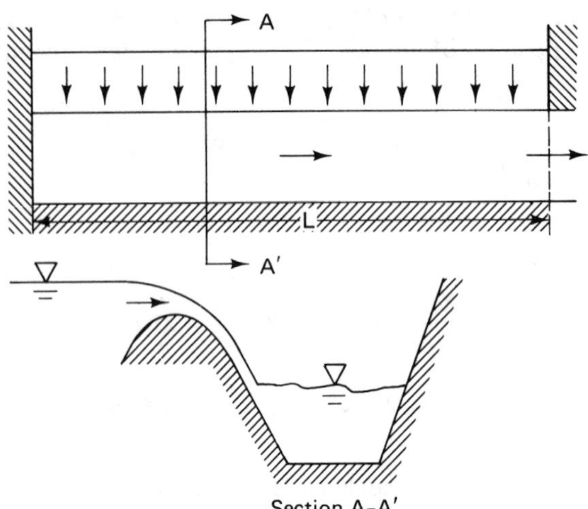

Figure 8.14 Side-channel spillway.

The side-channel spillway must provide a slope steep enough to carry away the accumulating flow in the channel, but a minimum slope and depth at each point along the channel is desired in order to minimize construction costs. For this reason, an accurate water surface profile for the maximum design discharge is important in the side-channel spillway design.

The flow profile in the side channel cannot be analyzed by the energy principle due to highly turbulent flow conditions that cause excess energy loss in the channel. However, an analysis based on the momentum principle has been validated by both model and prototype measurements.

The momentum principle considers the forces and change of momentum between two adjacent sections, dx distance apart, in the side-channel

$$\Sigma F = \rho(Q + \Delta Q)(V + \Delta V) - \rho Q V \tag{8.12}$$

where ρ is the density of water, V is the longitudinal velocity, and Q is the discharge at the upstream section. The symbol Δ signifies the incremental quantity at the adjacent downstream section.

The forces represented on the left-hand side of Equation (8.12) usually include the weight component of the water body between the two sections in the direction of the flow, $(\rho g A\, dx) \sin\theta$; the unbalanced hydrostatic forces,

$$\rho g A \bar{d} \cos\theta - \rho g (A + \Delta A)(\bar{d} + \Delta \bar{d}) \cos\theta$$

and a friction force, F_f, on the channel bottom. Here A is the water cross-sectional area, $\bar{d}$ is the distance between the centroid of the area and the water surface, and θ is the angle of the channel slope.

The momentum equation may thus be written as

$$\rho g A \Delta x \sin\theta + [\rho g A\bar{d} - \rho g(A + \Delta A)(\bar{d} + \Delta\bar{d})]\cos\theta - F_f = \rho(Q + \Delta Q)(V + \Delta V) - \rho Q V \tag{8.13}$$

Let $S_0 = \sin\theta$ for a reasonably small angle, and $Q = Q_1$, $V + \Delta V = V_2$, $A = (Q_1 + Q_2)/(V_1 + V_2)$, and $F_f = \gamma A S_f \Delta x$; the above equation may be simplified to

$$\Delta d = -\frac{Q_1(V_1 + V_2)}{g(Q_1 + Q_2)}\left(\Delta V + V_2\frac{\Delta Q}{Q_1}\right) + S_0\,\Delta x - S_f\,\Delta x \tag{8.14}$$

where Δd is the change in water surface elevation between the two sections. This equation is used to compute the water surface profile in the side-channel. The first term on the right-hand side represents the change in water surface elevation between the two sections due to the impact loss caused by the water overfalling into the channel. The second term represents the change due to friction in the channel. Relating the water surface profile to a horizontal datum, we may write

$$\Delta z = \Delta d - S_0\,\Delta x = -\frac{Q_1(V_1 + V_2)}{g(Q_1 + Q_2)}\left(\Delta V + V_2\frac{\Delta Q}{Q_1}\right) - S_f\,\Delta x \tag{8.15}$$

It may be noticed that when $Q_1 = Q_2$ or when $\Delta Q = 0$, Equation (8.15) reduces to

$$\Delta z = \left(\frac{V_2^2}{2g} - \frac{V_1^2}{2g}\right) - S_f\,\Delta x \tag{8.16}$$

which is the energy equation for constant discharge in an open channel as derived in Chapter 6.

PROBLEMS

8.7.1 The overflow spillway of Example 8.2 discharges into a side-channel spillway with a horizontal bottom slope. If the wall opposite the overflow spillway crest is vertical and the depth of water at the exit end of the side-channel is the critical depth, determine the depth of water at the beginning (the upstream end) of the channel. The width of the channel bottom is 10 m.

8.7.2 Solve Problem 8.7.1 if the channel bottom slope is 0.004.

8.7.3 A rectangular side-channel spillway 100 m long discharges 5.0 m^3/sec/m. The channel is 6 m wide and has a bottom slope of 0.0009. Define the water surface profile for a critical depth at the exit end of the channel as shown schematically in Figure P8.7.3.

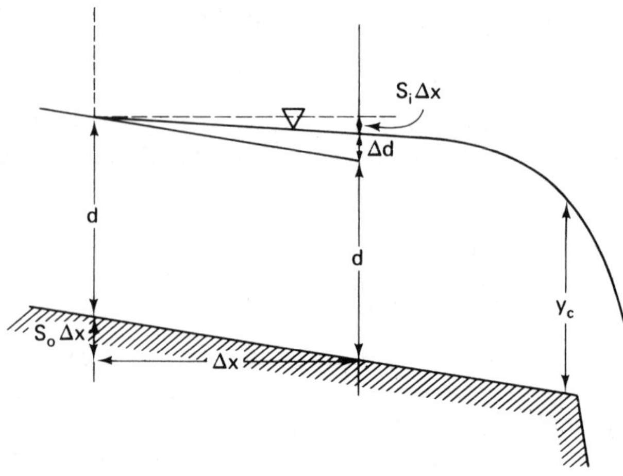

Figure P8.7.3

8.7.4 A 30-m-long overflow spillway drains into a 3-m-wide rectangular channel with bottom slope of 0.01. Determine the water surface profile for a total discharge of 36 m^3/sec if Manning's coefficient is $n = 0.022$.

8.8 SIPHON SPILLWAYS

When a closed conduit is elevated above the hydraulic gradient line (pressure line), the conduit experiences negative pressure. A spillway designed to discharge water in a closed conduit under negative pressure is known as a *siphon spillway*. A siphon spillway begins to discharge under negative pressure when the reservoir water level reaches a certain elevation that primes the conduit. At the beginning stage, the water overflows the spillway crest in the same manner as that of the overfall spillway described in Section 8.6. However, if the water that flows into the reservoir exceeds the capacity of the overfall, the water level at the crest will rise until it reaches and passes the level of the crown, C. At this point the conduit is primed and siphon action begins. Theoretically, the discharge head is increased by the amount equal to $(H - H_a)$, (Figure 8.15), and the discharge rate can thus be materially increased. The large head allows rapid discharge of the excess water until it drops to the near entrance elevation.

The portion of the spillway conduit rising above the hydraulic gradient line (HGL) is under negative pressure. Since the hydraulic gradient line represents zero atmospheric pressure, the vertical distance measured between the hydraulic gradient line and the conduit immediately above the HGL indicates the negative pressure head, $-P/\gamma$, at the location. The crown of a siphon is the highest point in the conduit, and, hence, it is subjected to the maximum negative pressure. The maximum negative pressure at a spillway crown must not be allowed to decrease below the vapor pressure of water at the temperature measured.

If the negative pressure at any section in the conduit drops below the water vapor pressure, the liquid vaporizes, and vast numbers of small vapor cavities

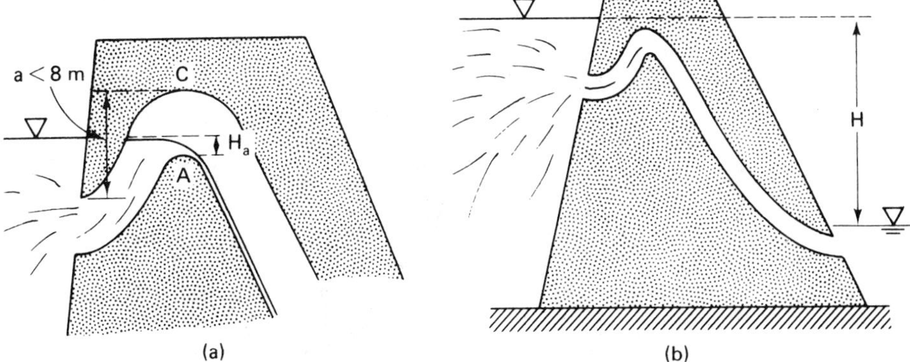

Figure 8.15 Schematic representation of a siphon spillway: (a) beginning stage, (b) siphoning stage.

form. These vapor bubbles are carried down the conduit with the flow. When a cavity bubble reaches the region of higher pressure, the vapor condenses again into liquid and sudden collapse takes place. As the bubble collapses, water surrounding the bubble cavity rushes in at great speed to fill the cavity. All this water collides in the cavity with a great deal of momentum, and potentially damaging pressure is created.

Since, under ordinary conditions, the atmospheric pressure is equivalent to a 10.3 m height of water column, the maximum distance between the crown (highest point in the siphon) and the water surface elevation in the reservoir is limited to approximately 8 m. This difference allows us to include the vapor pressure head, the velocity head, and the head losses between the reservoir and the crown.

Example 8.3

The rectangular siphon shown in Figure 8.16 has a constant cross section of 1 m × 1 m and is 40 m long. The distance between the entrance and the crown is 10 m, the friction factor $f = 0.025$, the inlet coefficient = 0.1, the bend loss coefficient = 0.8 at the crown, and the exit loss coefficient = 1.0. Determine the discharge and the pressure head at the crown section.

Solution

The energy relationship between point 1 (the upstream reservoir) and point 2 (the outlet) may be written as

$$\frac{V_1^2}{2g} + \frac{P_1}{\gamma} + z_1 = \frac{V_2^2}{2g} + \frac{P_2}{\gamma} + z_2 + \underset{\text{entrance loss}}{0.1\frac{V_2^2}{2g}} + \underset{\text{bend loss}}{0.8\frac{V_2^2}{2g}} + \underset{\text{friction loss}}{0.025\left(\frac{L}{D}\right)\frac{V_2^2}{2g}} + \underset{\text{exit loss}}{1.0\frac{V_2^2}{2g}}$$

Here, $V_1 = 0$ and $P_1/\gamma = P_2/\gamma = 0$; $z_1 = 6$ m and $z_2 = 0$. Thus, the above equation may be simplified, to

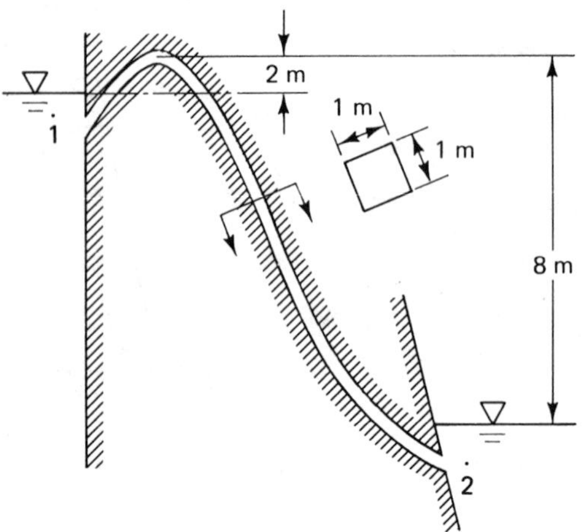

Figure 8.16

$$6 = \left[1 + 0.1 + 0.8 + 0.025\left(\frac{40}{1}\right)\right]\frac{V_2^2}{2g}$$

$$V_2 = 6.37 \text{ m/sec}$$

Hence, the discharge Q is

$$Q = AV = (1)^2(6.37) = 6.37 \text{ m}^3\text{/sec}$$

The energy relationship between point 1 (the reservoir) and point C (the crown) may be expressed as

$$\frac{V_1^2}{2g} + \frac{P_1}{\gamma} + 6 = \frac{V_c^2}{2g} + \frac{P_c}{\gamma} + 8 + 0.1\frac{V_c^2}{2g} + 0.025\left(\frac{10}{1}\right)\frac{V_c^2}{2g} + 0.8\frac{V_c^2}{2g}$$

This equation may be simplified since, $V_1 = 0$, $P_1/\gamma = 0$, and $V_c = V_2 =$ 6.37 m/sec.

$$6 = \frac{(6.37)^2}{2 \cdot 9.81}(1 + 0.1 + 0.8 + 0.25) + \frac{P_c}{\gamma} + 8$$

Therefore, the pressure head at the crown section is P_c/γ

$$\frac{P_c}{\gamma} = -6.45 \text{ m}$$

PROBLEMS

8.8.1 Water is discharged at the rate of 0.1 m^3/sec in a siphon 50 m long. The crown is located 1.2 m above the reservoir water elevation and 10 m away from the entrance. If the siphon diameter is 20 cm, the friction factor is 0.021, and the head loss coefficients are the same as in Example 8.3, determine the elevation difference between the crown section and the downstream reservoir. What is the pressure at the crown?

8.8.2 A siphon spillway has a throat section of 1 m high and a uniform width of 2 m. The siphon discharges at elevation 100 m. The siphon is primed when the headwater reaches an elevation of 110 m. The total energy loss in the conduit, exclusive of outlet velocity head, is 3.2 m. Determine the discharge.

8.8.3 For the reservoir in Problem 8.6.3, a rectangular siphon spillway is used instead of the overflow spillway. If the design elevation in the reservoir is 1 m above the crest of the siphon spillway and the total head loss is 5 $V^2/2g$, determine the width of the siphon.

8.8.4 A siphon spillway is designed to discharge 20 m^3/sec with a crest elevation of 30 m and outlet at 0 m. The radii of the siphon conduit at the crest are 1 m and 2 m, and the allowable gage pressure at the section is −8 m of water column. The crest section is followed, in order, by a vertical section, a 90° bend with centerline radius of curvature of 3 m, and a horizontal section at elevation 0 m. The distance from the entrance to the siphon crest is 3.2 m, and from the vertical section to the outlet is 15 m. If the siphon conduit has a Manning's coefficient $n = 0.025$ and the coefficients of entrance and of bend are, respectively, $K_e = 0.5$ and $K_b = 0.3$, determine the proper dimensions of the sections to satisfy the given requirements.

8.9 CULVERTS

Culverts are built at the points of lowest valley to pass water under the embankments of highways or railroads. The objective of the hydraulic design of culverts is to determine the most economic dimension that can provide the passage of a designed discharge without exceeding the allowable headwater elevation. The major components of a culvert usually include the inlet, the pipe barrel, the outlet, and an outlet energy dissipator, if necessary.

Inlet structures protect embankments from erosion and improve the hydraulic conditions of culverts. Outlet structures are designed to protect culvert outlets from scouring. Although simple in appearance, the hydraulics of culverts may involve that of both the pressure pipe and the open channel. Under various discharge conditions, the hydraulic operation of culverts may be classified into four categories

1. submerged inlet and outlet,
2. submerged inlet with full flow but free discharge at the outlet,

3. submerged inlet with partially full pipe,
4. unsubmerged inlet.

These categories are shown in Figure 8.17. The hydraulic conditions are discussed next.

1. Submergence of the culvert outlet may be the result of inadequate drainage downstream. In this case, the culvert discharge is primarily determined by the tail water elevation (TW) and the head loss of the culvert, regardless of the culvert slope. The culvert flow can be treated as a pressure pipe, and the head loss of the culvert, h_L, is the sum of

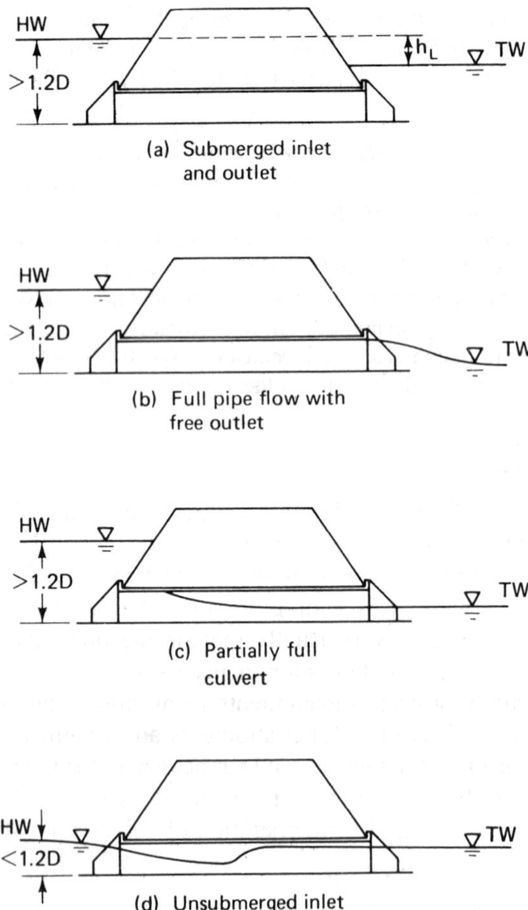

Figure 8.17 Hydraulic operation of culverts.

an entrance loss, h_{ent}, friction loss, h_f, and the velocity head in the barrel.

$$h_L = h_{ent} + h_f + \frac{V^2}{2g}$$

Substituting Equations (3.32) and (3.16) for h_{ent} and h_f, respectively, we have

$$h_L = k_{ent}\left(\frac{V^2}{2g}\right) + \frac{n^2V^2L}{R_h^{4/3}} + \frac{V^2}{2g} \tag{8.17}$$

The entrance coefficient, k_{ent}, is approximately 0.5 for a square-edged entrance and approximately 0.1 for a well-rounded entrance. Common values used for the Manning's roughness coefficient are $n = 0.012$ for concrete pipe and $n = 0.024$ for corrugated steel pipe. Equation (8.17) may be rearranged to express the direct relationship between the discharge and the dimensions of the culvert at any given elevation difference, h_L, between tail water and head water. For a circular culvert,

$$h_L = \left[K_{ent} + \left(\frac{n^2L}{R_h^{4/3}}\right)(2g) + 1\right]\frac{8Q^2}{\pi^2gD^4} \tag{8.18}$$

where Q is the discharge, D is the diameter, and R_h is the hydraulic radius ($R_h = D/4$) of the culvert barrel. For culverts with noncircular cross sections, the head loss may be calculated by Equation (8.17) with the corresponding hydraulic radius calculated by using the cross-sectional area, A, and the wetted perimeter, P.

2. If the discharge carried in a culvert has a normal depth that is larger than the barrel height, the culvert will flow full even if the tail water level drops below that of the outlet. In this case, the discharge is controlled by the head loss and the level of the head water (HW). The hydraulics are the same as discussed before.
3. If the normal depth is less than the barrel height, with the inlet submerged and free discharge at the outlet, a partially full pipe flow condition will normally result, as illustrated in Figure 8.17(c). The culvert discharge is controlled by the entrance condition, and the flow is said to be under *entrance* control. The discharge can be calculated by

$$Q = C_dA\sqrt{2gh} \tag{8.19}$$

where h is the hydrostatic head above the center of the orifice and A is the cross-sectional area. C_d is the coefficient of discharge; common values used in practice are $C_d = 0.62$ for a square-edged entrance and $C_d = 1.0$ for a well-rounded entrance.

4. When the hydrostatic head at the entrance is less than $1.2D$, air will break into the barrel and the culvert will flow under no pressure. In this case, the culvert slope and the barrel wall friction determine the flow condition in the culvert for open channel flow. Due to a sudden reduction of the water area at the entrance, the flow usually enters the culvert in a supercritical condition. The critical depth takes place at the entrance of the barrel. The friction of the barrel wall gradually dissipates the energy. If the rate of dissipation is higher than the flow could gain from the barrel slope, the depth of the flowing water will increase in the downstream direction. Depending on the tail water level, the supercritical flow may convert to subcritical flow through a hydraulic jump. The flow conditions can be computed by applying the water surface profiles developed for open channels.

Example 8.4

A corrugated steel pipe is used as a culvert that must carry a flow rate of 5.3 m^3/sec and discharge into the air. At the entrance, the maximum available water head is 3.2 m above the bottom as shown in Figure 8.18. The culvert is 35 m long and has a square-edged entrance and slope of 0.003. Determine the diameter of the pipe.

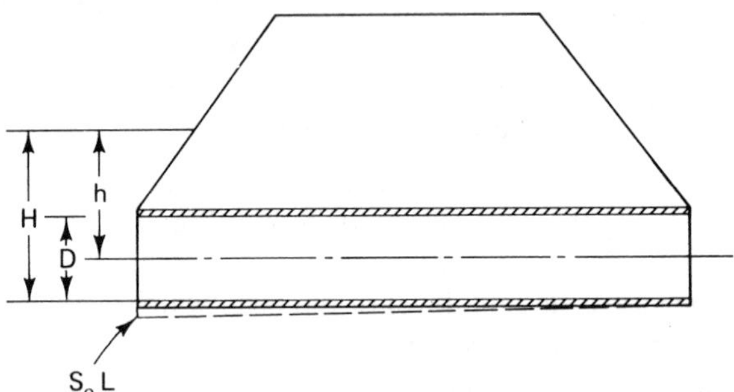

Figure 8.18

Solution

(a) Allowing full pipe flow, the energy balance of the culvert flow may be expressed as (see Figure 8.18),

$$h_L = H - D + S_o L = 3.2 - D + 0.003 \cdot 35$$
$$= 3.305 - D$$

where D is the diameter of the pipe. Also, from Equation (8.18), we have,

$$h_L = \left(k_{ent} + \frac{n^2 L}{R_h^{4/3}} \cdot 2g + 1\right)\frac{8Q^2}{\pi^2 \cdot gD^4}$$

$$= \left[0.5 + \frac{(0.024)^2 \cdot 35 \cdot 19.62}{(D/4)^{4/3}} + 1\right]\frac{8 \cdot (5.3)^2}{\pi^2 \cdot 9.81 \cdot D^4}$$

Combining both equations, we get

$$D + \left(1.5 + \frac{2.51}{D^{4/3}}\right) \cdot \left(\frac{2.321}{D^4}\right) = 3.305$$

By trial, it is found that $D = 1.395$ m.

(b) If the pipe flow is partially full, then the discharge is controlled by the entrance condition only. In this case, the head h, is measured above the centerline of the pipe, and we have,

$$h + \frac{D}{2} = 3.2$$

so that

$$h = 3.2 - \frac{D}{2}$$

and the orifice formula, Equation (8.19), shall be used to compute the discharge

$$Q = 5.3 = C_d A\sqrt{2gh} = C_d \cdot \pi(D/2)^2 \cdot \sqrt{2gh}$$

$$= 0.62 \cdot \pi(D/2)^2 \cdot \sqrt{2 \cdot 9.81 \cdot (3.2 - D/2)}$$

By trial, it is found that $D = 1.24$ m.

Because the rate of friction dissipation is higher than the barrel slope, it is evident that the adequate solution is the one in part (a).

PROBLEMS

8.9.1 A culvert is installed on a slope of 0.05. It is designed to discharge 7.5 m^3/sec at the maximum head water elevation of 4.8 m. Determine the diameter of the corrugated steel pipe to be used if the length of the culvert is 12 m.

8.9.2 A rectangular concrete conduit is to be used as a culvert on a slope of 0.02. The culvert is 15 m long and has a cross section of 2 m × 2 m. If the tail water elevation is 1.8 m above the crown at the outlet, determine the head water elevation necessary to pass a 10 m^3/sec discharge.

8.9.3 A circular corrugated metal highway culvert is 60 m long and is laid on a slope of 0.1, with free outlet. Assume that Manning's coefficient for the corrugated metal is $n = 0.022$ and that the inlet contraction coefficient is 0.6. The culvert is designed to pass a 2.5 m^3/sec discharge when the head water elevation is 0.5 m above the crown at the inlet. Determine the size of the pipe to be used. Neglect the approaching velocity head.

8.9.4 A culvert needs to be designed to handle a maximum discharge of 20 m^3/sec across a length of 100 m. The design elevation difference between the head water at the entrance and the tail water at the exit is 5.5 m. The design condition requires that both inlet and outlet be square-edged and flush with the walls. Assume that a concrete pipe of circular cross section is to be used that has a Manning's coefficient of $n = 0.013$. Select a pipe diameter based on the slope of your choice. Neglect the approaching velocity head.

8.9.5 Circular concrete pipes ($n = 0.013$) are used as culvert on a slope of 0.09. The culvert is 1.2 m in diameter, 42 m long. The entrance is square edged and flush with the wall. The tail water level is 0.6 m below the culvert crown at the outlet. Determine the discharge if the head water level is (a) 0.5 m above the crown at the inlet, (b) coincident with the crown, and (c) 0.5 m below the crown.

8.9.6 For the culvert in Problem 8.9.5, determine the head water elevation for (a) a discharge of 2 m^3/sec, (b) a discharge of 5 m^3/sec, and (c) a discharge of 10 m^3/sec.

8.10 STILLING BASINS

When the water velocity at an outlet is very high, the excessive amount of kinetic energy carried in the flow may be devastating to the outlet structure or its surroundings. A transitional energy dissipating structure is used. This situation often occurs at the end of a spillway where spilled water is highly accelerated and direct disposal in the downstream channel could produce enormous erosion. To avoid damage, a transition structure commonly known as a *stilling basin* may be constructed. The water flow changes from the supercritical state to the subcritical state, which contains much less damaging power. The stilling basin may be horizontal or inclined to match the slope of the channel. In any case, it should provide obstruction (or friction force) sufficient to overcome the gravitational forces so that the flow can be decelerated until a hydraulic jump is processed.

The relation between the energy to be dissipated and the depth of the flow in the basin can be associated to the Froude number, $V/\sqrt{gD}$, as will be discussed in Section 10.4. Generally speaking, no special stilling basin is needed when the outlet flow Froude number is less than 1.7. As the Froude number increases, *energy dissipators* such as baffles, sills, and blocks may be installed along the floor of the basin to enhance the energy dissipation within the limited length of the basin. The U.S. Bureau of Reclamation has obtained a comprehensive set of curves to define the dimensions of both the stilling basin and the various types of dissipators. These curves, which are based on extensive experimental data, are shown in Figures 8.19, 8.20, and 8.21.

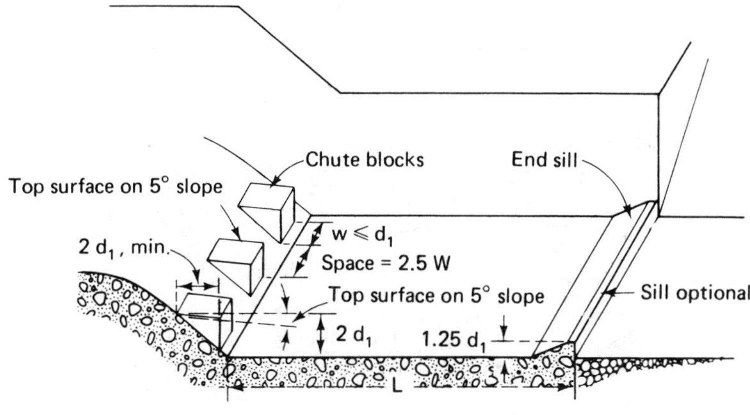

(a) Type IV basin dimensions

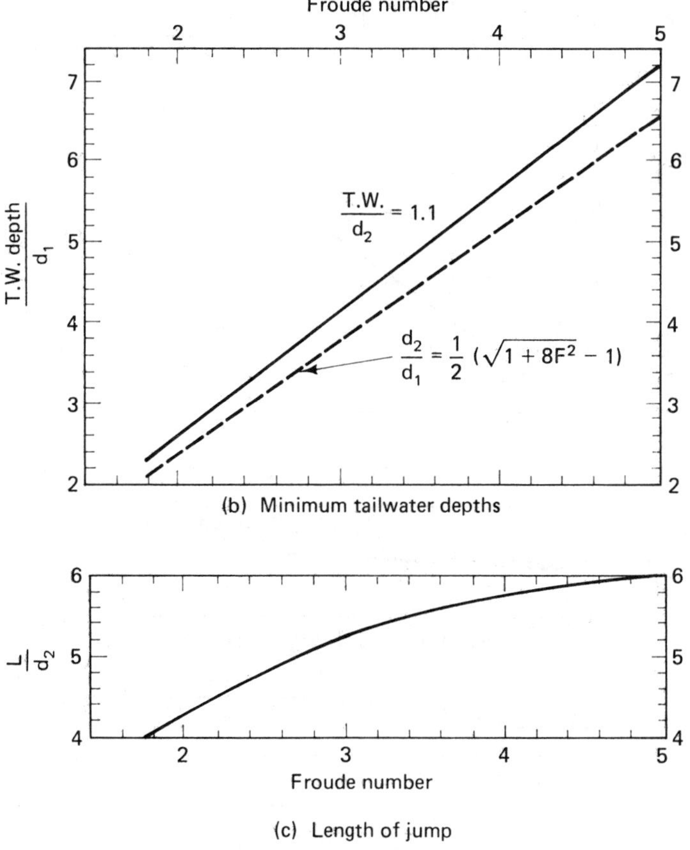

(c) Length of jump

Figure 8.19 Energy dissipator, the U.S. Bureau of Reclamation Type IV (for approaching Froude number between 2.5 and 4.5). (Courtesy of U.S. Bureau of Reclamation.)

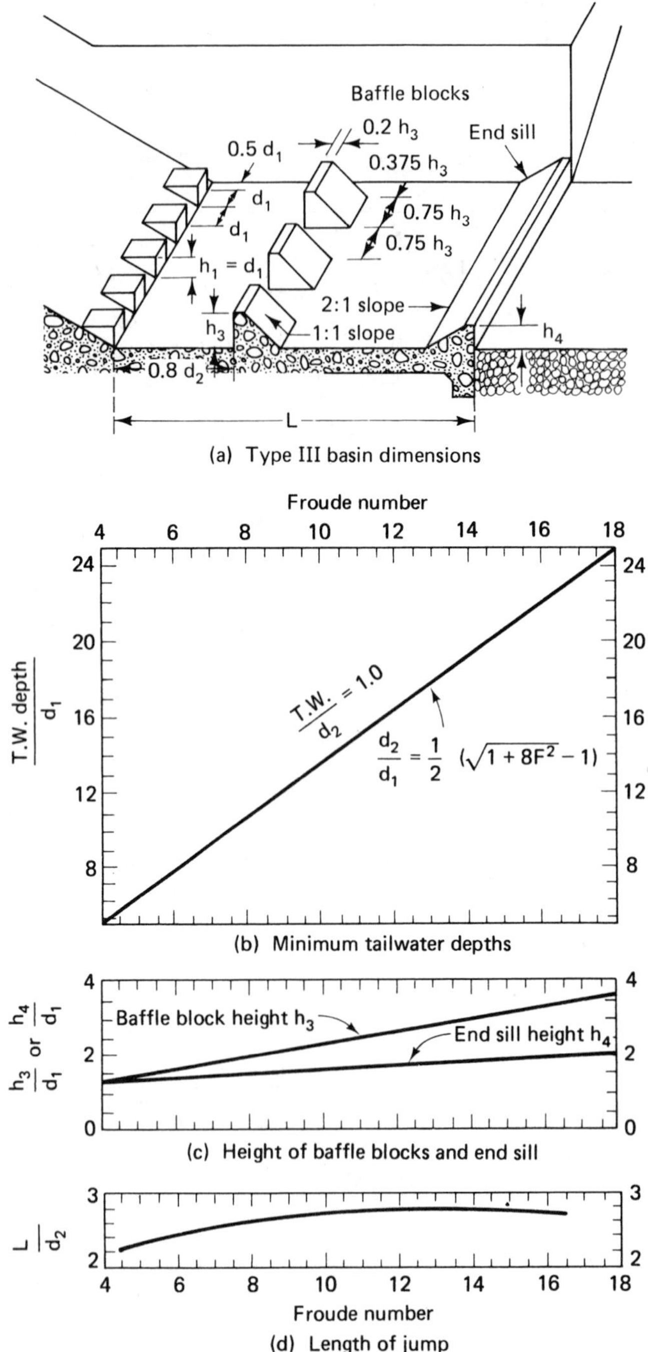

Figure 8.20 Energy dissipator, the U.S. Bureau of Reclamation Type III (for Froude number above 4.5 and approaching velocity less than 20 m/sec). (Courtesy of U.S. Bureau of Reclamation.)

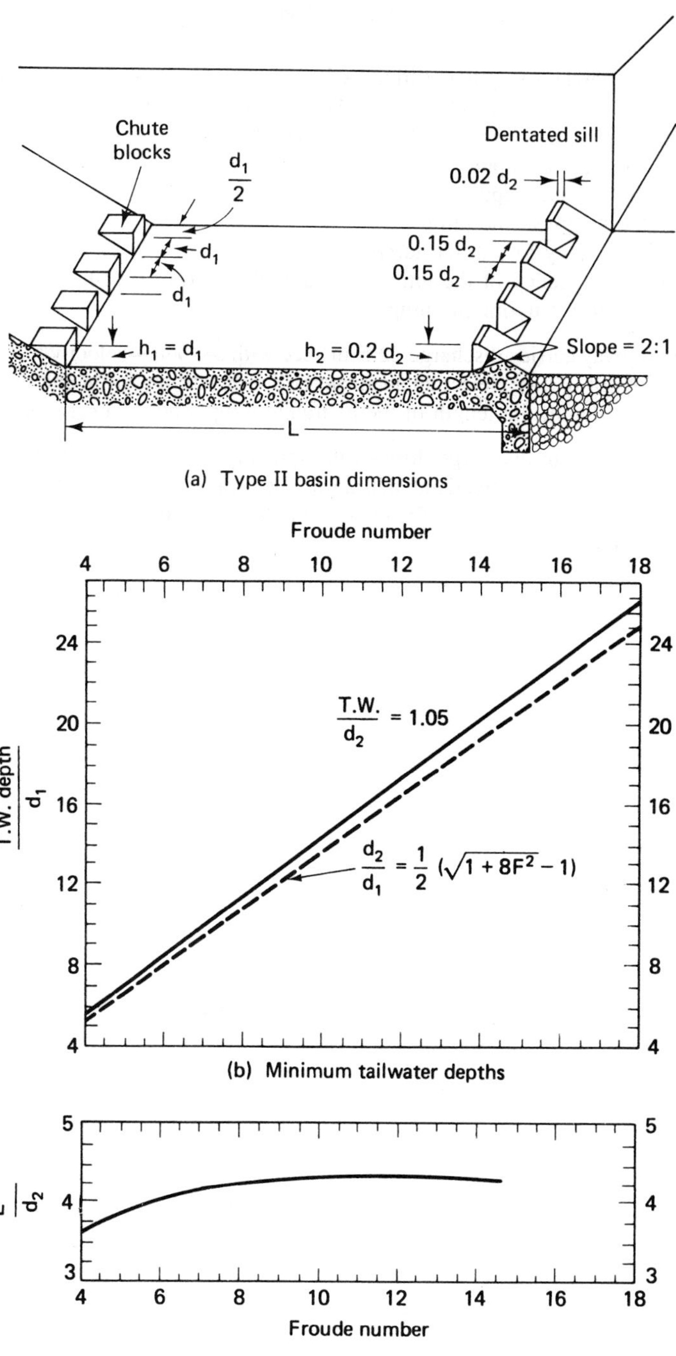

Figure 8.21 Energy dissipator, the U.S. Bureau of Reclamation Type II (for Froude number above 4.5). (Courtesy of U.S. Bureau of Reclamation.)

PROBLEMS

8.10.1 A horizontal rectangular stilling basin of U.S.B.R. Type III is used at the outlet of a spillway to dissipate energy. The spillway discharges 10 m^3/sec and has a uniform width of 10 m. At the point where the water enters the basin, the velocity is 10 m/sec. Compute
(a) the sequent depth of the hydraulic jump,
(b) the length of the jump,
(c) the energy loss in the jump,
(d) the efficiency of the jump defined as the ratio of specific energy after and before the hydraulic jump.

8.10.2 A spillway carries discharge 22.5 m^3/sec with the outlet velocity of 15 m/sec at a depth of 0.2 m. Select an adequate U.S.B.R. stilling basin and determine the sequent depth, the length of the hydraulic jump, and the energy loss.

8.10.3 An increase in discharge through the spillway in Problem 8.10.2 to 45 m^3/sec will increase the outlet depth to 0.25 m. Determine the sequent depth, the length, the energy loss, and the efficiency of the hydraulic jump if a U.S.B.R. Type II basin is used.

9

WATER MEASUREMENTS

Measurement of water flow provides fundamental data for analysis, design, and operation for every hydraulic system. A variety of devices and methods are available for measuring water flow in the field and in the laboratory. Hydraulic measurements are usually performed to measure velocity, pressure, or discharge and are based on the fundamental laws of physics and fluid mechanics. In general, each device is designed to perform under certain specific conditions; hence, there are limitations to each measurement device. Proper selection of a device for a particular application should be based on an understanding of the fundamental fluid mechanical principles, which are discussed in this chapter. Many details about installation and operation of a particular device can be found in specialized literature such as the fluid meter publications of the American Society of Mechanical Engineers (ASME).

9.1 PITOT TUBES AND CURRENT METERS—VELOCITY MEASUREMENTS

Water flow velocity in every conduit varies between near-zero values close to the stationary wall, to a maximum value in the midstream. It is often interesting to measure the velocity distribution in a conduit. This is done by making local measurements at several positions in a cross section. The measurements should only be made with velocity probes of small size, so that local flow patterns will not be disturbed by the presence of the probe in the field. Instruments commonly used for velocity measurement are *Pitot tubes* and *current meters*.

Pitot tubes* are hollow tubes bent to measure pressure in the flowing stream. The probe usually consists of two tubes that are bent in such a way that the open end of one tube is directly opposite the velocity vector and the other is parallel to the flow. The probe body is aligned in the direction parallel to the flow (Figure 9.1). In order to reduce the size of the probe, the two tubes are usually combined into one concentric construction, that is, a smaller tube is inside a larger one [Figure 9.1(b)].

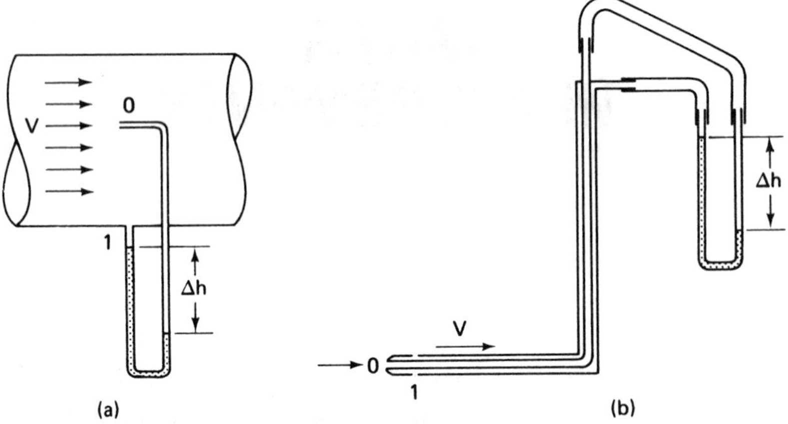

Figure 9.1 Schematic representation of a Pitot tube: (a) separated tubes, (b) combined tubes.

At the end opening, 0, the flow reaches a stagnant point where the velocity is zero. The pressure sensed at this opening is the stagnant point pressure, or *stagnation pressure*. At the side openings, 1, the flow velocity, v, is practically undisturbed. These openings sense the dynamic (or ambient) pressure at the site.

Applying Bernoulli's equation between the two positions, 0 and 1, and neglecting the small vertical distance in between, we may write

*Henri de Pitot (1695–1771) first used an open-ended glass tube with a 90° bend to measure velocity distribution in the Seine river. The rise of water elevation in the vertical part of the tube indicated static pressure. Pitot did not apply Bernoulli's principle to obtain the correct velocity as discussed in this section.

$$\frac{p_0}{\gamma} + 0 = \frac{p_1}{\gamma} + \frac{V^2}{2g}$$

Hence, the flow velocity can be determined

$$V^2 = 2g\left(\frac{p_0 - p_1}{\gamma}\right) = 2g\left(\frac{\Delta p}{\gamma}\right) = 2g(\Delta h)$$

or

$$V = \sqrt{2g\,\Delta h} \tag{9.1}$$

Notice that the quantity (Δh) indicates the height of the water column shown in the manometer. A different, immersible indicator liquid, such as mercury, is used for measuring the velocity of water. In such a case, the height of the indicator liquid column shown in the manometer must be converted into equivalent height of water column for computing the velocity using Equation (9.1).

Pitot tubes are widely used to measure the pressure and the velocity of water flow. They are both reliable and accurate, for they involve a simple physical principle and a simple setup. The Pitot tube is especially good for measuring water velocities under conditions in which the exact direction of the stream flow cannot be determined and in which misalignments of the probe to the stream are likely to occur. The Prandtl-type Pitot tube, as shown in Figure 9.1(b), has approximately a 1% error at an angle of 20° to the direction of the stream flow.

The outer diameter of a Pitot tube is typically small, say 5 mm. The two pressure tubes inside are much smaller. Because of the small diameters of these tubes, caution must be used to keep air bubbles from becoming trapped inside. Surface tension at the interface can produce a predominant effect in the small tubes that may totally inactivate the system.

Example 9.1

A Pitot tube is used to measure velocity at a certain point in a water pipe. The manometer indicates a column of 14.6 cm height. The indicator fluid has specific gravity of 1.95. Compute the velocity.

Solution

The column of 14.6 cm indicator fluid (sp. gr. 1.95) is equivalent to a water column of h.

$$h = 14.6 \cdot (1.95 - 1.0) = 13.87 \text{ cm} = 0.1387 \text{ m}$$

Applying Equation (9.1), we have

$$V = \sqrt{2gh} = \sqrt{2 \cdot 9.81 \cdot 0.1387} = 1.65 \text{ m/sec}$$

Current meters are frequently used to measure the speed of water current. There are two different types of current meter, the cup type and the propeller type.

The *cup-type* current meter usually consists of four to six evenly shaped cups mounted radially about a vertical axis of rotation [Figure 9.2(a)]. The current rotates the cups about the axis at a rate proportional to the current speed. Each revolution is registered by a conductor that transmits electrical contact made by the cup wheel to an earphone or a counter. Most cup-type current meters do not register speed below a few centimeters per second because of starting friction.

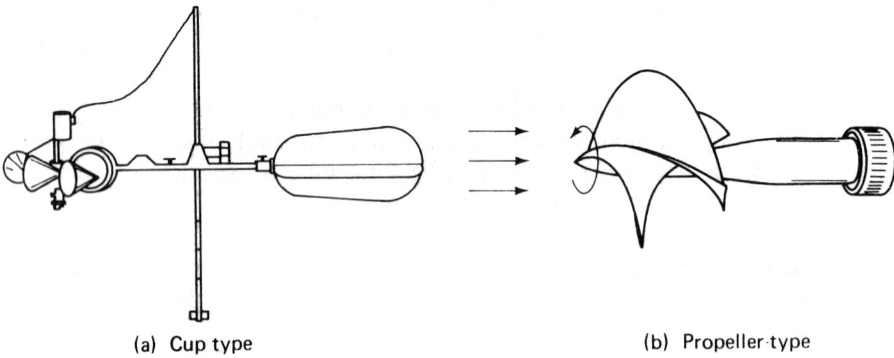

Figure 9.2 Current meters: (a) cup type, (b) propeller type.

The *propeller-type* current meter has a horizontal axis of rotation [Figure 9.2(b)]. It is more suitable for measuring higher ranges in the midstream regions.

Depending on the design and construction of the current meter, the speed of axial rotation may not be linearly proportional to the speed of the water current. For this reason, each current meter must be individually calibrated before it is used for field measurements. Calibrations may be carried out in a straight stretch of a towing tank by towing the probe through still water at constant speeds. A calibration curve covering the range of applicable speeds should be provided with each current meter.

PROBLEMS

9.1.1 A Pitot tube directed into a 4-m/sec stream of water has a difference of 14.2 cm in the column in a manometer. Determine the specific gravity of the indicator fluid.

9.1.2 A static tube (with the open end directly opposite to the velocity vector) is inserted into a small stream of flowing oil. Specific gravity of the oil is 0.85. There is a rise of the fluid, h, in the vertical tube. What is the velocity?

9.1.3 A Pitot tube (with both stagnant point pressure and dynamic pressure measured) is used in the stream in Problem 9.1.2 What is the equivalent height of the oil column?

9.1.4 Measurements in a water tunnel indicate that the stagnation pressure and the dynamic pressure differ by 12.2 cm of mercury column. What is the water velocity?

9.2 PRESSURE MEASUREMENTS

Pressure at any point in a liquid is defined as the normal force exerted by liquid on the unit area surface at the point. A common method used to measure this force is to measure the pressure made through the openings in the wall boundary. The liquid is connected through the openings to a bank of manometers in which the value of liquid pressure is determined by the height of liquid in a column. When the liquid is stationary, the manometer measures the hydrostatic pressure at the opening. If the liquid is moving past the boundary, the pressure at the opening decreases with the increase of flow velocity at the opening. The amount of pressure decrease can be calculated by Bernoulli's principle.

It is particularly important that the openings on the boundary satisfy certain characteristics so that the true ambient pressure can be registered. The openings must be flush with the surface and normal to the boundary. Figure 9.3(a) and (b) schematically demonstrates the various correct and incorrect connections of boundary openings for pressure measurement. The plus signs (+) indicate that the opening registers a higher than actual pressure value; the minus signs (−) indicate that the opening registers a lower than actual pressure value. In order to eliminate the irregularities and variations that might cause significant errors in measurement, a series of piezometric connections may sometimes be made at a given cross section in a closed conduit. The connections are in turn connected into a single manometer column so that the manometer registers an average pressure of all the piezometric openings in the cross section. This multiopening

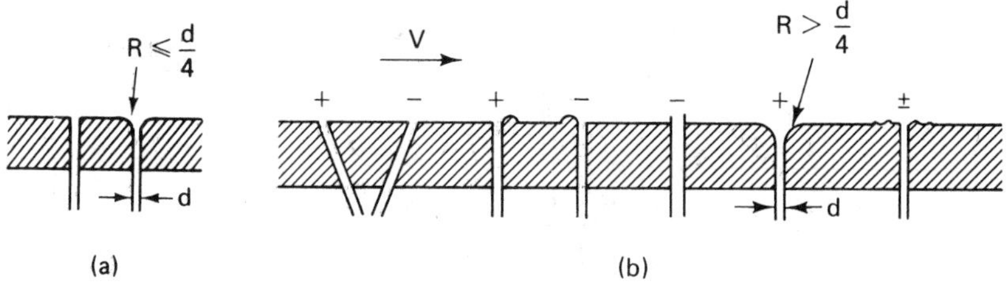

Figure 9.3 Pressure openings: (a) correct connections, (b) incorrect connections.

system is effective in sections of relatively straight pipe where the velocity profiles are reasonably symmetric and the pressure difference existing between one side of the pipe and the other is very small. If a large pressure difference exists between any of the openings, measurement error can develop because water may flow out of the higher-pressure side and into the lower-pressure openings. Caution must be taken to ensure that no net flow occurs through any piezometric openings.

To increase the sensitivity, inclined manometers may be used in which a small change in pressure can drive the indicator fluid a great distance along the sloped manometer tube (Figure 9.4). Other manometer configurations were discussed in Chapter 1.

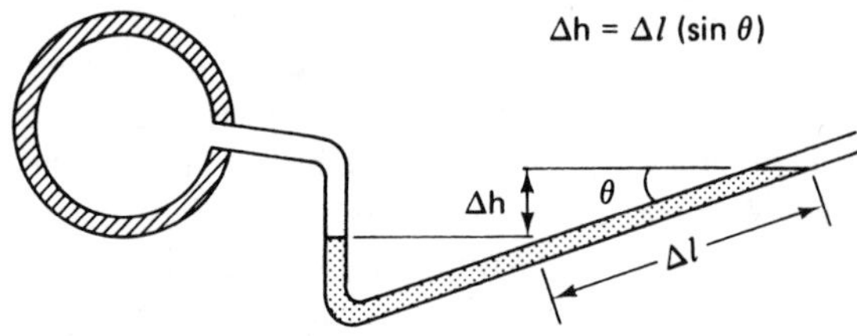

Figure 9.4 Inclined manometer.

The manometer systems may be used to measure pressure in both gases and liquid under steady flow conditions. They are not suitable for applications in time-varying flow fields that require high-frequency response both in probes and in recording systems. Electronic pressure cells are available commercially. The quartz crystal pressure cell, for example, can respond to flow field fluctuation from direct current up to 20,000 Hz over a wide range of pressure. Similar to the piezometric openings, the pressure cells must also be mounted flush with the boundary surface. Slight irregularity in alignment may cause significant error in measurement, as discussed previously.

Example 9.2

An open-ended U-tube (Figure 9.5) is filled with water. Oil is poured into one leg of the U-tube and causes the water surface in one leg to rise 6 cm above the other. The oil column measures 8.2 cm. What is the specific gravity of the oil?

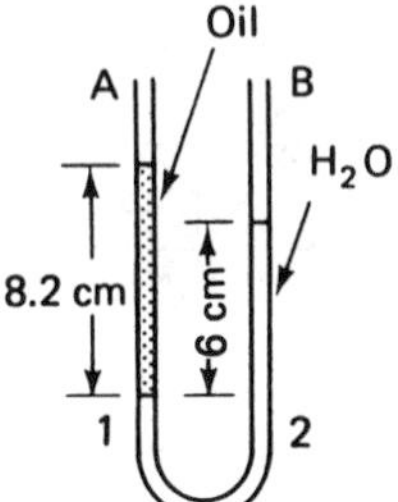

Figure 9.5

Solution

To balance the columns in the two legs of the U-tube, the pressure at levels 1 and 2 must be the same. That is, the pressure generated by the 6-cm water column must be the same as that by the 8.2-cm oil column.

$$p_{\text{water}} = 6 \text{ cm} \cdot (\gamma_{\text{water}}); \qquad p_{\text{oil}} = 8.2 \text{ cm} \cdot (\gamma_{\text{oil}})$$

$$6 \cdot (\gamma_{\text{water}}) = 8.2 \cdot (\gamma_{\text{oil}})$$

$$\text{specific gravity of oil} = \frac{\gamma_{\text{oil}}}{\gamma_{\text{water}}} = \frac{6}{8.2} = 0.732$$

PROBLEMS

9.2.1 If the U-tube in Example 9.2 is connected to a pressure vessel at leg B and if the column in leg A is mercury instead of oil, what is the pressure in the vessel? End A is open to atmosphere and the vessel contains pressurized air.

9.2.2 At one section a straight circular pipe of 1.2-m diameter has two pressure taps bored into the vessel wall with exactly the same type of connections. Open-ended mercury U-tubes are connected to the taps to read pressure. If the tap on top of the pipe reads 7.24 cm of mercury column and that on the bottom reads 16.06 cm, what is the specific gravity of the fluid inside the pipe? What is the mean pressure in the pipe?

9.2.3 A simple, open-ended glass tube is bent 90°. The tube is inserted into a flowing stream with one end facing the direction of the flow and the other end oriented vertically above the water. It is noted that the water surface in the vertical tube rises 4 cm above the stream surface. What is the stream velocity?

9.3 DISCHARGE MEASUREMENTS IN PIPE FLOWS—VENTURI METERS, NOZZLE METERS, ORIFICE METERS, AND BEND METERS

Although measurement of discharge in pipe flow can be accomplished by several different methods, the simplest and most reliable method is volumetric (or weight) determination. This method requires only a stopwatch and an open tank to collect the water flowing from the pipe. The discharge rate can be determined by measuring the water volume (or weight) collected per unit time. Because of its absolute reliability, this method is frequently used for calibration of various types of flow meters. It cannot be used in routine, operational applications because the flowing water must be totally diverted into a container tank every time a measurement is made.

The variation in pressure head, associated with localized changes in velocity and energy head loss through a sudden change in pipe cross-sectional geometry, may be correlated to the pipe flow discharge. This principle is usually used with Venturi meters, nozzle meters, and orifice meters.

A *Venturi meter* is a machine-cast section of pipe with a narrow throat. Two piezometric openings are installed at the entrance and at the throat, as shown in Figure 9.6. Applying the Bernoulli equation at sections 1 and 2, and neglecting head loss, we get

$$\frac{V_1^2}{2g} + \frac{p_1}{\gamma} + z_1 = \frac{V_2^2}{2g} + \frac{p_2}{\gamma} + z_2 \tag{9.2}$$

The continuity equation between the two sections is

$$A_1 V_1 = A_2 V_2 \tag{9.3}$$

where A_1 and A_2 are, respectively, the pipe and the throat cross-sectional areas. Substituting Equation (9.3) into Equation (9.2) and rearranging, we have the discharge, Q

$$Q = \frac{A_1}{\sqrt{\left(\frac{A_1}{A_2}\right)^2 - 1}} \sqrt{2g\left(\frac{p_1 - p_2}{\gamma} + z_1 - z_2\right)} \tag{9.4}$$

The equation may be simplified to

$$Q = C_d A_1 \sqrt{2g\left[\Delta\left(\frac{p}{\gamma} + z\right)\right]} \tag{9.5a}$$

where

$$C_d = \frac{1}{\sqrt{\left(\frac{A_1}{A_2}\right)^2 - 1}} \tag{9.5b}$$

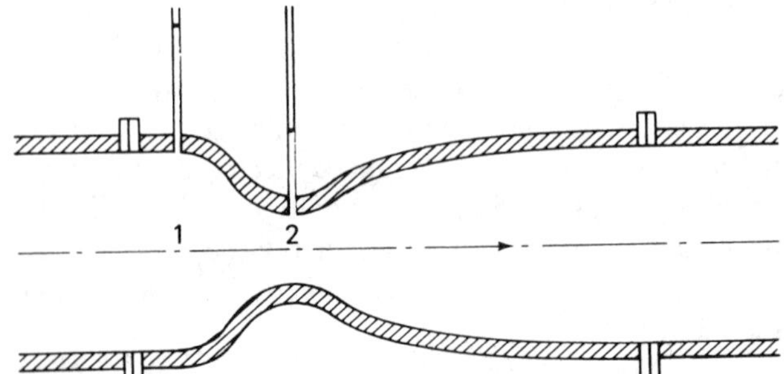

Figure 9.6 Venturi meter.

For Venturi meters installed in a horizontal position,

$$Q = C_d A_1 \sqrt{2g\left(\frac{\Delta p}{\gamma}\right)} \tag{9.5c}$$

For Venturi meters, the coefficient, C_d, can be directly computed from the values of A_1 and A_2. The difference between this theoretically computed value and that obtained from experiments should not exceed a few percent under most practical conditions.

For satisfactory operation, the meter should be installed in a section of the pipe where the flow is relatively undisturbed before it enters the meter. To ensure this, a section of straight and uniform pipe that is free from fittings and is at least 30 diameters in length must be provided before the meter installation.

Example 9.3

A 6-cm (throat) Venturi meter is installed in a 12-cm diameter horizontal water pipe. The differential manometer shows a reading of 15.2 cm of mercury column. Calculate the discharge.

Solution

$$A_1 = \frac{\pi}{4}\cdot(12)^2 = 113.1 \text{ cm}^2 \qquad A_2 = \frac{\pi}{4}\cdot(6)^2$$

The coefficient, C_d, can be calculated from the area ratio by using Equation (9.5b).

$$C_d = \frac{1}{\sqrt{\left(\frac{A_1}{A_2}\right)^2 - 1}} = 0.258$$

Applying Equation (9.5c) for a horizontally installed Venturi meter, the discharge, Q, can be calculated.

$$Q = 0.258\cdot\frac{113.1}{10{,}000}\sqrt{2\cdot 9.81\left(\frac{15.2\cdot 12.6}{100}\right)} = 0.0179 \text{ m}^3/\text{s}$$

Nozzle meters and *orifice meters* (Figure 9.7) are based on the same principle of energy head conversion and head loss across a concentric obstruction installed in the pipe conduit. The discharge equations for the nozzle meter and the orifice meter have the same form as that derived for the Venturi meter [Equation 9.5a)]. The main difference in application is that the value of the coefficient of discharge for the nozzle meters and orifice meters would be different from the theoretical value, C_d, calculated by using Equation (9.5b).

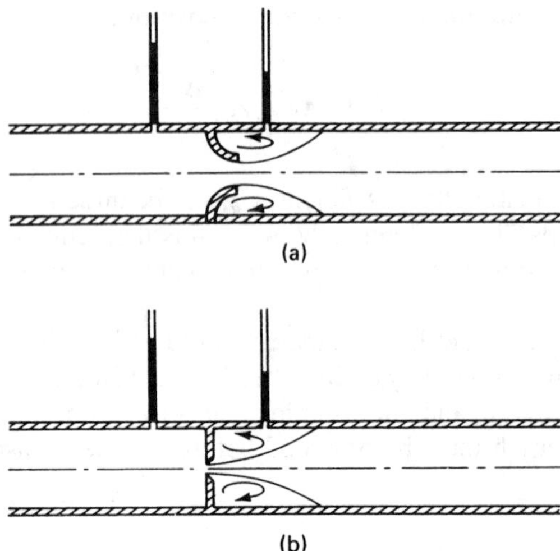

Figure 9.7 (a) Nozzle meter, (b) orifice meter.

This is primarily due to separation of the stream flow from the pipe wall boundary immediately downstream from the obstruction (*vena contracta*).

Nozzle meters and orifice meters encounter an especially significant amount of head loss since most of the pressure energy that converted to kinetic energy (to rush the fluid through the narrow opening) cannot be recovered. The coefficient of discharge may vary significantly from one meter to another. The value depends not only on the status of flow in the pipe (the pipe Reynolds number) but also on the area ratio between the nozzle (or orifice) and the pipe, the location of the pressure taps, and the upstream and downstream condition of the pipe flow. For this reason, on-site calibration is recommended for each meter installed. If installation of a pipe orifice meter is made without an on-site calibration, reference should be made to the manufacturer's data, and the detailed installation requirements should be followed.

The coefficient for nozzle meters and orifice meters cannot be directly computed from the area ratio, A_1/A_2. The discharge equations [Equations (9.5a), (9.5b), and (9.5c)] must be modified by an experimental coefficient, C_v.

$$Q = C_v C_d A_1 \sqrt{2g\, \Delta\left(\frac{P}{\gamma} + z\right)} \tag{9.6a}$$

where z is the difference in elevations between the two pressure taps. For horizontal installations,

$$Q = C_v C_d A_1 \sqrt{2g\, \Delta\left(\frac{P}{\gamma}\right)} \tag{9.6b}$$

Extensive research on the fluid nozzle has been sponsored by ASME and the International Standards Association (ISA) to standardize the nozzle geometry, installation, specification, and experimental coefficients. One of the typical

ASME flow nozzle installations commonly used in the United States and the corresponding experimental coefficients are shown in Figure 9.8.

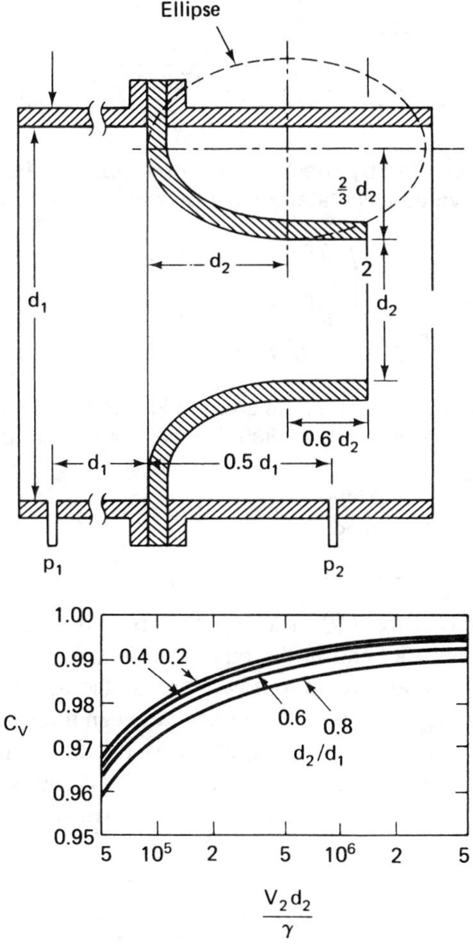

Figure 9.8 ASME nozzle dimensions and coefficients.

Compared to Venturi meters and nozzle meters, orifice meters are even more affected by flow conditions. For this reason, detailed instructions of installation and calibration curves must be provided by the manufacturer for each type and size. If a meter is not installed strictly according to the instructions, it should be calibrated individually on site.

Example 9.4

An ASME flow nozzle 6-cm in diameter is installed in a 12-cm waterline. The attached differential manometer reads a 15.2-cm column of mercury. The water in the pipeline is 20°C. Determine the discharge.

Solution

The diameter ratio is $d_2/d_1 = 6/12 = 0.5$, and $C_d = 0.258$ (from Example 9.3). Assume the experimental coefficient $C_v = 0.99$. The corresponding discharge can be calculated from Equation (9.6b)

$$Q = 0.99 \cdot 0.258 \cdot 0.01131 \sqrt{2 \cdot 9.81 \left(\frac{15.2 \cdot 12.6}{100}\right)} = 0.0177 \text{ m}^3/\text{sec}$$

This value must be verified by checking the corresponding Reynolds number of the nozzle. The N_{R_e} value calculated based on the discharge is

$$N_{R_e} = \frac{\left(\dfrac{0.0177}{\dfrac{\pi}{4}(0.06)^2} \cdot 0.06\right)}{1.007 \cdot 10^{-6}} = 3.5 \cdot 10^5$$

With this Reynolds number value, the chart in Figure 9.8 gives a better value of the experimental coefficient, $C_v = 0.986$. Hence, the correct discharge is

$$Q = \frac{0.986}{0.99} \cdot 0.0177 = 0.0176 \text{ m}^3/\text{sec}$$

A *bend meter* (Figure 9.9) measures the pressure difference between the outer and inner sides of a bend in the pipeline. Centrifugal force developed at a pipe bend forces the mainstream to flow closer to the outer wall of the pipe at the bend. Difference in pressure is developed between the inside and outside of the bend. The pressure difference increases as the flow rate increases. The relationship between the pressure difference measured and the discharge in the pipe can be calibrated for flow rate determinations. The discharge equation may be expressed as

$$Q = C_d A \sqrt{2g\left[\Delta\left(\frac{P_o}{\gamma} - \frac{P_i}{\gamma}\right)\right]} \qquad (9.7)$$

where A is the pipe cross-sectional area and P_i and P_o are the local pressure values

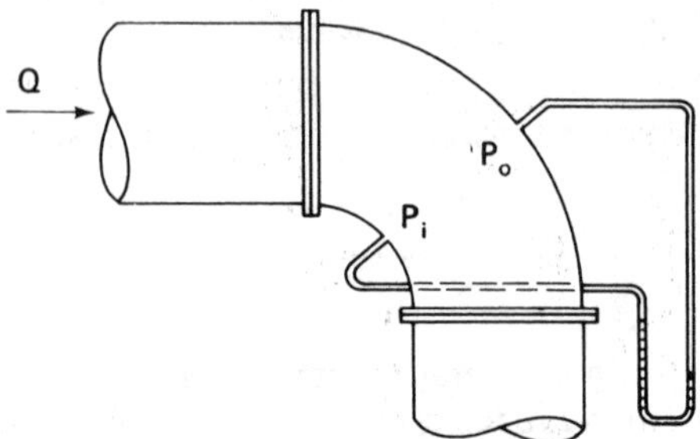

Figure 9.9 Bend meter.

registered at the inside and outside of the pipe bend, respectively. C_d is the discharge coefficient, which can be determined by calibration in place.

If the bend meter cannot be calibrated in place, the pipe discharge may still be determined with an accuracy of approximately 10% if the pipe flow Reynolds number is sufficiently large and if at least 30 diameters of straight pipe are provided upstream from the bend. In such a case, the discharge coefficient is approximated by

$$C_d = \frac{R}{2D} \tag{9.8}$$

where R is the center line radius at the bend and D is the pipe diameter.

Bend meters are inexpensive. An elbow already in the pipeline may be used without additional installation cost or added energy head loss.

PROBLEMS

9.3.1 The maximum flow rate in a 20-cm diameter waterline is 0.12 m^3/sec. If a 12-cm (throat) Venturi meter is installed in the pipe to measure the flow, estimate how long a vertical U-tube must be provided for the differential manometer if water and mercury are used in the manometer.

9.3.2 A 20-cm Venturi meter is installed in a 50-cm diameter waterline. If the pressure taps read 30 N/cm^2 and 20 N/cm^2 when the meter is installed in a horizontal position, what is the flow rate?

9.3.3 A 10-cm ASME flow nozzle is installed in a 20-cm waterline. The attached manometer contains mercury and water and registers a difference of a 42-cm column. Calculate the discharge of the pipe.

9.3.4 Determine the discharge in the 40-cm diameter waterline shown in Figure P9.3.4. The nozzle meter has a throat diameter of 32 cm and is built and installed according to the ASME standard.

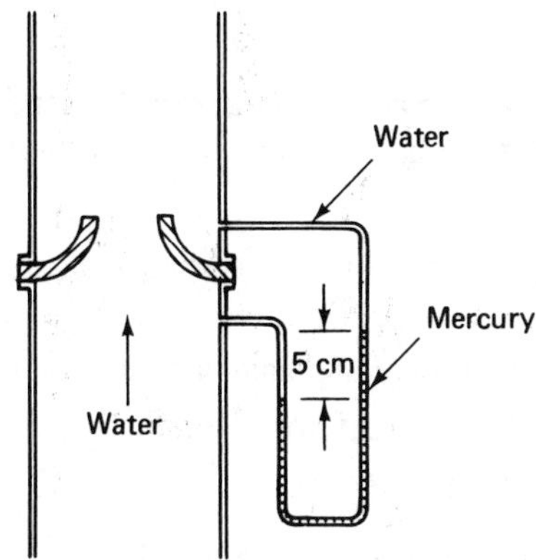

Figure P9.3.4

9.3.5 A 30-cm orifice plate is installed in a 50-cm-diameter, horizontal waterline. In a laboratory calibration, 6.78 m^3 of water were collected in 21.0 sec. The water mercury manometer reads a difference of 18.24 cm in a vertical mercury column. What is the discharge coefficient of the orifice meter?

9.3.6 A 5-cm orifice meter is installed in a 10-cm diameter horizontal pipeline. A differential mercury manometer shows a deflection of 9.0 cm that includes the head loss. Assuming that the pressure energy converted to kinetic energy at the orifice (to speed up the water flow) is totally lost through the meter, determine the discharge in the pipe. If a carefully streamlined venturi meter is used instead of the orifice meter, and it allows 60% of the pressure energy to recover, what would be the deflection of mercury column for the same discharge?

9.3.7 A bend meter installed in a 75-cm-diameter water pipe is shown in Figure 9.9. The installation delivered 48 m^3 of water in 1 min. Determine the deflection in cm of mercury when the meter is in a horizontal position.

9.3.8 Determine the coefficient of discharge for the bend meter in Problem 9.3.7 if the meter is installed in a vertical position for the same discharge. Note that in the vertical position the pressure taps are mounted along a 45° line from the horizontal datum.

9.4 DISCHARGE MEASUREMENTS IN OPEN CHANNELS—WEIRS, GATES, AND PARSHALL FLUMES

Essentially, a weir is a simple overflow structure extending across a channel and normal to the direction of the flow. Various types of weirs exist, and they are generally classified by shape. Weirs may be either sharp-crest or broad-crest.

Sharp-Crest Weirs Sharp-crest weirs (see Figure 9.10) include the following four basic types:

1. the horizontal weir without end contractions,
2. the horizontal weir with end contractions,
3. the V-notch weir,
4. the trapezoidal weir.

The *uncontracted horizontal weir* extends across the entire width of a uniform reach in the channel. A standard uncontracted, horizontal weir should meet the following requirements:

1. The crest of the weir should be horizontal, sharp-edged, and normal to the flow.
2. The weir plate should be vertical and have a smooth upstream surface.
3. The approach channel should be uniform and the water surface should be free from large surface waves.

The basic discharge equation for a standard, uncontracted, horizontal weir is

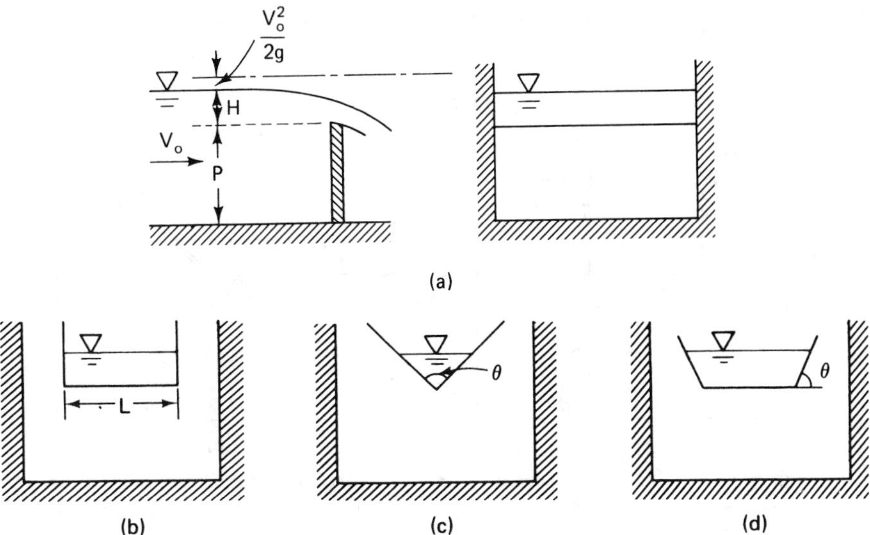

Figure 9.10 Common sharp-crest weirs: (a) uncontracted horizontal weir, (b) contracted horizontal weir, (c) V-notched weir, (d) trapezoidal weir.

$$Q = C_d L H^{3/2} \tag{9.9}$$

where L is the length of the crest, H is the head on the weir, and C_d is the discharge coefficient, which is derived from experimental data by the U.S. Bureau of Reclamation. Using the British units (H, L, p in feet, and Q in cubic feet per second), the discharge coefficient may generally be expressed as

$$C_d = 3.22 + 0.40\frac{H}{p} \tag{9.10}$$

where p is the weir height.

The *contracted horizontal weir* has a crest that is shorter than the width of the channel so that water must be contracted both horizontally and vertically in order to flow over the crest. The weir may be contracted at either one end or at both ends. The general discharge equation may be expressed as

$$Q = C_d\left(L - \frac{nH}{10}\right)H^{3/2} \tag{9.11}$$

where n is the number of contractions at the end [$n = 1$ for contraction at one end and $n = 2$ for contraction at both ends, as shown in Figure 9.10(b)]. The coefficient of discharge, C_d, can effectively be determined only by calibration in place.

A standard contracted horizontal weir is one whose crest and sides are so far removed from the bottom and sides of the channel that full contraction is developed. The dimension of a standard contracted horizontal weir is shown in

Figure 9.11. The discharge equation of the standard weir is given by the U.S.B.R.* as

$$Q = 3.33(L - 0.2H)H^{3/2} \tag{9.12}$$

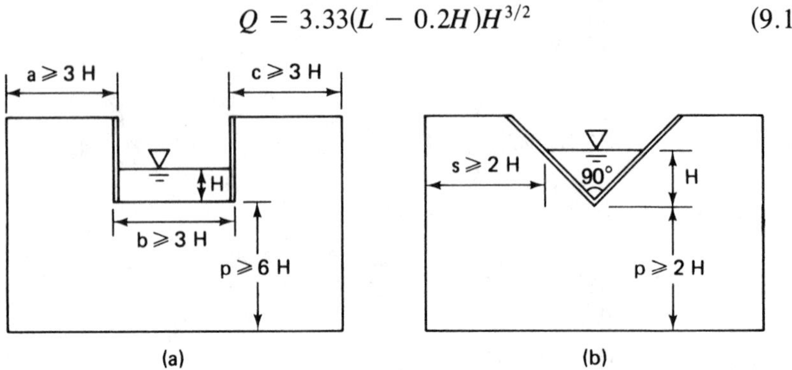

Figure 9.11 U.S.B.R. standard weirs: (a) contracted horizontal weir, (b) 90°-V-notch weir.

This formula is developed for the British measurement system, with L and H in feet and Q in the unit of cubic feet per second. The formula can be derived for the use of the S. I. units, as demonstrated in Example 9.5.

The triangular weir is especially useful when accuracy in measurement is desired for a large range of water elevations. The discharge equation for a triangular weir takes the general form of

$$Q = C_d\left(\tan\frac{\theta}{2}\right)H^{5/2} \tag{9.13}$$

where θ is the weir angle, as shown in Figure 9.10(c), and the discharge coefficient, C_d, is determined by calibration in place.

A standard U.S.B.R. *90°-V-notch weir* consists of a thin plate. The sides of the notch are inclined 45° from the vertical. The weir operates as a contracted horizontal weir and all conditions for accuracy stated for the standard contracted horizontal weir apply. The minimum distances of the sides of the weir from the channel banks should be at least twice the head on the weir, and the minimum distance from the crest to the channel bottom should be at least twice the head on the weir. The discharge equation of the standard 90°-V-notch weir is given by the U.S.B.R. as

$$Q = 2.49H^{2.48} \tag{9.14}$$

This formula, like Equation (9.12), is developed for the British measurement system. Units used in the formula are restricted to feet and to cubic feet per second.

The *trapezoidal weir* has hydraulic characteristics in between that of the

* *Water Measurement Manual,* U.S. Bureau of Reclamation, 1967.

contracted horizontal weir and the triangular weir. The general discharge equation developed for the contracted horizontal weir may be applied to the trapezoidal weir with an individually calibrated discharge coefficient.

A U.S.B.R. standard trapezoidal weir is also known as the *Cipolletti weir*. It has a horizontal crest and the sides incline outwardly at a slope of 1 (horizontal) to 4 (vertical). All conditions for accuracy stated for the standard weirs described above apply to the Cipolletti weir. The height of the weir crest should be at least twice the head of the approach flow above the crest, H, and the distances from the sides of the notch to the sides of the channel should also be at least twice the head. The discharge equation for the Cipolletti weir is given by the U.S.B.R. as

$$Q = 3.367LH^{3/2} \tag{9.15}$$

As in Equations (9.12) and (9.14), the units for the above formula are restricted to feet and cubic feet per second. The standardized dimension of the Cipolletti weir is given in Figure 9.12.

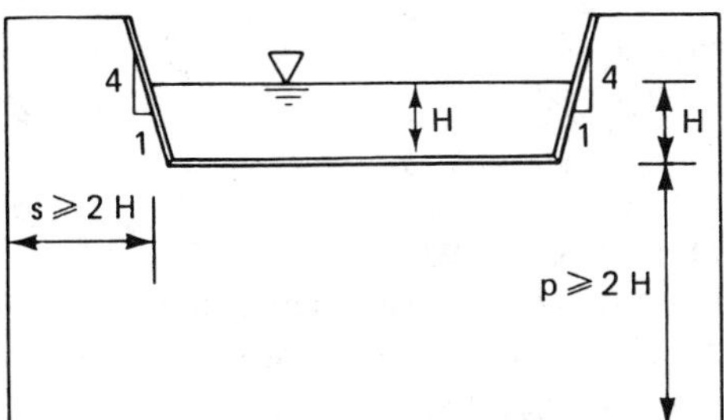

Figure 9.12 U.S.B.R. standard trapezoidal weir.

Example 9.5

Laboratory measurement made on a U.S.B.R. standard contracted horizontal weir shows a discharge of 0.25 m³/sec under a head of $H = 0.2$ m. Determine the discharge coefficient in the given units if $L = 1.56$ m.

Solution

Applying Equation (9.11) for the contracted horizontal weir, we get

$$Q = C_d\left(L - \frac{nH}{10}\right)H^{3/2}$$

Here, $L = 1.56$ m, $H = 0.2$ m, and $n = 2$ for contraction at both ends;

$$0.25 = C_d\left(1.56 - \frac{2 \cdot 0.2}{10}\right)(0.2)^{3/2}$$

$$C_d = 1.84$$

Broad-Crest Weirs A broad-crest weir provides a stretch of elevated channel floor over which the critical flow takes place (Figure 9.13). Depending on the height of the weir in relation to the depth of the approaching channel, the discharge equation may be derived from the balance of forces and momentum between the upstream approach section 1 and the section of minimum depth 2 on the crest of the weir. For unit width of the weir, the following equation may be written:

$$\rho q\left(\frac{q}{d_2} - \frac{q}{d_1}\right) = \frac{1}{2}\gamma d_1^2 - \frac{1}{2}\gamma d_2^2 - \frac{1}{2}\gamma h(2d_1 - h) \tag{9.16}$$

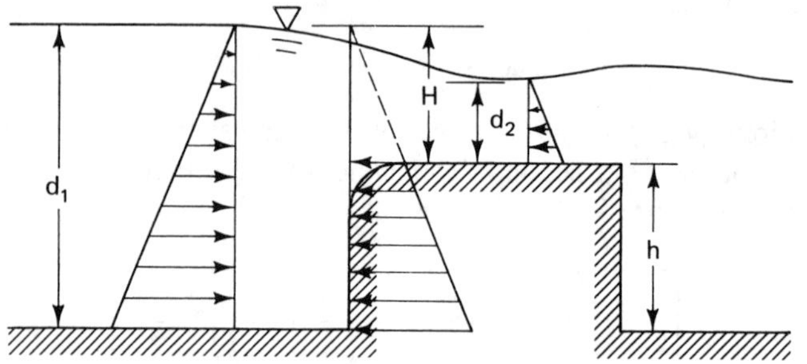

Figure 9.13 Broad-crest weir.

The conditions provided above are not sufficient for us to simplify Equation (9.16) into a one-to-one relationship between the approach water depth and the discharge. An additional equation was obtained from experimental measurements* for the average flow

$$d_1 - h = 2d_2 \tag{9.17}$$

Substituting Equation (9.17) into Equation (9.16) and simplifying, we have

$$q = 0.433\sqrt{2g}\left(\frac{d_1}{d_1 + h}\right)^{1/2} H^{3/2} \tag{9.18}$$

The total discharge over the weir is

$$Q = L \cdot q = 0.433\sqrt{2g}\left(\frac{d_1}{d_1 + h}\right)^{1/2} LH^{3/2} \tag{9.19}$$

where L is the length of the weir crest and H is the height of the approach water above the crest.

Considering the limit of a weir with zero height ($h = 0$) to infinity ($h \to \infty$), Equation (9.19) may vary from $Q = 1.92LH^{3/2}$ to $Q = 1.36LH^{3/2}$ with the units in meters and seconds.

*H. A. Doeringsfeld and C. L. Barker, "Pressure-momentum theory applied to the broad-crested weir", *Trans. ASCE*, 106 (1941), pp. 934–46.

Venturi Flumes The use of a weir is probably the simplest method for measuring discharge in open channel flows. The disadvantage, however, is the relatively high energy loss and the sedimentation deposited in the pool immediately upstream of the weir. These difficulties can be partially overcome by using a *critical flow flume* (*Venturi flume*).

A variety of Venturi flumes have been designed for field application. Most of the flumes operate with a submerged outflow condition and have a critical depth created at a contracted section (throat) and a hydraulic jump at the exit. The discharge through the flume can be calculated by reading the water depth from the observation wells located at the critical flow section and at another reference section.

The most extensively used critical flow flume in the United States is the *Parshall flume*, developed by R. L. Parshall* in 1920. The flume was experimentally developed for the British measurement system. It has fixed dimensions as shown in Figure 9.14 and Table 9.1. Empirical discharge equations were developed to correspond to each flume size. These equations are listed in Table 9.2.

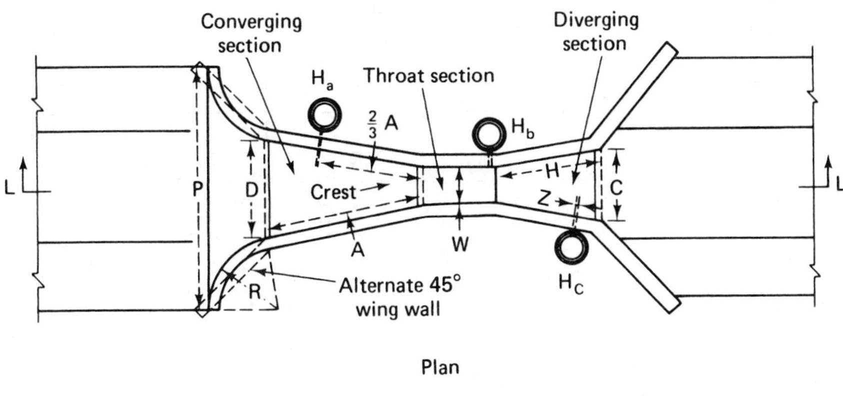

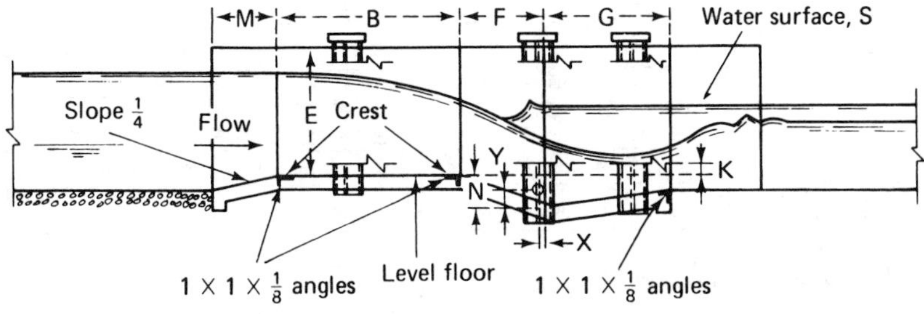

Figure 9.14 Parshall flume dimensions. (Courtesy of U.S. Bureau of Reclamation.)

*R. L. Parshall and C. Rohwer, *The Venturi Flume*, Colorado Agricultural Experimental Station Bulletin No. 265 (1921). R. L. Parshall, "The Improved Venturi Flume," *Trans. ASCE*, 89 (1926), 841–51.

TABLE 9.1 Parshall Flume Dimensions

	W		A		2/3A		B		C		D		E		F		G		H		K		M		N		P		R		X		Y		Z		FREE-FLOW CAPACITY MIN.	FREE-FLOW CAPACITY MAX.
	FT	IN	FT	IN	FT	IN	FT	IN	FT	IN	FT	IN	FT	IN	FT	IN	FT	IN	FT	IN	FT	IN	FT	IN	FT	IN	FT	IN	FT	IN	FT	IN	FT	IN	FT	IN	SEC-FT	SEC-FT
	0	1 1)	1	2 9/32	0	9 17/32	1	2	0	3 21/32	0	6 19/32	0	6 to 9	0	3	0	8	0	8 1/8	0	3/4	–	–	0	1 1/8		–	–		0	5/16	0	1/2	0	1/8	0.01	0.19
2)		2 1)	1	4 5/16		10 7/8	1	4		5 5/16		8 13/32		6 to 10		4 1/2		10		10 1/8		7/8		–		1 11/16		–	–			5/8		1		1/4	0.02	0.47
		3 1)	1	6 3/8	1	1/4	1	6		7		10 3/16	1 to 1 1/2			6	1	0	1	5/32		1	–			2 1/4		–	–			1		1 1/2		1/2	0.03	1.13
	0	6	2	7/16	1	4 5/16	2	0	1	3 1/2	1	3 5/8	2	0	1	0	2	0	–		0	3	1	0	0	4 1/2	2	11 1/2	1	4	0	2	0	3	–		0.05	3.9
		9	2	10 5/8	1	11 1/8	2	10	1	3	1	10 5/8	2	6	1	0	1	6	–			3	1	0		4 1/2	3	6 1/2	1	4		2		3	–		0.09	8.9
	1	0	4	6	3	0	4	4 7/8	2	0	2	9 1/4	3	0	2	0	3	0	–			3	1	3		9	4	10 3/4	1	8		2		3	–		0.11	16.1
	1	6	4	9	3	2	4	7 7/8	2	6	3	4 3/8	3	0	2	0	3	0	–			3	1	3		9	5	6	1	8		2		3	–		0.15	24.6
	2	0	5	0	3	4	4	10 7/8	3	0	3	11 1/2	3	0	2	0	3	0	–			3	1	3		9	6	1	1	8		2		3	–		0.42	33.1
3)	3	0	5	6	3	8	5	4 3/4	4	0	5	1 7/8	3	0	2	0	3	0	–			3	1	3		9	7	3 1/2	1	8		2		3	–		0.61	50.4
	4	0	6	0	4	0	5	10 5/8	5	0	6	4 1/4	3	0	2	0	3	0	–			3	1	6		9	8	10 3/4	2	0		2		3	–		1.3	67.9
	5	0	6	6	4	4	6	4 1/2	6	0	7	6 5/8	3	0	2	0	3	0	–			3	1	6		9	10	1 1/4	2	0		2		3	–		1.6	85.6
	6	0	7	0	4	8	6	10 3/8	7	0	8	9	3	0	2	0	3	0	–			3	1	6		9	11	3 1/2	2	0		2		3	–		2.6	103.5
	7	0	7	6	5	0	7	4 1/4	8	0	9	11 3/8	3	0	2	0	3	0	–			3	1	6		9	12	6	2	0		2		3	–		3.0	121.4
	8	0	8	0	5	4	7	10 1/8	9	0	11	1 3/4	3	0	2	0	3	0	–			3	1	6		9	13	8 1/4	2	0		2		3	–		3.5	139.5
	10	0	–		6	0	14	0	12	0	15	7 1/4	4	0	3	0	6	0	–		0	6	–		1	1 1/2	–		–		0	9	1	0	–		6	200
	12	0	–		6	8	16	0	14	8	18	4 3/4	5	0	3	0	8	0	–			6	–		1	1 1/2	–		–			9	1	0	–		8	350
	15	0	–		7	8	25	0	18	4	25	0	6	0	4	0	10	0	–			9	–		1	6	–		–			9	1	0	–		8	600
	20	0	–		9	4	25	0	24	0	30	0	7	0	6	0	12	0	–		1	0	–		2	3	–		–			9	1	0	–		10	1000
4)	25	0	–		11	0	25	0	29	4	35	0	7	0	6	0	13	0	–		1	0	–		2	3	–		–			9	1	0	–		15	1200
	30	0	–		12	8	26	0	34	8	40	4 3/4	7	0	6	0	14	0	–		1	0	–		2	3	–		–			9	1	0	–		15	1500
	40	0	–		16	0	27	0	45	4	50	9 1/2	7	0	6	0	16	0	–		1	0	–		2	3	–		–			9	1	0	–		20	2000
	50	0	–		19	4	27	0	56	8	60	9 1/2	7	0	6	0	20	0	–		1	0	–		2	3	–		–			9	1	0	–		25	3000

1) Tolerance on throat width(w) ± 1/64 inch; tolerance on other dimensions ±1/32 inch. Sidewalls of throat must be parallel and vertical

2) From Colorado State University Technical Bulletin No. 61

3) From US Department of Agriculture Soil Conservation Circular No. 843

4) From Colorado State University Bulletin No. 426-A

TABLE 9.2

Throat Width	Discharge Equation	Free Flow Capacity (cfs)	
3 in.	$Q = 0.992H_a^{1.547}$	0.03–1.9	(9-20)
6 in.	$Q = 2.06H_a^{1.58}$	0.05–3.9	(9-21)
9 in.	$Q = 3.07H_a^{1.53}$	0.09–8.9	(9-22)
1 to 8 ft	$Q = 4WH_a^{(1.522W^{0.026})}$	Up to 140	(9-23)
10 to 50 ft	$Q = (3.6875W + 2.5)H_a^{1.6}$	Up to 2000	(9-24)

In the equations above, Q is the discharge in cubic feet per second (cfs), W is the throat width, and H_a is the water level reading from the observation well, a, measured in feet. Readers should notice that these are derived strictly for the units stated and for the dimensions specified in Figure 9.14 and Table 9.1. These equations do not have an equivalent version in the metric system.

When the ratio of gage reading H_b (from observation well b) to gage reading H_a (from observation well a) exceeds the following values:

0.50 for flumes 1, 2, and 3 in. wide
0.60 for flumes 6 in. and 9 in. wide
0.70 for flumes 1 ft to 8 ft wide
0.80 for flumes 8 ft to 50 ft wide

the flow is said to be submerged. The effect of the downstream submergence is to reduce the discharge through the flume. In this case, the discharge computed by the above equations should be corrected by considering both readings H_a and H_b.

Figure 9.15 shows the rate of submerged flow through a 1-ft Parshall

Flume Size (ft)	1.0	2.0	3.0	4.0	6.0	8.0
Correction factor	1.0	1.8	2.4	3.1	4.3	5.4

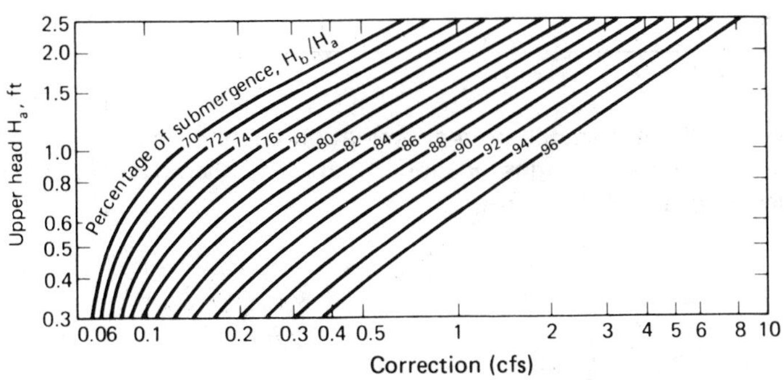

Figure 9.15 Flow-rate correction for 1-ft submerged Parshall flume. [After R. L. Parshall, *Measuring Water in Irrigation Channels with Parshall Flumes and Small Weirs*, U.S. Soil Conservation Service, Circular 843 (1950). R. L. Parshall, *Parshall Flumes of Large Size*, Colorado Agricultural Experimental Station Bulletin No. 426A (1953).]

flume. The diagram is made applicable to larger flumes (up to 8 ft) by multiplying the corrected discharge with a factor given for the particular size selected.

Figure 9.16 shows the rate of submerged flow through a 10-ft Parshall flume. The diagram is also made applicable to larger size flumes (up to 50 ft) by multiplying the corrected discharge for the 10-ft flume by a factor given for the particular size used.

Flume Size (ft)	10	15	20	30	40	50
Correction factor	1.0	1.5	2.0	3.0	4.0	5.0

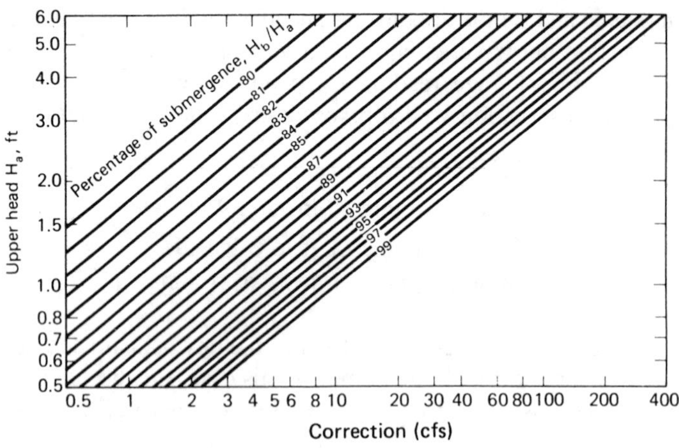

Figure 9.16 Flow-rate correction for 10-ft submerged Parshall flume [After R. L. Parshall, *Measuring Water in Irrigation Channels with Parshall Flumes and Small Weirs,* U.S. Soil Conservation Service, Circular 843 (1950). R. L. Parshall, *Parshall Flumes of Large Size,* Colorado Agricultural Experimental Station Bulletin No. 426A (1953).]

Example 9.6

A 4-ft Parshall flume is installed in an irrigation channel to monitor the rate of flow. If the readings at gages H_a and H_b are 2.5 ft and 2.0 ft, respectively. Determine the channel discharge.

Solution

$$H_a = 2.5 \text{ ft} \qquad H_b = 2.0 \text{ ft}$$

Submergence, $H_b/H_a = 80\%$.

Equation (9.23) gives the value of unsubmerged discharge

$$Q_0 = 4 \times 4(2.5)^{1.522 \times 4^{0.026}} = 67.92 \text{ cfs}$$

Under the given conditions, the flume is operating at 80% submergence, and the value should be corrected accordingly.

From Figure 9.15, we found the flowrate correction for a 1-ft Parshall flume to be 1.9 cfs. For the 4-ft flume, the correction flowrate is Q_c.

$$Q_c = 3.1 \times 1.9 = 5.89 \text{ cfs}$$

The correct discharge of the channel is

$$Q = Q_0 - Q_c = 62.03 \text{ cfs}$$

PROBLEMS

9.4.1 A contracted horizontal weir with a crest 4 m long has a head 1.1 m above the crest. Determine the discharge if the height of the crest is 3 m.

9.4.2 A rectangular sharp-crest weir with end contractions has a crest 2.0 m long. How high should it be placed in a channel to maintain an upstream depth of 2.25 m for a discharge of 1 m^3/sec?

9.4.3 The discharge in a rectangular channel 4 m wide is constant. A depth of 2.3 m is maintained by a 1.7-m high rectangular contracted weir with a 1-m-long horizontal crest. This weir is to be replaced by an uncontracted horizontal weir that will maintain the same upstream depth. Determine the height of the weir.

9.4.4 Flow occurs over a sharp-crest, uncontracted horizontal weir 1 m high under a head of 0.3 m. If the weir is replacing a weir just one-half of its height, what change in depth of water can be expected in the upstream channel?

9.4.5 Laboratory tests on a 60°-V-notch weir gave the following results: $H = 0.3$ m, $Q = 0.022$ m^3/sec, $H = 0.6$ m, and $Q = 0.132$ m^3/sec. Determine the formula of weir discharge.

9.4.6 A broad-crest rectangular weir is 1 m high and 3 m long. The weir is constructed with a well-rounded upstream corner and a smooth surface. What is the discharge if the head is 0.4 m?

9.4.7 Calculate the flow rate over a U.S.B.R. standard contracted horizontal weir 3 m long that has a head of 0.6 m.

9.4.8 What length Cipoletti weir (U.S.B.R. standard trapezoidal weir) is needed for a flow up to 0.8 m^3/sec if the maximum head is limited to 0.25 cm?

9.4.9 Determine the discharge in cubic meters per second measured by a 10-ft Parshall flume if the gage reading H_a is 1 m at a free-flow condition.

9.4.10 Determine the discharge through a 4-ft Parshall flume if the gage reading from well *a* is 1.0 m and from well *b* is 0.8 m.

9.4.11 A 40-ft Parshall flume measures $H_a = 1.0$ m and $H_b = 0.95$ m; determine the discharge.

10

HYDRAULIC SIMILITUDE AND MODEL STUDIES

Use of small models for studying the prototype hydraulic design can be dated at least to Leonardo da Vinci.* But the method developed for using the results of experiments conducted on a *scaled model* to predict quantitatively the behavior of a full-size hydraulic structure (or prototype) was realized only after the turn of this century. The principle on which the model studies are based comprises the theory of *hydraulic similitude*. The analysis of the basic relationship of the various physical quantities involved in the static and dynamic behaviors of water flow in a hydraulic structure is known as *dimensional analysis*.

All important hydraulic structures are now designed and built after certain preliminary model studies have been completed. Such studies may be conducted for any one or more of the following purposes:

* Leonardo da Vinci (1452–1519), a genius, Renaissance scientist, engineer, architect, painter, sculptor, and musician.

1. The determination of the discharge coefficient of a large measurement structure, such as an overflow spillway or a weir.
2. The development of an effective method for energy dissipation at the outlet of a hydraulic structure.
3. The reduction of energy loss at the intake structure or at the transition section.
4. The development of an efficient, economic spillway or other type of flood-releasing structure for a reservoir.
5. The determination of the average time of travel in a temperature control structure, for example, in a cooling pond in a power plant.
6. The determination for the best cross section, location, and dimensions of the various components of the structure, such as the breakwater, the docks, and the locks, etc., in harbor and waterway design.
7. The determination of the dynamic behaviors of the floating, semi-immersible, and floor-installed structures in transportation and installation of offshore structures.

River models have also been extensively used in hydraulic engineering to determine

1. The pattern a flood wave travels through a river channel.
2. The effect of artificial structures, such as bends, levees, dikes, jetties, and training walls, on the sedimentation movements in the channel reach, as well as in the upstream and downstream channels.
3. The direction and force of currents in the channel or harbor and their effect on navigation and marine life.

10.1 DIMENSIONAL HOMOGENEITY

When a physical phenomenon is described by an equation or a set of equations, all terms in each of the equations must be kept dimensionally homogeneous. In other words, all terms in an equation must be expressed in the same units.

In fact, to derive a relationship among several parameters involved in a physical phenomenon, one should always check the equation for homogeneity of units. If all terms in an equation do not appear to have the same unit, then one can be sure that certain important parameters may be missing or misplaced.

Based on the physical understanding of the phenomenon and the concept of dimensional homogeneity, the solution of many hydraulic problems may be obtained. For example, we understand that the speed of surface wave propagation on water surface, C, is related to the gravitational acceleration, g, and the water depth, d. Generally, we may write

$$C = f(g, d) \tag{10.1}$$

The units of the physical quantities involved are indicated in the brackets,

$$C = [LT^{-1}]$$
$$g = [LT^{-2}]$$
$$d = [L]$$

Since the left-hand side of Equation (10.1) has the units of $[LT^{-1}]$, those units must then appear explicitly on the right-hand side. Thus, d and g must combine as a product and the function, f, must be the square root. We have

$$C = \sqrt{gd}$$

as discussed in Chapter 6, (Equation 6.11).

The dimensions of the physical quantities commonly used in hydraulic engineering are listed in Table 10.1.

TABLE 10.1 Dimensions of Physical Quantities Commonly Used in Hydraulic Engineering

Quantity	Dimension	Quantity	Dimension
Length	L	Force	MLT^{-2}
Area	L^2	Pressure	$ML^{-1}T^{-2}$
Volume	L^3	Shear Stress	$ML^{-1}T^{-2}$
Angle (radians)	None	Specific Weight	$ML^{-2}T^{-2}$
Time	T	Modulus of Elasticity	$ML^{-1}T^{-2}$
Discharge	L^3T^{-1}	Coefficient of Compressibility	$M^{-1}LT^2$
Linear Velocity	LT^{-1}	Surface Tension	MT^{-2}
Angular Velocity	T^{-1}	Momentum	MLT^{-1}
Acceleration	LT^{-2}	Angular Momentum	ML^2T^{-1}
Mass	M	Torque	ML^2T^{-2}
Moment of Inertia	ML^2	Energy	ML^2T^{-2}
Density	ML^{-3}	Power	ML^2T^{-3}
Viscosity	$ML^{-1}T^{-1}$	Kinematic Viscosity	L^2T^{-1}

10.2 PRINCIPLES OF HYDRAULIC SIMILITUDE

Similarity between hydraulic models and prototypes may be achieved in three basic forms

1. geometric similarity,
2. kinematic similarity,
3. dynamic similarity.

Geometric similarity implies similarity of form. The model is a geometric reduction of the prototype and is accomplished by maintaining a fixed ratio for all homologous lengths between the model and the prototype.

The physical quantities involved in geometric similarity are length, L, area, A, and volume, Vol. The ratio of homologous lengths in prototype and model is a constant and can be expressed as

$$\frac{L_p}{L_m} = L_r \tag{10.2}$$

An area, A, is the product of two homologous lengths; hence, the ratio of the homologous area is also a constant and can be expressed as

$$\frac{A_p}{A_m} = \frac{L_p^2}{L_m^2} = L_r^2 \tag{10.3}$$

A volume, Vol, is the product of three homologous lengths; the ratio of the homologous volume can be expressed as

$$\frac{\text{Vol}_p}{\text{Vol}_m} = \frac{L_p^3}{L_m^3} = L_r^3 \tag{10.4}$$

Example 10.1

A geometrically similar open channel model is constructed with a 5:1 scale. If the model measures a discharge of 0.2 m³/sec, determine the corresponding discharge in the prototype.

Solution

The velocity ratio between the prototype and the model is

$$\frac{V_p}{V_m} = \frac{\frac{L_p}{T}}{\frac{L_m}{T}} = \frac{L_p}{L_m} = L_r = 5$$

Since in geometric similarity, time in model and prototype remains unscaled. The area ratio between the prototype and the model is

$$\frac{A_p}{A_m} = \frac{L_p^2}{L_m^2} = L_r^2 = 25$$

Accordingly, the discharge ratio is

$$\frac{Q_p}{Q_m} = \frac{A_p V_p}{A_m V_m} = (25)(5) = 125$$

Thus, the corresponding discharge in the prototype is

$$Q_p = 125 Q_m = 125 \cdot 0.2 = 25 \text{ m}^3/\text{sec}$$

Kinematic similarity implies similarity in motion. Kinematic similarity between a model and the prototype is attained if the homologous moving particles have the same velocity ratio along geometrically similar paths. The kinematic similarity involves the scale of time as well as length. The ratio of times required for homologous particles to travel homologous distances in a model and its prototype is

$$\frac{T_p}{T_m} = T_r \tag{10.5}$$

The velocity V is defined in terms of distance per unit time; thus, the ratio of velocities can be expressed as

$$\frac{V_p}{V_m} = \frac{\dfrac{L_p}{T_p}}{\dfrac{L_m}{T_m}} = \frac{\dfrac{L_p}{L_m}}{\dfrac{T_p}{T_m}} = \frac{L_r}{T_r} \tag{10.6}$$

The acceleration, a, is defined in terms of length per unit time square; thus, the ratio of homologous acceleration is

$$\frac{a_p}{a_m} = \frac{\dfrac{L_p}{T_p^2}}{\dfrac{L_m}{T_m^2}} = \frac{\dfrac{L_p}{L_m}}{\dfrac{T_p^2}{T_m^2}} = \frac{L_r}{T_r^2} \tag{10.7}$$

The discharge Q, is expressed in terms of volume per unit time; thus,

$$\frac{Q_p}{Q_m} = \frac{\dfrac{L_p^3}{T_p}}{\dfrac{L_m^3}{T_m}} = \frac{\dfrac{L_p^3}{L_m^3}}{\dfrac{T_p}{T_m}} = \frac{L_r^3}{T_r} \tag{10.8}$$

Kinematic models constructed for hydraulic machinery may frequently involve angular displacement, θ, expressed in radians, which is equal to the tangential displacement, L, divided by the length of radius, R, of the curve at the point of tangency. The ratio of angular displacements may be expressed as

$$\frac{\theta_p}{\theta_m} = \frac{\dfrac{L_p}{R_p}}{\dfrac{L_m}{R_m}} = \frac{\dfrac{L_p}{L_m}}{\dfrac{R_p}{R_m}} = \frac{\dfrac{L_p}{L_m}}{\dfrac{L_p}{L_m}} = 1 \tag{10.9}$$

The angular velocity, N, in revolutions per minute, is defined as angular displacement per unit time; thus, the ratio

$$\frac{N_p}{N_m} = \frac{\dfrac{\theta_p}{T_p}}{\dfrac{\theta_m}{T_m}} = \frac{\dfrac{\theta_p}{\theta_m}}{\dfrac{T_p}{T_m}} = \frac{1}{T_r} \tag{10.10}$$

The angular acceleration, α, is defined as angular displacement per unit time square,

$$\frac{\alpha_p}{\alpha_m} = \frac{\dfrac{\theta_p}{T_p^2}}{\dfrac{\theta_m}{T_m^2}} = \frac{\dfrac{\theta_p}{\theta_m}}{\dfrac{T_p^2}{T_m^2}} = \frac{1}{T_r^2} \tag{10.11}$$

Example 10.2

A 10:1 scale model is constructed to study the flow motions in a cooling pond. If the designed discharge from the power plant is 200 m^3/sec and the model can accommodate a maximum flow rate of 0.1 m^3/sec, determine the time ratio.

Solution

The length ratio between the prototype and the model is

$$L_r = \frac{L_p}{L_m} = 10$$

and the discharge ratio is $Q_r = \dfrac{200}{0.1} = 2000$, and

$$Q_r = \frac{Q_p}{Q_m} = \frac{\dfrac{L_p^3}{T_p}}{\dfrac{L_m^3}{T_m}} = \left(\frac{L_p}{L_m}\right)^3 \cdot \left(\frac{T_m}{T_p}\right) = L_r^3 \cdot T_r^{-1}$$

Substituting the length ratio into the discharge ratio gives the time ratio

$$T_r = \frac{T_p}{T_m} = \frac{L_r^3}{Q_r} = \frac{(10)^3}{2000} = 0.5$$

and

$$T_m = 2T_p$$

A unit time period measured in the model is equivalent to two periods of time in the prototype pond.

Dynamic similarity implies similarity in forces involved in motion. Dynamic similarity between a model and its prototype is attained if the ratio of homologous forces in the model and prototype is kept at a constant value, or

$$\frac{F_p}{F_m} = F_r \tag{10.12}$$

Many hydrodynamic phenomena may involve several different kinds of forces in action. Since models are usually built to simulate the prototype on reduced scales, they usually are not capable of simulating all the forces simultaneously. In practice, a model is designed to study the effects of only a few *dominant forces*. Dynamic similarity requires that the ratios of these forces be kept the same between the model and the prototype. Hydraulics phenomena governed by the various types of force ratios are discussed in Sections 10.3, 10.4, 10.5, and 10.6. Since force is equal to mass, M, multiplied by acceleration, a, and since mass is equal to density, ρ, times volume, Vol, the force ratio may be expressed as

$$\begin{aligned}\frac{F_p}{F_m} &= \frac{M_p a_p}{M_m a_m} = \frac{\rho_p \,\mathrm{Vol}_p \, a_p}{\rho_m \,\mathrm{Vol}_m \, a_m} \\ &= \frac{\rho_p}{\rho_m} \cdot \frac{L_p^3}{L_m^3} \cdot \frac{\dfrac{L_p}{T_p^2}}{\dfrac{L_m}{T_m^2}} = \frac{\rho_p}{\rho_m} \cdot \frac{L_p^4}{L_m^4} \cdot \frac{1}{\dfrac{T_p^2}{T_m^2}} \\ &= \rho_r L_r^4 T_r^{-2}\end{aligned} \tag{10.13}$$

and the ratio of homologous masses, force divided by acceleration, may be expressed as

$$\frac{M_p}{M_m} = \frac{\dfrac{F_p}{a_p}}{\dfrac{F_m}{a_m}} = \frac{\dfrac{F_p}{F_m}}{\dfrac{a_p}{a_m}} = F_r T_r^2 L_r^{-1} \tag{10.14}$$

Since work is equal to force multiplied by distance, the ratio of homologous work in dynamic similarity is

$$\frac{\overline{W}_p}{\overline{W}_m} = \frac{F_p L_p}{F_m L_m} = F_r L_r \tag{10.15}$$

Power is the time rate of doing work; thus, the ratio of homologous powers in the model and prototype is

$$\frac{P_p}{P_m} = \frac{\dfrac{\overline{W}_p}{T_p}}{\dfrac{\overline{W}_m}{T_m}} = \frac{\overline{W}_p}{\overline{W}_m} \cdot \frac{1}{\dfrac{T_p}{T_m}} = \frac{F_r L_r}{T_r} \tag{10.16}$$

Example 10.3

A 59,680-w (80-hp) pump is used to power a water supply system. The model constructed to study the system has an 8:1 scale. If the velocity ratio is 2:1, determine the power needed for the model pump.

Solution

By substituting the length ratio into the velocity ratio, the time ratio is obtained

$$V_r = \frac{L_r}{T_r} = 2; \qquad L_r = 8$$

$$T_r = \frac{L_r}{2} = \frac{8}{2} = 4$$

The same fluid is used in the model and the prototype, unless otherwise specified; then $\rho_r = 1$, and the force ratio can be calculated, from Equation (10.13)

$$F_r = \rho_r L_r^4 T_r^{-2} = \frac{(1)(8)^4}{(4)^2} = 256$$

The power ratio is, from Equation (8.15)

$$P_r = \frac{F_r L_r}{T_r} = \frac{(256)(8)}{(4)} = 512$$

And the power required for the model pump is

$$P_m = \frac{P_p}{P_r} = \frac{59{,}680}{512} = 116.6 \text{ w} = 0.156 \text{ hp}$$

Example 10.4

The model designed to study the prototype of a hydraulic machine must

1. be geometrically similar,
2. have the same discharge coefficient defined as $Q/A\sqrt{2gH}$,
3. have the same ratio of peripheral speed to the water discharge velocity, $\omega D/(Q/A)$.

Determine the scale ratios in terms of discharge Q, head H, diameter D, and rotational angular velocity ω.

Solution

It is important to recognize that although the energy head, H, is expressed in units of length, it is not necessarily modeled as a linear length dimension. To have the same ratio of peripheral speed to the water discharge velocity, we have,

$$\frac{\omega_p D_p}{Q_p/A_p} = \frac{\omega_m D_m}{Q_m/A_m}$$

or,

$$\frac{\omega_r D_r A_r}{Q_r} = \frac{T_r^{-1} L_r L_r^2}{L_r^3 T_r^{-1}} = 1$$

To have the same discharge coefficient, we have,

$$\frac{Q_p/(A_p\sqrt{2gH_p})}{Q_m/(A_m\sqrt{2gH_m})} = \frac{Q_r}{A_r\sqrt{(gH)_r}} = 1$$

or,

$$\frac{L_r^3 T_r^{-1}}{L_r^2 (gH)_r^{1/2}} = 1$$

From which, we get,

$$(gH)_r = \frac{L_r^2}{T_r^2}$$

Since the gravitational acceleration, g, is the same for model and prototype, we may write,

$$H_r = L_r^2 T_r^{-2}$$

The other ratios asked for are

$$\text{Discharge ratio: } Q_r = L_r^3 T_r^{-1}$$
$$\text{Diameter ratio: } D_r = L_r, \text{ and}$$
$$\text{Angular speed ratio: } \omega_r = T_r^{-1}$$

PROBLEMS

10.2.1 A 1-m-long, 1:30 model is used to study the wave force on a prototype of a sea wall structure. If the total wave force measured on the model is 2.27 N and the velocity scale is 1:10, determine the force per unit length of the prototype.

10.2.2 The moment exerted on a gate structure is studied in a laboratory water tank with a 1:125 scale model. If the moment measured on the model is 1.5 N · m on the 1-m long gate arm, determine the moment exerted on the prototype.

10.2.3 A 1:100 scale model is constructed to study a gate prototype that is designed to drain a reservoir. If the model reservoir is drained in 5.2 min, how long should it take to drain the prototype?

10.2.4 An overflow spillway is designed to have a 100-m long crest to carry the discharge of 1150 m^3/sec under a permitted maximum head of 3 m. The operation of the prototype spillway is studied by a 1:50 scale model in a hydraulic laboratory.

(a) Determine the model discharge based on Equation (8.9).

(b) The model velocity measured at the end (toe) of the spillway is 3 m/sec. Determine the corresponding velocity in the prototype.

(c) What are the model and prototype Froude numbers at the toe?

(d) If the force on a bucket energy dissipator at the toe of the spillway is measured to be 37.5 N, determine the corresponding force on the prototype.

10.2.5 If a 1:5 scale model, at 1200 rpm, is used to study the prototype of a centrifugal pump that produces 1 m^3/sec at 30-m head when rotating at 400 rpm, determine the model discharge and head.

10.2.6 A 1:25 scale model is being designed to study a prototype hydraulic structure. The velocity ratio between the model and the prototype is 1:5, and the measurement accuracy is required to be within 1% of the total force. Determine the accuracy of the force measurement in the model if the expected force on the prototype is 45,000 N. [Hint: use Equation (8.9)].

10.2.7 Sedimentation in a river section 5 km long is to be studied in a lab channel 20 m long. A time ratio of 4 will be used and the prototype discharge is 75 m^3/sec. Select a suitable length scale and determine the model discharge.

10.2.8 A 1:20 scale model of a prototype energy dissipation structure is constructed to study force distribution and water depths. A velocity ratio of 7.75 is used.

Determine the force ratio and prototype discharge if the model discharge is 300 ℓ/sec.

10.2.9 A 1:50 scale model is used to study the power requirements of a prototype submarine. The model will be towed at a speed 50 times greater than the speed of the prototype in a tank filled with sea water. Determine the conversion ratios from the prototype to the model for the following quantities: (a) time, (b) force, (c) energy, (d) power.

10.3 PHENOMENA GOVERNED BY VISCOUS FORCE—REYNOLDS NUMBER LAW

Water in motion always involves inertial forces. When the inertial forces and the viscous forces can be considered to be the only forces that govern the motion, the ratio of these forces acting on homologous particles in a model and its prototype is defined by the Reynolds number law

$$N_R = \frac{\text{inertial force}}{\text{viscous force}} \tag{10.17}$$

The inertial force defined by Newton's second law of motion, $F = ma$, can be expressed by the ratio in Equation (10.13):

$$F_r = M_r\frac{L_r}{T_r^2} = \rho_r L_r^4 T_r^{-2} \tag{10.13}$$

The viscous force defined by Newton's law of viscosity,

$$F = \mu\left(\frac{dV}{dL}\right)\cdot A$$

may be expressed by

$$F_r = \frac{\mu_p\left(\dfrac{dV}{dL}\right)_p A_p}{\mu_m\left(\dfrac{dV}{dL}\right)_m A_m} = \mu_r L_r^2 T_r^{-1} \tag{10.18}$$

where μ is the viscosity and V denotes the velocity.

Equating values of F_r from Equations (10.13) and (10.18), we get

$$\rho_r L_r^4 T_r^{-2} = \mu_r L_r^2 T_r^{-1}$$

from which

$$\frac{\rho_r L_r^4 T_r^{-2}}{\mu_r L_r^2 T_r^{-1}} = \frac{\rho_r L_r^2}{\mu_r T_r} = \frac{\rho_r L_r V_r}{\mu_r} = 1 \tag{10.19}$$

Rearranging the above equation, we may write

$$\frac{\left(\dfrac{\rho_p L_p V_p}{\mu_p}\right)}{\left(\dfrac{\rho_m L_m V_m}{\mu_m}\right)} = 1$$

or

$$\frac{\rho_p L_p V_p}{\mu_p} = \frac{\rho_m L_m V_m}{\mu_m} = N_R \tag{10.20}$$

Equation (10.20) states that when the inertial force and the viscous force are considered to be the only forces governing the motion of the water, the Reynolds number of the model and the prototype must be kept at the same value.

If the same fluid is used in both the model and the prototype, the scale ratios for many physical quantities can be derived based on the Reynolds number law. These quantities are listed in Table 10.2.

TABLE 10.2 Scale Ratios for Reynolds Number Law (for Water in Both Model and Prototype, $\rho_r = 1$, $\mu_r = 1$)

Geometric Similarity		*Kinematic Similarity*		*Dynamic Similarity*	
Length	L_r	Time	L_r^2	Force	1
Area	L_r^2	Velocity	L_r^{-1}	Mass	L_r^3
Volume	L_r^3	Acceleration	L_r^{-3}	Work	L_r
		Discharge	L_r	Power	L_r^{-1}
		Angular Velocity	L_r^{-2}		
		Angular Acceleration	L_r^{-4}		

Example 10.5

In order to study a transient process, a model is constructed at a 10:1 scale. Water is used in the prototype, and it is known that viscous forces are the dominant ones. Compare the time and force ratios, if the model uses

(a) water,

(b) oil 5 times more viscous than water, with $\rho_{oil} = 0.8\rho_{water}$.

Solution

(a) From Table 10.2

$$T_r = L_r^2 = (10)^2 = 100$$

$$V_r = L_r^{-1} = (10)^{-1} = 0.1$$

$$F_r = 1$$

(b) From the Reynolds number law,

$$\frac{\rho_p L_p V_p}{\mu_p} = \frac{\rho_p L_m V_m}{\mu_m}$$

we have

$$\frac{\rho_r L_r V_r}{\mu_r} = 1$$

Since the ratios of viscosity and density are, respectively,

$$\mu_r = \frac{\mu_p}{\mu_m} = \frac{\mu_{\text{water}}}{\mu_{\text{oil}}} = \frac{\mu_{\text{water}}}{5\mu_{\text{water}}} = 0.2$$

$$\rho_r = \frac{\rho_p}{\rho_m} = \frac{\rho_{\text{water}}}{\rho_{\text{oil}}} = \frac{\rho_{\text{water}}}{0.8\rho_{\text{water}}} = 1.25$$

From the Reynolds number law

$$V_r = \frac{\mu_r}{\rho_r L_r} = \frac{(0.2)}{(1.25)(10)} = 0.016$$

The time ratio is

$$T_r = \frac{L_r}{V_r} = \frac{\rho_r L_r^2}{\mu_r} = \frac{(1.25)(10)^2}{(0.2)} = 625$$

The force ratio, then, is

$$F_r = \frac{\rho_r L_r^4}{T_r^2} = \frac{(\rho_r L_r^4)}{\left(\dfrac{\rho_r^2 L_r^4}{\mu_r^2}\right)} = \frac{\mu_r^2}{\rho_r} = \frac{(0.2)^2}{1.25} = 0.032$$

This example demonstrates the importance of selecting the model fluid. The properties of the liquid used in the model, especially the viscosity, greatly affect the performance in the Reynolds number models.

PROBLEMS

10.3.1 A Reynolds number scale model is used to study the operation of a prototype hydraulic device. The model is built on a 1:5 scale and uses water at 20°C. The prototype discharges 11.5 m^3/sec of water at 90°C temperature. Determine the model discharge.

10.3.2 The moment exerted on a ship's rudder is studied with a 1:20 scale model in a water tunnel using the same temperature as the river water. If the torque measured on the model is 10 N·m for a water tunnel velocity of 20 m^3/sec, determine the corresponding torque and speed for the prototype.

10.3.3 A 1:10 scale model of a water supply piping system is to be tested at 20°C to determine the total head loss in the prototype that carries water at 85°C. The prototype is designed to carry 5.0 m^3/sec discharge with 1-m diameter pipes. Determine the model discharge and model velocity. Discuss how losses determined from the model are converted to the prototype losses.

10.3.4 A submerged vehicle moves at 5 m/sec in the ocean. At what theoretical speed must a 1:10 model be towed for there to be dynamic similarity between the model and the prototype? Assume that the sea water and towing tank water are the same.

10.3.5 A structure is built underwater on the ocean floor where a strong current of 5 m/sec is measured. The structure is to be studied by a 1:25 model in a water tunnel using sea water at the same temperature as that measured in the ocean. What speed must the water tunnel provide in order to study the force load on the structure due to the current? If the required tunnel velocity is judged to be impractical, can the study be performed in a wide tunnel using air at 20°C? What would the corresponding air speed in the tunnel need to be?

10.4 PHENOMENA GOVERNED BY GRAVITY FORCE—FROUDE NUMBER LAW

When inertial force and gravity force are considered to be the only dominant forces in the fluid motions, the ratio of the inertial forces acting on the homologous elements of the fluid in the model and prototype can be defined by Equation (10.13)

$$\frac{F_p}{F_m} = \rho_r L_r^4 T_r^{-2} \tag{10.13}$$

and the ratio of gravity forces, which is determined by the weight of the homologous fluid elements involved,

$$\frac{F_p}{F_m} = \frac{M_P g_P}{M_m g_m} = \frac{\rho_P g_P L_p^3}{\rho_m g_m L_m^3} = \rho_r g_r L_r^3 \qquad (10.21)$$

Equating the values from Equations (10.3) and (10.21), we get

$$\rho_r L_r^4 T_r^{-2} = \rho_r g_r L_r^3$$

Rearranging, we get

$$g_r L_r = \frac{L_r^2}{T_r^2} = V_r^2$$

or

$$\frac{V_r}{g_r^{1/2} L_r^{1/2}} = 1 \qquad (10.22)$$

From which

$$\frac{\left(\dfrac{V_p}{g_P^{1/2} L_P^{1/2}}\right)}{\left(\dfrac{V_m}{g_m^{1/2} L_m^{1/2}}\right)} = 1$$

Hence,

$$\frac{V_P}{g_P^{1/2} L_P^{1/2}} = \frac{V_m}{g_m^{1/2} L_m^{1/2}} = N_F \qquad \text{(Froude number)} \qquad (10.23)$$

In other words, when the inertial force and the gravity force are considered to be the only forces that dominate the fluid motions, the Froude number of the model and the prototype should be kept at the same value.

If the same fluid is used in both the model and the prototype, and they are both subjected to the same gravitational force field, many physical quantities can be derived based on the Froude number law. These quantities are listed in Table 10.3.

TABLE 10.3 Ratios for the Froude Number Law ($g_r = 1, \rho_r = 1$)

Geometric Similarity		*Kinematic Similarity*		*Dynamic Similarity*	
Length	L_r	Time	$L_r^{1/2}$	Force	L_r^3
Area	L_r^2	Velocity	$L_r^{1/2}$	Mass	L_r^3
Volume	L_r^3	Acceleration	1	Work	L_r^4
		Discharge	$L_r^{5/2}$	Power	$L_r^{7/2}$
		Angular Velocity	$L_r^{-1/2}$		
		Angular Acceleration	L_r^{-1}		

Example 10.6

An open channel model 30 m long is built to satisfy Froude number law. What is the flow in the model for a prototype flood of 700 m³/sec if the scale used is 20:1? Determine also the force ratio.

Solution

From Table 10.3, the discharge ratio is

$$Q_r = L_r^{5/2} = (20)^{2.5} = 1789$$

Thus, the model flow should be

$$Q_m = \frac{Q_P}{Q_r} = \frac{700 \text{ m}^3/\text{sec}}{1789} = 0.391 \text{ m}^3/\text{sec} = 391 \ \ell/\text{sec}$$

The force ratio is

$$F_r = \frac{F_P}{F_m} = L_r^3 = (20)^3 = 8000$$

PROBLEMS

10.4.1 An overflow spillway with a 300-m crest is designed to discharge 3600 m³/sec. A 1:20 model of the cross section of the dam is built in the laboratory flume 1 m wide. Calculate the required laboratory flow rate. Neglect viscosity and surface tension effects.

10.4.2 If a 1:1000 scale tidal basin model is used to study the operation of a prototype satisfying the Froude number law, what length of time in the model represents the period of one day in the prototype?

10.4.3 A ship 100 m long designed to travel at a top speed of 1 m/sec is to be studied in a towing tank with a 1:50 scale model. Determine what speed the model must be towed for (a) the Reynolds number law and (b) the Froude number law.

10.4.4 An overflow spillway is designed to be 100 m high and 120 m long, carrying a discharge of 1200 m³/sec under an approaching head of 2.75 m. The spillway operation is to be analyzed by a 1:50 model in a hydraulic laboratory. Determine

(a) The model discharge,

(b) If the discharge coefficient at the model crest measures 2.12, what is the prototype crest discharge coefficient?

(c) If the velocity at the outlet of the model spillway measures 25 m/sec, what is the prototype velocity?

(d) If the U.S.B.R. Type II stilling basin, 50 m wide is used to dissipate the energy at the toe of the spillway, what is the energy dissipation in the model and in the prototype as measured in units of kilowatts?

(e) What is the efficiency of the dissipator in the prototype?

10.4.5 An energy dissipator is being designed to force a hydraulic jump at the end of a spillway channel discharging 400 m^3/sec. The initial depth in the 20-m wide prototype is expected to be 0.8 m. Determine the discharge of the 1:10 scale model and the velocity and force ratios between prototype and model.

10.4.6 A 1:25 model is built to study a stilling basin at the outlet of a steep spillway chute. The stilling basin consists of a horizontal floor (apron) with U.S.B.R. Type II baffles installed to stabilize the location of the hydraulic jump. The prototype has a rectangular cross section 25 m wide designed to carry a 75-m^3/sec discharge. The velocity immediately before the jump is 10 m/sec. Determine the following:

(a) The model discharge.

(b) The depth downstream of the jump in the prototype if the dynamic force measured on the model baffles is 16.2 N.

(c) The force on the baffles per unit width of the prototype channel.

(d) The energy dissipated in the basin.

10.5 PHENOMENA GOVERNED BY SURFACE TENSION—WEBER NUMBER LAW

Surface tension is a measure of energy level on the surface of a liquid body. The force is of primary importance in hydraulic engineering practice in the study of the motion of small surface waves or control of evaporation from a large body of water, such as a water storage tank or reservoir.

Surface tension, denoted by σ, is measured in terms of force per unit length. Hence, the force is $F = \sigma L$. The ratio of analogous surface tension forces in prototype and in model is

$$F_r = \frac{F_p}{F_m} = \frac{\sigma_p L_p}{\sigma_m L_m} = \sigma_r L_r \tag{10.24}$$

Equating the surface tension force ratio to the inertial force ratio [Equation (10.13)] gives

$$\sigma_r L_r = \rho_r \frac{L_r^4}{T_r^2}$$

Rearranging gives

$$T_r = \left(\frac{\rho_r}{\sigma_r}\right)^{1/2} L_r^{3/2} \tag{10.25}$$

By substituting for T_r the basic relationship of $V_r = L_r/T_r$, the Equation 10.25 may be rearranged to give

$$V_r = \frac{L_r}{\left(\frac{\rho_r}{\sigma_r}\right)^{1/2} L_r^{3/2}} = \left(\frac{\sigma_r}{\rho_r L_r}\right)^{1/2}$$

or

$$\frac{\rho_r V_r^2 L_r}{\sigma_r} = 1 \qquad (10.26)$$

Hence,

$$\frac{\rho_p V_p^2 L_p}{\sigma_p} = \frac{\rho_m V_m^2 L_m}{\sigma_m} = N_w \qquad \text{(Weber number)} \qquad (10.27)$$

In other words, the Weber number must be kept at the same value in the model and in the prototype for studying phenomena governed by surface tension force. If the same liquid is used in both model and prototype, then $\rho_r = 1.0$, and $\sigma_r = 1.0$, and Equation 10.27 can be simplified to

$$V_r^2 L_r = 1$$

or

$$V_r = L_r^{1/2} \qquad (10.28)$$

Since $V_r = L_r/T_r$, we may also write

$$\frac{L_r}{T_r} = L_r^{1/2}$$

Thus,

$$T_r = L_r^{3/2} \qquad (10.29)$$

PROBLEMS

10.5.1 A model is built to study the surface tension phenomenon in a reservoir. Determine the conversion ratios between the model and the prototype for the following quantities if the model is built with a 1:100 scale: (a) rate of flow, (b) energy, (c) pressure, (d) power. The same fluid is used in the model and the prototype.

10.5.2 A measuring device includes certain small glass tubes of a given geometry. To study the surface tension effect, a 5:1 scale model (larger than prototype) is built. Determine the discharge and force ratios.

10.5.3 Determine the surface tension of a liquid in the prototype if a time ratio of 2 is established with a 1:10 scale model. The surface tension of the liquid in the model is 150 dyn/cm. What is the force ratio?

10.6 PHENOMENA GOVERNED BY BOTH GRAVITY AND VISCOUS FORCES

In the case of surface vessels moving through water or the propagation of shallow water waves in open channels, both gravity and viscous forces may be important. The study of these phenomena requires that both the Froude number and Reynolds number laws be satisfied simultaneously. That is,

$$\frac{\rho_r L_r V_r}{\mu_r} = \frac{V_r}{(g_r L_r)^{1/2}}$$

Assuming that both the model and the prototype are affected by the earth's gravitational field, $g_r = 1$ and since $\nu = \mu/\rho$, the above relationship may be simplified to

$$\nu_r = L_r^{3/2} \tag{10.30}$$

This requirement can only be met by choosing a special model fluid with a kinematic viscosity ratio to water equal to the three-half power of the scale ratio. In general, this requirement is difficult to meet. For example, a 1:10 scale model would require that the model fluid have a kinematic viscosity of 30 times less than that of water, which is obviously impossible.

However, two expedients may be available depending on the relative importance of the two forces in the particular phenomenon. In the case of ship resistance, the ship model may be built according to the Reynolds model law and may operate in a towing tank in accordance with the Froude number law. In the case of shallow water waves in open channels, empirical relationships such as Manning's formula [Equation (3.28)] may be used as an auxiliary condition for the wave measurements, according to the Froude number law.

10.7 MODELS FOR FLOATING AND SUBMERGED BODIES

Model studies for floating and submerged bodies are performed in order to obtain information on

1. the friction drag along the boundary of the moving vessel,
2. the form drag resulting from flow separation from the vessel boundary due to the boundary shape,

3. the force expended in the generation of gravity waves,
4. the stability of the body in withstanding the water waves and the wave forces on the body.

The first two forces are strictly viscous phenomena, and, therefore, the models should be designed according to the Reynolds number law. The third force is a gravity force governed phenomenon and, hence, must be analyzed by applying the Froude number law. All three measurements may be performed simultaneously in water in a towing tank. In analyzing the data, however, the friction forces and the form drag forces are first computed from the measurements by using known formulas and drag coefficients. The remaining force measured in towing the vessel through the water surface is the force expended in generating the gravity waves (wave resistance), and it is scaled up to the prototype values by the Froude number law.

The analysis procedure is demonstrated in Example (10.7). For subsurface vessels, such as submarines, the effect of surface waves on the vessels may be neglected. Hence, the Froude number model is not needed. To study the stability and wave force on stationary offshore structures, the effect of inertial force must be taken into consideration. The inertial force, defined as $F_i = M' \cdot a$, can be calculated directly from the prototype dimensions. Here, M' is the mass of water displaced by the portion of the structure immersed below the waterline (also known as the *virtual mass*), and a is the acceleration of the water mass.

Example 10.7

A ship model with a maximum cross-sectional area of 0.78 m^2 immersed below the waterline, has a characteristic length of 0.9 m. The model is towed in a wave tank at the speed of 0.5 m/sec. For the particular shape of the vessel, it is found that the drag coefficient can be approximated by $C_D = (0.06/N_R^{0.25})$ for $10^4 \leq N_R \leq 10^6$, and $C_D = 0.0018$ for $N_R > 10^6$. The Froude number law is applied for the 1:50 model. During the experiment, a total force of 0.40 N is measured. Determine the total resistance force on the prototype vessel.

Solution

Based on the Froude number law, we may determine the velocity ratio,

$$V_r = L_r^{1/2} = (50)^{1/2} = 7.07$$

Hence, the corresponding velocity of the vessel is

$$V_p = V_r \cdot V_m = 0.5 \cdot 7.07 = 3.54 \text{ m/sec}$$

The Reynolds number for the model is

$$N_R = \frac{V_m \cdot L_m}{\nu} = \frac{0.5 \cdot 0.9}{1.003 \cdot 10^{-6}} = 4.49 \cdot 10^5$$

and the drag coefficient for the model is

$$C_{D_m} = \frac{0.06}{(4.49 \cdot 10^5)^{1/4}} = 0.0023$$

The drag force on a vessel is defined as $D = C_D(\frac{1}{2}\rho A \cdot V^2)$, where ρ is the water density, and A is the projected area of the immersed portion of the vessel on a plane normal to the direction of the motion. Thus, the model drag force can be calculated as

$$D_m = C_{D_m}(\tfrac{1}{2}\rho_m A_m \cdot V_m^2) = 0.0023 \cdot \tfrac{1}{2} \cdot 1000 \cdot 0.78 \cdot 0.5^2 = 0.2243 \text{ N}$$

The model wave resistance is the difference between the measured towing force and the drag force

$$F_w = 0.4 - 0.2243 = 0.1757 \text{ N}$$

For the prototype, the Reynolds number is

$$N_R = \frac{V_p \cdot L_p}{\nu_p} = \frac{V_p \cdot L_r \cdot L_m}{\nu_p} = 1.59 \cdot 10^8$$

The drag coefficient of the prototype vessel is $C_{D_p} = 0.0018$, and the drag force is

$$D_p = C_{D_p}(\tfrac{1}{2}\rho_p A_p \cdot V_p^2) = C_{D_p}(\tfrac{1}{2}\rho_p A_m \cdot L_r^2 \cdot V_p^2) = 21937 \text{ N}$$

The wave resistance on the prototype vessel is calculated by applying Froude number law (see Table 10.3)

$$F_{w_p} = F_{w_r} \cdot F_{w_m} = L_r^3 \cdot F_{w_m} = 21963 \text{ N}$$

Hence the total resistance force on the prototype is

$$F = D_p + F_{w_p} = 21937 + 21963 = 43900 \text{ N}$$

PROBLEMS

10.7.1 A ship 100 m long moves at 1.5 m/sec in freshwater at 15°C. A 1:100 scale model of the prototype ship is to be tested in a towing tank containing a liquid of specific gravity 0.9. What viscosity must this liquid have for both Reynolds and Froude number laws to be satisfied?

10.7.2 A 1:250 ship model is towed in a wave tank and a wave resistance of 10.7 N is measured. Determine the corresponding prototype wave resistance on the prototype.

10.7.3 A barge model 1 m long is tested in a towing tank at the speed of 1 m/sec. Determine the prototype velocity if the prototype is 150 m in length. The model has 2-cm draft and is 10 cm wide. The drag coefficient is $C_D = 0.25$ for $N_R > 5 \cdot 10^4$, and the towing force required to tow the model is 0.3 N. What force is required to tow the barge in waterways?

10.7.4 A concrete caisson 60 m wide, 120 m long, and 12 m high is to be towed in sea water in the longitudinal direction to an offshore construction site where it will be sunk. The calculated floating depth of the caisson is 8 m, with 4 m remaining above the water surface. A 1:100 model is built to study the operation of the prototype. If the model is towed in a wave tank using sea water, what is the model speed that corresponds to the prototype speed of 1.5 m/sec? The model study considers both the skin drag and form drag force (Reynolds number) and the resistance due to the generation of gravity waves in motion (Froude number).

10.8 OPEN CHANNEL MODELS

Open channel models may be used to study either the velocity discharge-slope relationship or the effect of flow patterns on the changes in bed configuration. For the former applications, relatively long reaches of the river channel can be modeled. A special example is the U.S. Army Engineers Waterways Experiment Station in Vicksburg, Mississippi, where the Mississippi River is modeled on one site. In these applications, where the influence of changes in bed configuration are only of secondary concern, a fixed-bed model may be used. Basically, this model is used in studying the velocity-slope relationship in a particular channel; therefore, the effect of bed roughness is important.

An empirical relation, such as the Manning equation [Equation (3.28)], may be used to assume the similarity between the prototype and the model

$$V_r = \frac{V_p}{V_m} = \frac{\dfrac{1}{n_p} R_{h_p}^{2/3} S_p^{1/2}}{\dfrac{1}{n_m} R_{h_m}^{2/3} S_m^{1/2}} = \frac{1}{n_r} R_{h_r}^{2/3} S_r^{1/2} \tag{10.31}$$

If the model is built with the same scale ratio for the horizontal dimensions ($\overline{X}$) and vertical dimensions ($\overline{Y}$), known as the *undistorted model,* then

$$R_{h_r} = \overline{X}_r = \overline{Y}_r = L_r \quad \text{and} \quad S_r = 1$$

and

$$V_r = \frac{1}{n_r} L_r^{2/3} \tag{10.32}$$

Since the Manning's roughness coefficient $n \propto R_h^{1/6}$ [Equation (6.3)], we may write

$$n_r = L_r^{1/6} \tag{10.33}$$

This model may frequently result in a model velocity so small (or, conversely, the model roughness will be so large) that realistic measurements cannot be made; or a model water depth may be so shallow that the physical characteristics of the flow may be altered. Such situations may be resolved by using a distorted model in which the vertical scale and the horizontal scale do not have the same value; usually, a smaller vertical scale ratio, $\overline{X}_r > \overline{Y}_r$. This means that

$$S_r = \frac{S_p}{S_m} = \frac{\overline{Y}_r}{\overline{X}_r} < 1$$

Hence, $S_m > S_p$, and the result is a larger slope for the model. The use of the Manning equation requires that the flow be fully turbulent in both model and prototype.

Open channel models involving problems of sediment transportation, erosion, or deposit require movable bed models. A movable channel bed consists of sand or other loose material that can be moved in response to the forces of the current at the channel bed. Normally, it is impractical to scale the bed material down to the model scale. A vertical scale distortion is usually employed on a movable bed model in order to provide a sufficient tractive force to induce bed material movement. Quantitative similarity is difficult to attain in movable bed models. For any sedimentation studies performed, it is important that the movable bed model be quantitatively verified by a number of field measurements.

Example 10.8

An open channel model is built to study the effects of tidal waves on sedimentation movement in a 10-km river reach (the reach meanders in an area 7 km long). The mean depth and width of the reach are 4 m and 50 m, respectively, and the discharge is 850 m^3/sec. Manning's roughness coefficient $n_p = 0.035$. If the model is to be constructed in a laboratory room 18 m long, determine a convenient scale and the model discharge.

Solution

In surface wave phenomenon the gravitational forces are dominant. The Froude number law will be used for the modeling. The laboratory length will limit the horizontal scale

$$\overline{X}_r = \frac{L_p}{L_m} = \frac{7000}{18} = 389$$

We will use $L_r = 400$ for convenience.

It is judged reasonable to use a vertical scale of $\overline{Y}_r = 80$ (enough to measure surface gradients). Recall that the hydraulic radius is the characteristic dimension in open channel and that for a large width-to-depth ratio the hydraulic radius is roughly equal to the water depth. Thus, we can make the following approximation:

$$R_{h_r} = \overline{Y}_r = 80$$

Since

$$N_F = \frac{V_r}{g_r^{1/2}\, R_{h_r}^{1/2}} = 1 \tag{10.22}$$

Then

$$V_r = R_{h_r}^{1/2} = \overline{Y}_r^{\,1/2} = (80)^{1/2}$$

Using Manning's formula [or Equation (10.31)],

$$V_r = \frac{V_p}{V_m} = \frac{1}{n_r} R_{h_r}^{2/3} S_r^{1/2} \qquad S_r = \frac{\overline{Y}_r}{\overline{X}_r}$$

we have

$$n_r = \frac{R_{h_r}^{2/3} S_r^{1/2}}{V_r} = \frac{\overline{Y}_r^{\,2/3}\left(\dfrac{\overline{Y}_r}{\overline{X}_r}\right)^{1/2}}{\overline{Y}_r^{\,1/2}} = \frac{\overline{Y}_r^{\,2/3}}{\overline{X}_r^{\,1/2}} = \frac{(80)^{2/3}}{(400)^{1/2}} = 0.928$$

Hence,

$$n_m = \frac{n_p}{n_r} = \frac{0.035}{0.928} = 0.038$$

The discharge ratio is

$$Q_r = A_r \cdot V_r = \overline{X}_r \overline{Y}_r \overline{V}_r = \overline{X}_r \overline{Y}_r^{\,3/2} = (400)(80)^{3/2} = 286{,}217$$

Thus, the model discharge required is

$$Q_m = \frac{Q_p}{Q_r} = \frac{850}{286{,}217} = 0.003 \text{ m}^3/\text{sec} \equiv 3\ \ell/\text{sec}$$

In order to use the Manning formula, turbulent flow must be ensured in the model. To verify the turbulent flow condition in the model, it is necessary to calculate the value of the model Reynolds number.

The horizontal prototype velocity is

$$V_p = \frac{850 \text{ m}^3/\text{sec}}{4 \text{ m} \cdot 50 \text{ m}} = 4.25 \text{ m/sec}$$

Hence,

$$V_m = \frac{V_p}{V_r} = \frac{4.25}{(80)^{1/2}} = 0.475 \text{ m/sec}$$

The model Reynolds number is

$$N_R = \frac{V_m \overline{Y}_m}{\nu} = \frac{0.475 \cdot 0.05}{1.1 \cdot 10^{-6}} = 21{,}598$$

which is much greater than the critical Reynolds number (2,000). Hence, the flow is turbulent in the model.

PROBLEMS

10.8.1 A new laboratory site is available for modeling the channel of Example 10.8 so that the length is no longer a restriction, but the roughness coefficient of the material to be used in the movable bed is $n_m = 0.018$. Determine a convenient scale and the corresponding model velocity.

10.8.2 A ship channel model is built to study sedimentation control in the prototype which is 40 m wide and 7.5 m deep and carries a discharge of 300 m^3/sec. For a vertical scale of 1:65 and a roughness coefficient of $n_p = 0.03$, determine all the other needed ratios for the study if $n_m = 0.02$.

10.8.3 A 1:100 scale model is constructed to study the pattern of flow in a river reach. If the reach has a Manning's coefficient $n = 0.025$, what should be the corresponding value of n in the model? Discuss the errors that may result from the study if the value of n were not modeled.

10.8.4 Determine the value of n in the model in Problem 10.8.3 if the vertical scale is exaggerated to 1:25 distortion.

10.8.5 A 1:300 scale model is constructed to study the discharge-depth relationship in a river reach with Manning's coefficient $n = 0.031$. If the model discharges 52 ℓ/sec and has Manning's coefficient $n = 0.033$, determine an adequate vertical scale ratio and the flowrate for the prototype.

A

GRAPHICAL FLOW NETS: ELECTRIC ANALOG AND NUMERICAL ANALYSIS

Two-dimension flow nets are constructed with sets of streamlines (ψ-lines) and equipotential lines (ϕ-lines). The streamlines are tangential to the velocity vector at all points in the flow field, while the equipotential lines represent constant velocity potential levels. Usually, the streamlines are chosen so as to divide the flow into a number of channels conveying equal flow rates, and the spacing of the equipotential lines is selected so that it is equal to the spacing between the streamlines everywhere in the flow field.

Some general characteristics of flow nets can be listed as follows:

1. Flow nets represent the flow patterns of irrotational flow* within given boundaries. It is not necessary that the flow be steady.
2. There is only one possible pattern of flow for a given set of boundary conditions.

* Flow of zero vorticity

3. The ϕ-lines intersect the ψ-lines and the fixed boundaries, at 90°.
4. Each unit grid resulting from the ψ-lines and the ϕ-lines approximate to squares, having equal median lines and 90° angles (except at points of zero velocity or boundary discontinuity).
5. In regions of uniform flow the squares are of equal size. In diverging flows they increase, and in converging flows they decrease in size in the direction of the flow. The size of the square is inversely proportional to the velocity at any point in the flow field.

The graphical procedure for the construction of a flow net may be summarized in the following steps:

1. Decide upon the number of streamlines required to yield the necessary details of the flow pattern. Sketch the streamlines in regions where the velocity distributions are evident, such as in parallel or radial flow to a well.
2. Lightly sketch the remaining portions of the streamlines with smooth curves. The spacing should decrease with decrease of radius on curves. The product of velocity and the radius is always constant in irrotational flows.
3. Sketch the equipotential lines; they should normally intersect all streamlines including the boundaries and form squares with the streamlines. It will be necessary to adjust the initial streamline locations from time to time until the above conditions are satisfied. The process is a trial-and-error one in which the operator improves the sketches with practice.
4. At the stagnation points (points of zero velocity), five-sided units occur. In this case, the equipotential lines adjacent to the stagnation point should be spaced nearly equally about the stagnation point.
5. Free water table surface must be located by trial such that when the flow net is drawn the velocity along the free surface is in accord with the boundary conditions of the problem. Since the free surface of the water table is affected by gravity, the spacing of the ϕ-lines along the water table should be such that the velocity head at any point on the water table equals the distance of the point below the total head line, h_0 (Figure A.1).

$$V = \sqrt{2gh_v}\,\frac{\Delta\phi}{\Delta s}$$

where $\Delta\phi$ is the spacing between the ϕ-lines and Δs is the spacing between the ψ-lines, h_v is the velocity head at the point

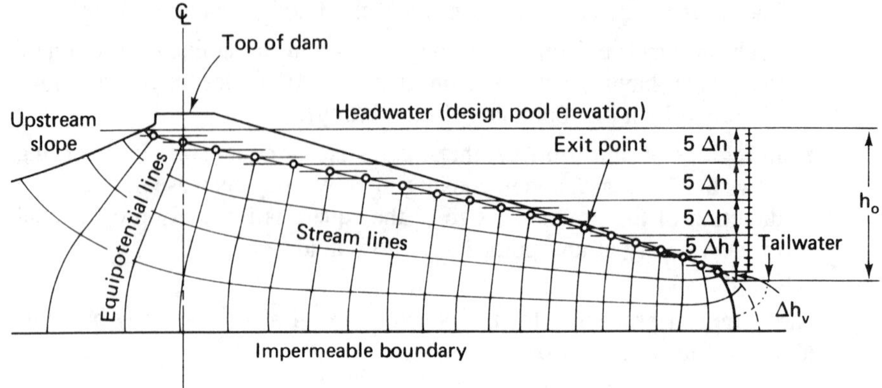

Figure A.1 Gravity-effected water table.

$$h_v = h_0 - h$$

6. The diagonals of the "squares" should form smooth curves which intersect each other normally. This fact is used to check the accuracy of the final flow pattern.

A.1 ELECTRIC ANALOG

The flow of an electric current in a two-dimensional conductor is analogous to irrotational flow. The electrical potential (voltage) is the counterpart of the velocity potential, and the electric current (I) is analogous to the fluid flow. A sheet of uniform, homogeneous conducting material or a shallow bath of electrolyte of uniform depth is shaped so that its boundary is geometrically similar to that of the prototype flow field to be investigated [for example, Figures A.2(a) and A.2(b)]. If a voltage drop is established along the conductor representing the boundary equipotential lines, a potential pattern geometrically similar to that in the flow field is traced [Figure A.2(a)]. The electric potential p is the counterpart of the velocity potential function ϕ, and the components of the electric field strength, E_x and E_y, are the counterparts of velocity components V_x and V_y, respectively. The corresponding equations for the electric field are

$$\frac{\partial^2 p}{\partial x^2} + \frac{\partial^2 p}{\partial y^2} = 0 \tag{1}$$

and

$$E_x = \frac{\partial p}{\partial x}, \qquad E_y = \frac{\partial p}{\partial y} \tag{2}$$

If a voltage drop is established along the boundary streamlines, then the ψ-line pattern can be obtained (Figure A.2b).

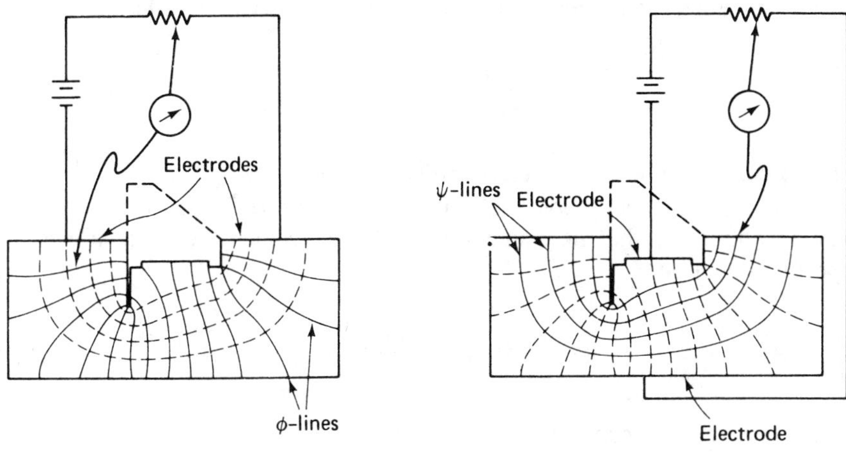

Figure A.2 Electrode arrangements in an electrical analog model for determining: (a) equipotential lines, and (b) streamlines of seepage beneath a masonry dam.

A voltmeter or potentiometer probe is used to locate the lines of constant potential. By fixing a constant voltage drop between the boundary voltage, each ϕ-line and ψ-line can be accurately located. The electrical analogy arrangement is shown in Figure A.2.

A.2 NUMERICAL ANALYSIS

Since the streamlines are everywhere parallel to the velocity vector in the flow field, a stream function, ψ, may be defined such that

$$\frac{\partial \psi}{\partial y} = V_x, \qquad \frac{\partial \psi}{\partial x} = -V_y \tag{3}$$

where V_x and V_y are the velocity components in the x- and y-directions respectively.

In a flow field free from vortex motion, or where the effect of vortices can be neglected, the flow is said to be *irrotational*. Flow nets may be constructed in irrotational flow fields where the *rotation of flow* is zero. The rotation of flow is defined as

$$\omega = \frac{1}{2}\left(\frac{\partial V_y}{\partial x} - \frac{\partial V_x}{\partial y}\right)$$

For irrotational flow, then

$$\frac{\partial V_y}{\partial x} - \frac{\partial V_x}{\partial y} = 0 \tag{4}$$

Substituting Equation (3) into Equation (4) and simplifying, we have,

$$\frac{\partial^2 \psi}{\partial x^2} + \frac{\partial^2 \psi}{\partial y^2} = 0 \tag{5}$$

which is commonly known as the *Laplace Equation*.

Similarly, a velocity potential function, ϕ, may be defined for the irrotational flow field such that the gradient of ϕ gives the velocity component in each respective direction.

$$\frac{\partial \phi}{\partial x} = V_x \qquad \frac{\partial \phi}{\partial y} = V_y \tag{6}$$

In the meantime, the continuity condition requires that,

$$\frac{\partial V_x}{\partial x} + \frac{\partial V_y}{\partial y} = 0 \tag{7}$$

Substituting Equation (6) into Equation (7) and simplifying, we have

$$\frac{\partial^2 \phi}{\partial x^2} + \frac{\partial^2 \phi}{\partial y^2} = 0 \tag{8}$$

A streamline (ψ-line) is defined as a line for which the stream function ψ has a constant value. An *equipotential line* (ϕ-line) is defined as a line for which the velocity potential function ϕ has a constant value. Flow net is constructed with a set of streamlines and its corresponding set of equipotential lines.

Impervious boundaries are streamlines since flow velocity is parallel and no flow is allowed to cross the boundary lines. In a certain flow field (for example, Figure A.4), an arbitrary value may be assigned to each boundary stream function. For example, one boundary may have the stream function value of 0 and the other boundary the stream function value of 100. Equation (5) may be solved by numerical integration in a quadrangular grid (for example, Figure A.4b) for given boundary conditions. The discretization of the Laplace equation may be expressed as follows:

A ψ-value is first assigned to each node in the quadrangular grid (Figure A.3), and the gradient of ψ at the midpoint of each grid arm (A, B, C, and D) may be expressed as, respectively,

$$\left.\frac{\partial \psi}{\partial y}\right|_A = \frac{\psi_1 - \psi_0}{L_1} \tag{9}$$

$$\left.\frac{\partial \psi}{\partial x}\right|_B = \frac{\psi_2 - \psi_0}{L_2} \tag{10}$$

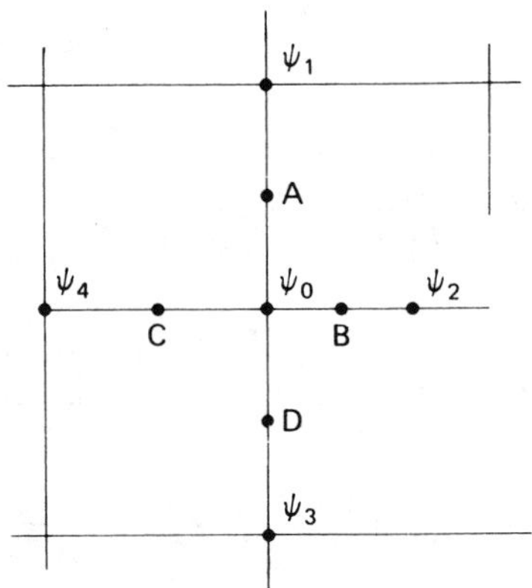

Figure A.3

$$\left.\frac{\partial \psi}{\partial y}\right|_C = \frac{\psi_0 - \psi_3}{L_3} \tag{11}$$

$$\left.\frac{\partial \psi}{\partial x}\right|_D = \frac{\psi_0 - \psi_4}{L_4} \tag{12}$$

where the L's are the length of each arm measured between the respective nodes. The arms may or may not have the same length depending on the boundary geometry. Combining Equations (9) and (11), we may write the second derivative of ψ at node 0:

$$\left.\frac{\partial^2 \psi}{\partial y^2}\right|_0 = \frac{\left.\frac{\partial \psi}{\partial y}\right|_A - \left.\frac{\partial \psi}{\partial y}\right|_C}{\left(\frac{L_1 + L_3}{2}\right)} = \frac{2\left(\frac{\psi_1}{L_1} + \frac{\psi_3}{L_3}\right)}{L_1 + L_3} - \frac{2\psi_0}{L_1 \cdot L_3} \tag{13}$$

Using the same approach, we may also write,

$$\left.\frac{\partial^2 \psi}{\partial x^2}\right|_0 = \frac{2\left(\frac{\psi_2}{L_2} + \frac{\psi_4}{L_4}\right)}{L_2 + L_4} - \frac{2\psi_0}{L_2 \cdot L_4} \tag{14}$$

Substituting Equations (13) and (14) into the Laplace equation (Equation 5), we have:

$$\frac{\frac{\psi_2}{L_2} + \frac{\psi_4}{L_4}}{L_2 + L_4} + \frac{\frac{\psi_1}{L_1} + \frac{\psi_3}{L_3}}{L_1 + L_3} - \frac{\psi_0}{L_2 \cdot L_4} - \frac{\psi_0}{L_1 \cdot L_3} = 0 \tag{15}$$

or,

$$\psi_0\left(\frac{1}{L_2 \cdot L_4} + \frac{1}{L_1 \cdot L_3}\right) = \frac{\dfrac{\psi_2}{L_2} + \dfrac{\psi_4}{L_4}}{L_2 + L_4} + \frac{\dfrac{\psi_1}{L_1} + \dfrac{\psi_3}{L_3}}{L_1 + L_3} \tag{16}$$

when the arms are of the same length, $L_1 = L_2 = L_3 = L_4 = a$, which is the length of the basic grid selected, the above relationship may be simplified to

$$\psi_0 = \frac{a^2}{2}\left(\frac{\psi_2 + \psi_4}{2a^2} + \frac{\psi_1 + \psi_3}{2a^2}\right) = \frac{\psi_1 + \psi_2 + \psi_3 + \psi_4}{4} \tag{17}$$

When a digital computer is used, a grid size is first adopted and the following information included: the initial value of stream function for every node in the mesh, the length of each arm according to the boundary geometry given, and a code specifying if a node is on a boundary or if it has one or more than one arms shorter than the basic grid size a. Generally, the smaller the grid, the more accurate the solution. However, the resolution is always limited by the capacity of the computer and the cost of operation.

Equipotential lines may be constructed in exactly the same way using assumed values for equipotential lines far upstream and downstream from the object.

Example A.1

Determine the pattern of streamlines for the seepage through the foundation of a dam as shown in Figure A.4.

Solution

A basic grid with arms length of 8 m appears to fit the boundary geometry fairly well. The dam bottom surface, as a boundary, is assigned a stream function value of 0; and the impervious layer deep in the foundation is assigned the value of 100.

Far from the dam, the direction of the seepage flow is assumed small and parallel to the horizontal direction. At these stations, same value as that of the impervious boundary is assumed. The effect of these assumptions diminishes with distance. It becomes negligible near the base of the dam.

To completely define the boundaries, values are also assigned to the water-foundation interface. It is very unlikely that the right values can be guessed at the first trial. Since the streamlines must intersect the boundary at right angles, for the interface we also have,

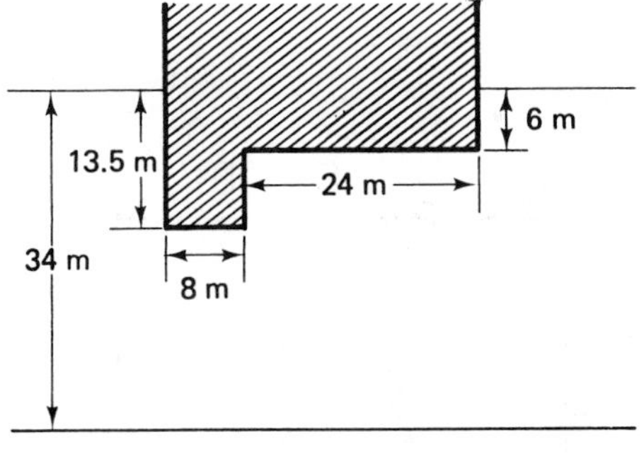

(a)

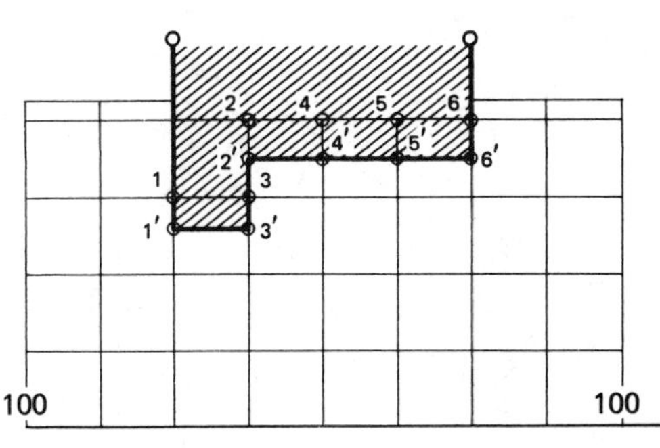

(b)

Figure A.4

$$\frac{\partial \psi}{\partial y} = 0, \quad \text{or,} \quad \frac{\psi_1 - \psi_0}{L_1} = 0, \quad \text{then,} \quad \psi_1 = \psi_0$$

A computer program is written to solve the Laplace equation of stream function and the result is plotted in Figure A.5 with the streamlines proportioned in as shown.

If more detailed flow patterns are needed for a particular region, the computer program may be applied again to the region of interest using smaller grid and boundary conditions obtained from the previous computation. For example, the rectangular region bounded by ABCD in Figure A.5 may be reprocessed by the same computer program using a grid of one half the original size and the boundary conditions presented in the result of the above example.

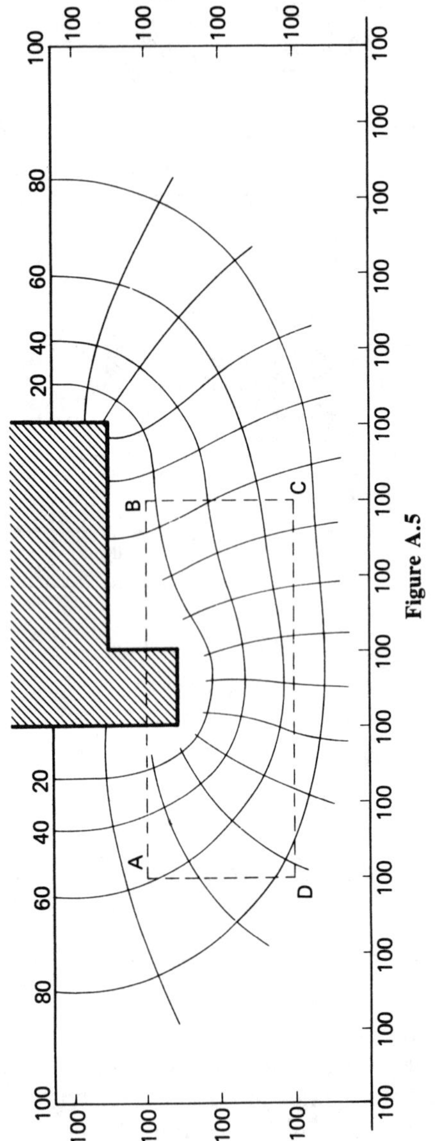

Figure A.5

SYMBOLS

A	area
b	width, aquifer penetration depth
C	Chezy's coefficient, celerity (surface wave speed)
C_d	discharge coefficient
C_{HW}	Hazen-Williams coefficient
°C	degree Celsius
D	diameter, hydraulic depth
d	water depth
d_1, d_2	initial, sequent depth
d_n, d_c	normal depth, critical depth
E	energy per unit weight of water (energy head)
e	roughness height, efficiency
E_b	modulus of elasticity (bulk)
E_c	composite modulus of elasticity
E_p	modulus of elasticity (pipe material)
E_s	specific energy
EGL	energy gradient line
F	force
f	friction factor
F_s	specific force
g	gravitational acceleration
H	total head, approaching head
H_a	approaching head
H_s	static pressure head
H_P	pump head

H_v	velocity head
h	elevation head
h_b	bend loss
h_c	contraction loss
h_d	discharge loss
h_e	entrance loss
h_E	enlargement loss
h_L	head loss
h_f	friction loss
Σh_{fc}	friction loss in clockwise direction
Σh_{fcc}	friction loss in counter-clockwise direction
HGL	hydraulic gradient line
I	moment of inertia, linear impulse
I_0	moment of inertia about the axis of rotation
K	coefficient of permeability, coefficient of energy loss
L	length
ℓ	liters
M	moment, momentum, mass
m	mass, model
m	meter
M, N	exponents of the gradually varied flow function
N	newtons
n	Manning's coefficient
N_F	Froude number
N_R	Reynolds number
N_s	specific speed of pump
N_W	Weber number
NPSH	net positive suction head
P	wetted perimeter, pressure, power
P_i, P_o	input, output power
p	pressure variable
Q	flow rate (discharge)
q	discharge per unit width
R	radius
r	variable radius, homologous ratio
r_i, r_o	inside, outside diameter
r_o, r_w	depression cone diameter, well diameter
R_h	hydraulic radius
S	slope, shape number
S_f	energy slope
S_c	critical slope, aquifer storage constant
S_0	channel slope
S_w	water surface slope
sp. gr.	specific gravity
T	temperature, transmissibility of aquifer
T	torque
t	time
th	thickness of pipe wall
T_r	return period
x, y, z	coordinate axes
u, v, w	velocity in x, y, z direction
V	mean velocity
V_i, V_o	inlet, outlet velocity
Vol	volume
W	weight
$\overline{W}$	work
z	elevation, side slope of cross section
σ	surface tension, cavitation parameter
α	porosity, energy coefficient, angle
β	vane angle, momentum coefficient
ρ	density
μ	viscosity, friction coefficient
ν	kinematic viscosity
γ	specific weight
τ	shear stress
τ_0	wall shear stress
ω	angular velocity
ϵ	Poisson's ratio
θ	angle, wall deflection angle
ω_s	specific speed

ANSWERS TO SELECTED PROBLEMS

CHAPTER ONE

1.2.1 7.99×10^7 Cal
1.2.3 29.38°C
1.2.5 73.84 KCal
1.3.1 927.6 kg/m
1.3.3 0.9830, 1157.16 kN
1.3.5 8.31 lb
1.3.7 1.6×10^5 Cal
1.3.9 0.0000685 slugs
1.4.1 68.7 poise
1.4.3 188.50 N
1.4.5 4.86 cP
1.4.7 929 stokes
1.4.9 15.8 ft^2/sec
1.5.1 20.4% of H_1
1.5.3 1.455 mm
1.6.1 −0.099%
1.6.3 (a) 1,157,161 N (b) 1,157,165 N

CHAPTER TWO

2.2.1 92.031 kN/m^2, 8.02%

2.2.3 5.08 m

2.2.5 40.3 m

2.4.1 1.051 bars

2.4.3 7.35 cm

2.4.5 126,549 N/m^2

2.4.7 87,427 N/m^2

2.4.11 29.95 m

2.5.1 (a) 132.3 kN at 1 m above bottom, (b) 485.1 kN at 1.45 m above bottom

2.5.5 963 D^3

2.5.7 (a) 6.67 m below water surface, (b) 6.62 m below water surface

2.5.9 1,250 kN

2.5.11 8.78 ft from ground

2.6.1 3.04×10^6 N @ 14.45°

2.6.3 $\gamma_2 = 1.334\ \gamma_1$

2.6.5 4,580,930 N

2.6.7 $2 R^2 \gamma$

2.6.9 F_H = 98 kN/ft width @ 1.47 m from bottom
F_V = 26.7 kN/ft width

2.6.11 338.6 N

2.8.1 1.58 m

2.8.3 $\gamma_A = \gamma_W, \qquad \gamma_B = 0.5\ \gamma_W$

2.8.5 40.7°

2.8.7 72,217 N-m, 144,082 N-m, 215,244 N-m

CHAPTER THREE

3.3.1 $949.9\ \vec{j} - 957.6\ \vec{i}$ or 1348.86 kN @ − 44.77°

3.3.3 $64.0\ \vec{i}$ kN

3.5.1 0.057 m^3/sec

3.5.3 5.67 m/km or 55,590 (N/m^2)/km

3.5.5 74.27 m^3/sec

3.5.7 0.0003

3.5.9 2.64%

3.5.11 524.1 ℓ/sec

3.5.13 8.0 ℓ/sec

3.6.1 (a) 330 m, (b) 720 m, (c) 397 m

3.6.3 (a) 7.71 m or 69,615 N/m^2, (b) 3.17 m or 31,496 N/m^2

3.6.7 (a) 136.83 ℓ/sec, (b) 94.2 ℓ/sec

3.6.9 17.6%

3.10.1 0.048 m^3/sec

3.10.3 155 m or 1,520 kN/m^2

3.10.5 244.4 ℓ/sec

3.10.7 (a) 0.18 m, (b) 0.43 m

3.10.9 844.37 ℓ/sec

CHAPTER FOUR

4.1.3 11.2 mm

4.1.5 (a) 0.44 m, (b) 147.7 m

4.1.7 (a) 1.85 m or 18.081 kN/m^2, (b) 6.74 m or 66.000 kN/m^2, (c) 6.66 m or 65.284 kN/m^2, (d) −0.9586 m or −9.30 kN/m^2, (e) 1.4413 m or 14.124 kN/m^2

4.1.9 89.6 kN/m^2

4.1.11 53.74 ℓ/sec

4.2.1 5 m or 49.03 kN/m^2

4.2.3 72 m

4.2.5 29.22 m or 286.34 kN/m^2

4.2.7 7.12 m

4.2.9 1.35 m

4.3.1 108.5 m (C below A)

4.3.3 1.41 m^3/sec (reservoir A)
1.05 m^3/sec (reservoir B)
0.40 m^3/sec (reservoir C)

4.3.4 3.53 m^3/sec (reservoir A)
2.174 m^3/sec (reservoir B)
1.36 m^3/sec (reservoir C)

4.4.1 4.53 ℓ/sec, 7.47 ℓ/sec

4.4.3 0.4247 m^3/sec (pipe 1)
0.3753 m^3/sec (pipe 2)
0.2024 m^3/sec (pipe 3)

4.5.1 15 m of water

4.5.3 (a) 1000 m, (b) 1023 m, (c) 1032 m

4.5.5 1.6 cm

4.6.1 9.91 m

4.6.3 5.7 m

CHAPTER FIVE

5.1.1 774 hp

5.1.3 64.7%

5.1.5 195 rpm, 58%

5.1.7 5.89 m^3/sec, 414 m

5.4.1 Pump II @ 3850 rpm with 50% efficiency

5.4.3 Pump II @ 3850 rpm with 48% efficiency

5.5.1 47.5 ℓ/sec

5.5.3 Two Pumps IV's in parallel @ 3850 rpm
Q total = 350 ℓ/sec, Q/pump = 175 ℓ/sec
H = 36 m @ 61% efficiency

5.5.5 (a) Series Q = 173 ℓ/sec @ P = 84.9 kW,
(b) Parallel Q = 143 ℓ/sec @ P = 142 kW

5.6.1 4.84 m

5.6.3 (a) 0.35 m (series), (b) 2.12 m (parallel)

5.6.5 (a) −1.59 m (20°C), (b) −2.61 m (50°C)

5.7.1 12.76 m^3/sec, 4693 kW

5.7.3 9.54 m^3/sec, 13,000 kW

CHAPTER SIX

6.1.1 (a) Unsteady varied flow, (b) Varied steady flow,
(c) Varied steady flow, (d) Varied unsteady flow

6.2.1 2.0 m

6.2.3 d = 4.9 m, b = 9.8 m

6.2.5 n = 0.0247

6.2.7 z = 3.34

6.2.9 4.37 m

6.3.1 0.5

6.4.1 (a) 1.38 m, (b) 1.96 m, (c) 3.65 m

6.4.3 (a) 0.00567, (b) 1.37 m, (c) 0.72

6.4.5 1.897 m, 0.00265

6.4.7 2.07 m

6.4.9 z_i = 2.94 m
at x = 0 d = 1.96 m
x = 10 d = 2.86 m
x = 20 d = 3.69 m
x = 30 d = 4.46 m

6.5.1 0.026 m, 30.96 m^3/sec

6.8.1 18 m

6.8.3 324.4 m

6.8.5 at d = 5.96 m L = 0 m (L = distance from initial point)
5.05 m 564 m
4.15 m 1125 m
3.25 m 1686 m
2.95 m 1872 m
2.65 m 2068 m
2.60 m 2099 m

6.8.7 at d = 0.19 m L = 0.0 m (L = distance from initial point)
0.33 m 74.46 m

	0.47 m	158.02 m
	0.61 m	333.59 m
6.8.9	at d = 6.011 m	L = 0 m (L = distance from initial point)
	5.20 m	8681 m
	4.30 m	19,322 m
	3.75 m	27,460 m

6.9.3 33.06°, 3.9°

6.10.1 Use curve $b_2/b_1 = 2.0$ with parameters $b/2.8$ against $x/7.93$

6.10.3 3.62°, 15.85 m

CHAPTER SEVEN

7.1.1 (a) 0.0265 cm/sec, (b) 1.326 cm/min

7.1.3 20,800 cm^3/sec

7.3.1 1,462 cm/min

7.3.3 0.0141 cm/min

7.4.1	for r = 0	S_d = 14.2 m	
	10	28.3 m	r measured from midpoint
	20	11.1 m	between wells ($\lvert r \pm 10 \rvert \geq r_w$)

7.4.3 h_w = 24.78 m, $h(500)$ = 44.68 m (x measured from lake)

7.4.5 0.659 m/hr

7.7.1 5.35 m^3/day/m width

7.7.3 6 K

7.8.1 0.314 m^3/day/m width

7.8.3 88.87 m^3/day/m width

CHAPTER EIGHT

8.4.1 P(heel) = 385,240 N/m^2
P(toe) = 558,970 N/m^2

8.4.3 Dam is stable against sliding.
P(heel) = 518,164 N/m^2, P(toe) = 281,743 N/m^2
Dam is stable against overturning.

8.5.1 1.35

8.5.3 2.57 m^3/sec, 1.45 m

8.6.1 0.6204 m

8.6.3 25 m

8.7.1 8.84 m

8.7.3 8.912 m

8.8.1 −2.31 m

8.8.3 6.52 m

8.9.1 (a) 1.244 m, (b) 1.307 m

8.9.3 (a) 0.834 m, (b) 1.084 m

8.9.5 (a) 3.26 m^3/sec, (b) 0.36 m

CHAPTER NINE

9.1.1 5.743
9.2.1 10,352 N/m^2
9.2.3 0.89 m/sec
9.3.1 0.396 m
9.3.3 0.45 m^3/sec
9.3.5 0.64
9.3.7 0.045 m Hg
9.4.1 8.11 m^3/sec
9.4.3 0.22 m
9.4.5 0.81 ($H^{2.58}$)
9.4.7 2.46 m^3/sec
9.4.9 7.46 m^3/sec
9.4.11 19.82 m^3/sec

CHAPTER TEN

10.2.1 6,810 N/m^2
10.2.3 52 min
10.2.5 24 ℓ/sec, 69.3 cm
10.2.7 L_r = 250, 0.0012 ℓ/sec
10.2.9 (a) 2500, (b) 1, (c) 50, (d) 0.02,
10.3.1 123 ℓ/sec
10.3.3 1450 ℓ/sec, 185 m/sec
10.3.5 125 m/sec
10.4.1 134 ℓ/sec
10.4.3 50 m/sec, 0.141 m/sec
10.5.1 (a) 10^5, (b) 10^7, (c) 10^2, (d) 10^7
10.5.3 150 dyn/cm, 10
10.7.1 1,025 kN · sec/m^2
10.7.3 12.25 m/sec, 1,012.5 kN
10.8.1 1:80, 0.475 m/sec
10.8.5 65.63, 8,295 m^3/sec

INDEX